Specialist Techniques
in Engineering Mathematics

Specialist Techniques in Engineering Mathematics

A.C. BAJPAI

L.R. MUSTOE

D. WALKER

Department of Engineering Mathematics
Loughborough University of Technology

JOHN WILEY & SONS

Chichester · New York · Brisbane · Toronto

British Library Cataloguing in Publication Data:

Bajpai, Avinash Chandra
Specialist techniques in engineering mathematics.
1. Engineering mathematics
I. Title
II. Mustoe, Leslie R.
III. Walker, Dennis
510'.2'462 TA330 80-41274

ISBN 0 471 27907 2 (Cloth)

ISBN 0 471 27908 0 (Paper)

Printed in Great Britain

Preface

This book is the third in a series of books written specifically for engineers and scientists. The first two books in the series, *Engineering Mathematics* and *Advanced Engineering Mathematics*, written by the same authors, have received a favourable response both from the United Kingdom and from abroad. They were designed for first and second year undergraduate students respectively. This volume is directed towards third year undergraduate and postgraduate science and engineering students in universities, polytechnics and colleges in all parts of the world. Whilst it would be helpful if the reader is familiar with both the *Engineering Mathematics* and *Advanced Engineering Mathematics* texts, this third volume is self-contained and can be read by students who have covered the earlier material by an alternative approach.

As the title of this book suggests, the topics covered are specialist techniques of which only some will be required by individual students. They are all topics which have come to be used more and more over recent years and which are necessary for the study of modern branches of engineering and science. The whole philosophy behind the way in which the text has been written is that a *readable* introduction should be given to these advanced topics. It is not intended that each of these should be comprehensively covered but rather that the reader will be given a good grounding and can then go on to study the more advanced volumes which exist on each topic.

Both final year undergraduates as well as postgraduate students starting their research will find an advantage in such an approach, for in our experience some real feeling for a topic is necessary before reading the standard texts which are often difficult for the inexperienced student to understand.

We make no apology for approaching the material from this standpoint. For example, we do not pretend that someone having read the chapter on Finite Elements will then become an expert in Finite Element techniques but it is hoped that he or she will then have an understanding of what the method involves.

Chapters one to four should be read in that order but other chapters are self-contained and do not rely on material covered earlier. A Bibliography and Reference section is included towards which the reader is directed throughout the chapters. Our earlier work *Advanced Engineering Mathematics* is referred to as A.E.M. in the text. Worked examples are given wherever possible and supplementary problems are provided at the end of most sections which illustrate both the theory and its application. Answers to most of the problems are given at the end of the book.

The authors believe that some proofs are necessary where basic principles are involved and have tried to keep a balance of rigour and applications throughout the text.

A debt of gratitude to the following is acknowledged with pleasure:

Staff and students of Loughborough University of Technology and other institutions who have participated in the development of this text; in particular *Dr. G.W. Irwin* for his help with the chapters on systems, control theory and random processes and *D. Storry* for assistance with the chapter on functional analysis.

S.J. Kelly and *B.M. Thompson* for providing the computer-based solutions in the Finite Element chapter.

Mrs. G. Anthony for drawing the diagrams.

Mrs. Barbara Bell for her great patience and co-operation in typing the entire manuscript and *Gordon Bell* for his help in general administration.

John Wiley and Sons for their help and co-operation.

Contents

CHAPTER FOUR: INTRODUCTION TO OPTIMAL CONTROL

CHAPTER FIVE: RANDOM PROCESSES

CHAPTER SIX: CARTESIAN TENSORS

CHAPTER SEVEN: THE FINITE ELEMENT METHOD

CHAPTER EIGHT: DESIGN OF EXPERIMENTS

CHAPTER NINE: FUNCTIONAL ANALYSIS

Chapter One

Introduction to Systems

1.1 INTRODUCTION

The word 'system' is part of our everyday vocabulary. Yet its widespread use may mean that we suffer a handicap when trying to model and predict the behaviour of systems in a mathematical way. Can we reasonably expect a theory to cope with engineering systems, biological systems and economic systems? In this book we shall concentrate on engineering systems, but many of our results will have general application. For the moment we shall say that by a system we mean an assembly of interrelated objects in which a particular input gives rise to a particular output. This is an intuitive statement. Engineering systems may be on a very large scale: for example a chemical plant or a reservoir network and the understanding of system behaviour is then very important. To help that understanding a mathematical model is needed.

A natural requirement is to be able to control the behaviour of the system in some way so that, for example, the desired output is maintained, or very nearly maintained. Consequently much of the work on systems covered in this book is concerned with aspects of control, although the precise nature of the control need not always be specified in the early stages of discussion.

Sometimes a system may be **discrete** insofar as the number of states it may adopt is finite. For example traffic lights go through a cycle of four states: red and amber, green, amber, red. By **state** is meant the description of the system behaviour. In this example the system undergoes a **transition** from one state to the next, which is well-defined. In general, discrete systems are modelled by finite-difference equations whereas **continuous** systems are often modelled by differential equations. Since most systems are sampled-data systems, i.e. measurements are taken at specified particular times, they are not perhaps truly continuous. An example might be the oscillations of a simple pendulum. Here the transition from one state to the next is continuous; we may define the state of the system by the position and velocity of the pendulum, since this knowledge together with the governing second-order differential equation is sufficient to determine the state at any later time.

We have so far mentioned two examples of **deterministic systems**. If an element of uncertainty is introduced then we deal with **probabilistic systems** in which we estimate unknown constants by probability models and in which we use the theory of **stochastic processes** to describe those time-varying quantities whose behaviour is not known. In the example of the traffic lights the state of the system may be altered by the volume of traffic on one or more roads leading to the junction and such volumes will not be predictable.

The concept of **control** of a system is well-known to most people: thermostats control central heating systems, switches control whether electric lights are on or off. There are several kinds of control and these are examined in Section 1.3.

The **stability** of a system is clearly important. If a system in steady-state is subjected to small disturbances it is vital that such disturbances are damped out. An unstable system may be made stable by introducing a control. We discuss stability in Chapter 3.

One important class of systems is that of **linear systems.** A system is said to be linear if, when an input x_1 produces an output y_1 and an input x_2 produces an output y_2, then the input $(\alpha x_1 + \beta x_2)$ produces the output $(\alpha y_1 + \beta y_2)$, where α and β are scalars. Linear systems can be analysed in detail. Since they possess many characteristics in common with non-linear systems (for example, steady-state, transients, stability) their quantitative analysis makes it possible to understand qualitatively similar but non-linear systems.

Chapter 4 examines briefly the theory of **optimal control.** It is assumed that the fundamental aspects of the performance of a system under control can be measured by a single number, called a **performance criterion.** The design of the system is then concerned with selecting that control which optimises the performance criterion.

1.2 SOME EXAMPLES OF SYSTEM MODELS

We now examine four examples of engineering systems to bring out basic features of systems theory and to illustrate the modelling process.

(i) Thermal system

Figure 1.1 is a schematic representation of a simple oven. The air outside the oven is assumed to be at a temperature T_0, the oven wall at a uniform temperature T_1 and the interior of the oven at a temperature T_2.

The internal temperature can be altered by an input current u from a heating coil.

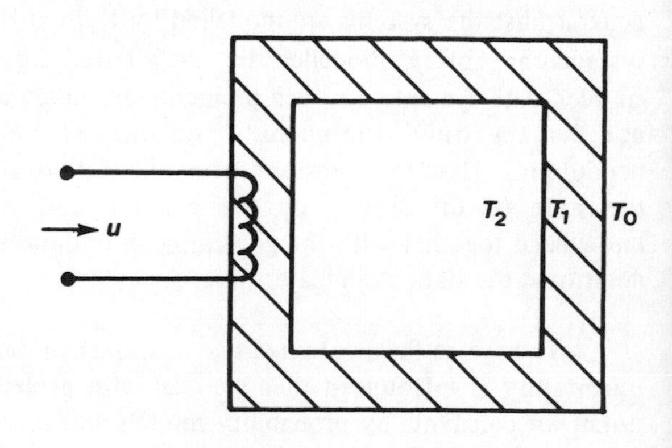

Figure 1.1

Let the surface areas of the outer and inner jackets be A_0, A_2 respectively and the coefficients of heat transfer of the surfaces be h_0 and h_2 respectively; the heat capacities of the oven space and the walls are C_2 and C_1 respectively. We **assume** that the temperatures are uniform throughout their respective regions and that temperature changes are distributed instantaneously. We further **assume** that the rate of **loss of heat** from a surface is proportional to its area and to the excess of that temperature over the temperature of the surroundings. (If your engineering discipline embraces the subject of heat transfer then you can justify these assumptions; if not, then you may accept them at face value.)

We now set up two **heat balance** equations.

(a) For the oven walls, consideration of the rate of heat transfer leads to

$$C_1 \dot{T}_1 = -A_0 h_0 (T_1 - T_0) + A_2 h_2 (T_2 - T_1) + u \tag{1.1}$$

(b) For the interior

$$C_2 \dot{T}_2 = A_2 h_2 (T_1 - T_2) \tag{1.2}$$

Outside the oven

$$\dot{T}_0 = 0 \tag{1.3}$$

A • over a symbol represents its time derivative.

(ii) Hydraulic System

Consider the system depicted in Figure 1.2. The purpose of the system is to maintain the head H of water in the second tank as nearly constant as possible despite the outflow Q varying. In this way a constant-pressure supply is maintained even though the demand varies.

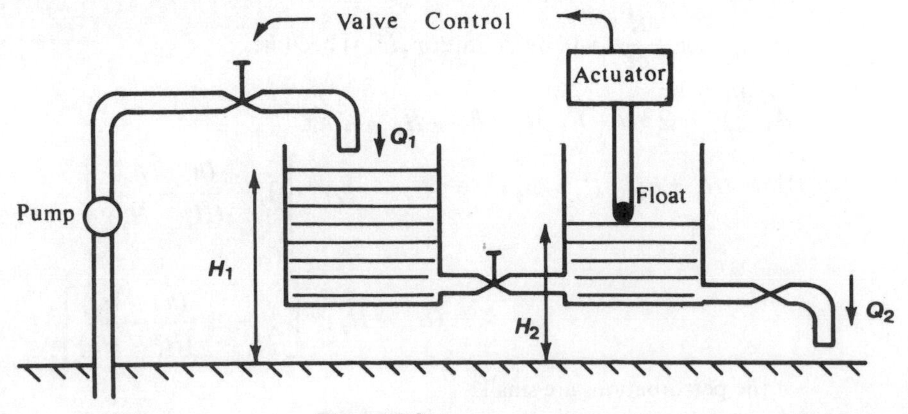

Figure 1.2

The second tank is supplied from a primary tank in which the head is $h_1(t)$. As the demand Q_2 increases, the level in the second tank falls. The float then communicates this information to an actuator which is a device that converts the information into action which causes more water to enter the second tank. In this example, the actuator operates a valve which pumps more water into the first tank and since this action provides a greater head in this tank, the flow rate into the second tank is increased. Precisely how the actuator operates is not considered.

The aim is to keep the head H_2 as close to some prescribed constant value $\bar{H}_2$ as possible. If at some instant, $Q_2 > Q_1$, then it is necessary to increase Q_1, hence the input is varied by an amount u, where u is a control variable (the volume of water per second) which will be positive if $H_2 < \bar{H}_2$ and negative if $H_2 > \bar{H}_2$. Assume that the inflow to the second tank is $K_2(H_1 - H_2)^{1/2}$ where K_2 is a constant for the first tank and that $Q_2 = K_3 H_2^{1/2}$ where K_3 is a constant for the second tank.
We now consider volume rates of change of water in each tank. Let the (constant) cross-sectional area of the first tank be A_1 and that of the second tank be A_2.

Using continuity of flow for the first tank we obtain

$$A_1 \frac{dH_1}{dt} = Q_1 - K_2(H_1 - H_2)^{1/2} \tag{1.4}$$

For the second tank

$$A_2 \frac{dH_2}{dt} = K_2(H_1 - H_2)^{1/2} - Q_2 \tag{1.5}$$

What is meant by 'steady- state' in this context? Consider the case where $Q_2 = Q$, a constant (and hence so is Q_1), $H_2 = \bar{H}_2$ and $H_1 = \bar{H}_1$, which is a constant. Then any other state can be regarded as a perturbation about this state. It follows that $u = 0$ in the steady state. Let $Q_1 = Q + u$, $Q_2 = Q + q_2$, $H_1 = \bar{H}_1 + h_1$, $H_2 = \bar{H}_2 + h_2$.

Then, since $\dfrac{d\bar{H}_1}{dt} = 0$ by definition, (1.4) becomes

$$A_1 \frac{dh_1}{dt} = Q + u - K_2(\bar{H}_1 + h_1 - \bar{H}_2 - h_2)^{1/2} \tag{1.6}$$

$$\text{Now } (\bar{H}_1 + h_1 - \bar{H}_2 - h_2)^{1/2} = (\bar{H}_1 - \bar{H}_2)^{1/2}\left[1 + \frac{(h_1 - h_2)}{(\bar{H}_1 - \bar{H}_2)}\right]^{1/2}$$

$$\simeq (\bar{H}_1 - \bar{H}_2)^{1/2}\left[1 + \tfrac{1}{2}\frac{(h_1 - h_2)}{(\bar{H}_1 - \bar{H}_2)}\right]$$

if the perturbations are small.

Furthermore, the 'steady-state' case of (1.4) is

$$0 = A_1 \frac{d\bar{H}_1}{dt} = Q - K_2(\bar{H}_1 - \bar{H}_2)^{1/2}$$

Hence (1.6) reduces to

$$A_1 \frac{dh_1}{dt} = u - \frac{K_2(h_1 - h_2)}{2(\bar{H}_1 - \bar{H}_2)^{1/2}} \tag{1.7}$$

Similarly the steady-state case of (1.5) is

$$0 = A_2 \frac{d\bar{H}_2}{dt} = K_2(\bar{H}_1 - \bar{H}_2)^{1/2} - Q$$

and (1.5) becomes

$$A_2 \frac{dh_2}{dt} = \frac{K_2(h_1 - h_2)}{2(\bar{H}_1 - \bar{H}_2)^{1/2}} - q_2 \tag{1.8}$$

Notice that equations (1.7) and (1.8) are linear and therefore we should now be able to analyse the system fairly easily. How responsive the system is when it is disturbed about its steady-state (or equilibrium state) is the problem of stability.

(iii) Electro-mechanical example

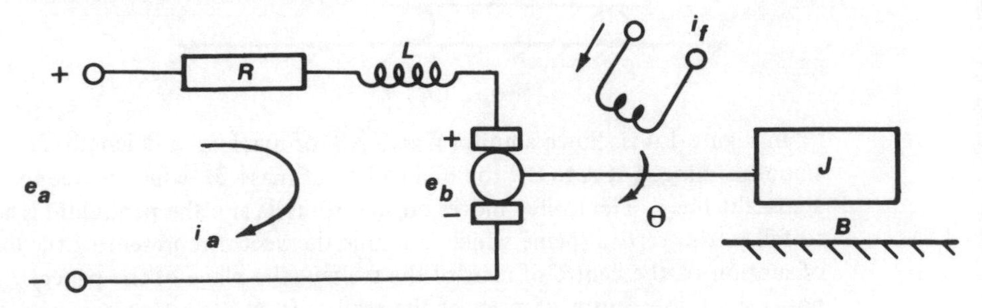

Figure 1.3

Figure 1.3 is a schematic representation of a system in which a d.c. motor rotates in a fixed field. The motor drives a load of inertia J subject to a viscous loading. Let e_a be the armature voltage, e_b the back e.m.f. developed, i_a the current in the armature of the motor and i_f the current in the fixed field. R and L are the resistance and inductance respectively of the armature, B a coefficient of viscous damping and θ the angle turned through by the load. We assume that (i) the torque output is $K_t\, i_a$ where K_t is a constant, (ii) the back e.m.f., $e_b = K_g\,\dot{\theta}$ where K_g is a constant. (Again, if you are able, justify these assumptions on physical grounds.) The equations governing the system are

(a) Newton's law of motion:

$$J\ddot{\theta} = -B\dot{\theta} + K_t\, i_a \tag{1.9}$$

(b) Kirchhoff's law:

$$L\frac{\mathrm{d}i_a}{\mathrm{d}t} + Ri_a + e_b = e_a \tag{1.10}$$

For purposes of control assume that i_f is fixed and it is required to select the e.m.f. supplied, e_a, to produce a desired angle θ (an example is the control of a gun turret). For simplification write $e_a = u$, $f_1 = -B/J$, $f_2 = K_t/J$, $f_3 = -K_g/L$, $f_4 = -R/L$, $C = 1/L$ so that (1.9) and (1.10) become, respectively,

$$\ddot{\theta} = f_1\dot{\theta} + f_2 i_a \tag{1.11}$$

$$\frac{\mathrm{d}i_a}{\mathrm{d}t} = f_3\dot{\theta} + f_4 i_a + Cu \tag{1.12}$$

(iv) **Inverted pendulum**

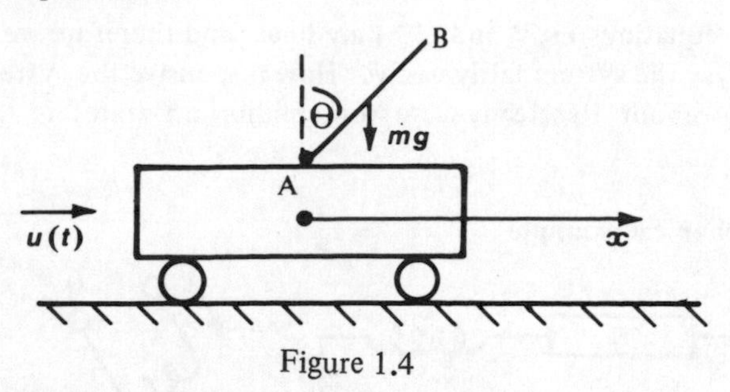

Figure 1.4

In Figure 1.4 is shown a uniform rod AB of mass m and length $2l$ which is smoothly hinged at A to the top of a trolley of mass M which is free to move in a straight line. The trolley moves on smooth rails and the pendulum is assumed to fall in the vertical plane which contains the vector representing the direction of motion of the centre of mass of the trolley (the plane of the paper). Let the position of the centre of mass of the trolley from a starting point be denoted $x(t)$. The angle with the vertical made by rod AB is $\theta(t)$.

At $\theta = 0$ the system is clearly in unstable equilibrium. The problem is to determine what force $u(t)$ to apply to the (centre of mass of the) trolley in order to maintain the rod in that vertical position. (A practical application is to control the position of a rocket on a thrust platform.)

Figure 1.5 shows the forces acting on the rod.

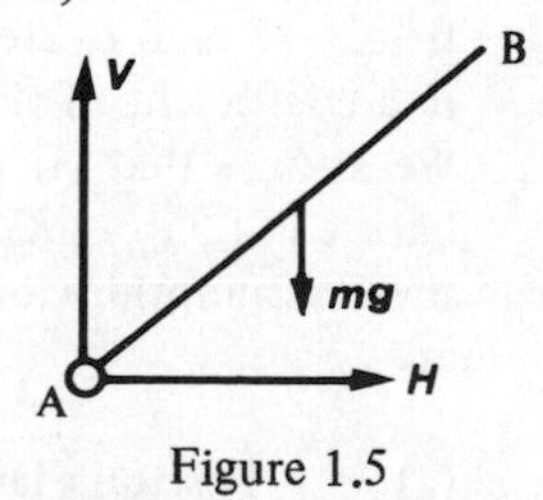

Figure 1.5

Resolving vertically,

$$V - mg = m \frac{d^2(l \cos \theta)}{dt^2} \tag{1.13}$$

Resolving horizontally,

$$H = m \frac{d^2(x + l \sin \theta)}{dt^2} \tag{1.14}$$

Taking moments about the centre of mass of the rod

$$I\ddot{\theta} = \frac{1}{3} ml^2 \ddot{\theta} = V\, l \sin \theta - H\, l \cos \theta \tag{1.15}$$

For the centre of mass of the trolley

$$u - H = M\ddot{x} \tag{1.16}$$

We have four unknowns – V, H, θ and x – and a solution is required for the last two.

Eliminate V and H to obtain

$$(m + M)\ddot{x} + ml(\ddot{\theta} \cos \theta - \dot{\theta}^2 \sin \theta) = u \tag{1.17}$$

$$\cos \theta\, \ddot{x} + \frac{4}{3} l\ddot{\theta} = g \sin \theta \tag{1.18}$$

We now **linearise** the equations by **assuming** θ to be **small** so that $\sin \theta \cong \theta$ and $\cos \theta \cong 1$. We neglect $\dot{\theta}^2 \theta$ where it occurs, since it is of second order of smallness. This can be justified physically. Then we obtain

$$(m + M)\ddot{x} + ml\ddot{\theta} = u \tag{1.19}$$

$$\ddot{x} + \frac{4}{3} l\ddot{\theta} = g\theta \tag{1.20}$$

Finally, we may eliminate $\ddot{x}$ and obtain

$$\frac{l}{3}(m + 4M)\ddot{\theta} - (m + M)g\theta = -u \tag{1.21}$$

Until the nature of the control force $u(t)$ is known we cannot proceed further. In the next section the concept of control is examined more closely.

Problems

1. A satellite revolves about its axis of symmetry. The angular position of the satellite at time t is given by $\phi(t)$ and the moment of inertia of the satellite is J about the symmetry axis. By using gas jets, a variable torque $\mu(t)$ can be exerted: this is the input variable to the system. Friction is assumed to be absent. Show that the differential equation of the system is

$$\ddot{\phi} = \frac{1}{J}\mu(t)$$

2. A tank is fed by two incoming flows with time-varying rates $F_1(t)$ and $F_2(t)$. Each feed contains dissolved material with constant concentrations c_1 and c_2 respectively. The tank is well stirred so that the concentration of the outgoing flow, at a rate $F(t)$, is equal to the concentration $c(t)$ in the tank. Let $V(t)$ be the volume of fluid in the tank. Let S be the constant cross-section area of the tank. Then

$$F(t) = k\left[\frac{V(t)}{S}\right]^{1/2}$$

where k is an experimentally determined constant.

Let the steady-state conditions be defined by F_{10}, F_{20}, F_0, V_0 and c_0. Assume that only small deviations from steady-state occur and that $F_1(t) = F_{10} + \mu_1(t)$, $F_2(t) = F_{20} + \mu_2(t)$, $V(t) = V_0 + x_1(t)$, $c(t) = c_0 + x_2(t)$.

First form the mass balance equations for the tank; then produce the 3 steady-state equations. Then derive equations for x_1 and x_2 in terms of the "input variables" μ_1, μ_2.

1.3 IDEAS ON CONTROL

A fundamental classification of control systems is that of whether the system is open-loop or closed-loop.

An **open-loop** system is one in which the control operates without taking into account the state of the system; for example, those traffic lights which change according to a pre-set pattern. See Figure 1.6(a).

If, however, there is some **feedback** of the states of the system to the control so that the control operates according to the state of the system then the system is referred to as **closed-loop**. See Figure 1.6(b). The hydraulic system of the last section is an example of a closed-loop system.

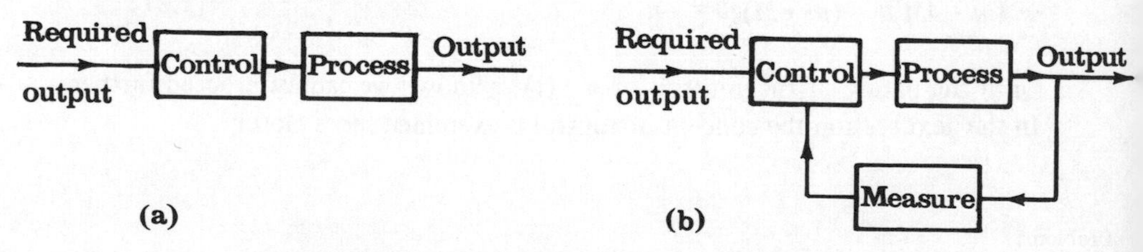

Figure 1.6

The aim of control is to produce a desired system response. The form of control chosen will depend on the nature of the system, the accuracy required for the output and the cost and ease of implementation of each control that is considered.

Sometimes the control may be **manual**: a human observer notes a measurement and adjusts some component of the system accordingly. This is obviously a closed-loop system and it is referred to often as a **sampled-data** system if the operator makes observations or measurements at specified (usually equally spaced) intervals of time. Often the control is **automatic** and the measurements are converted into a signal which alters the system. A particular class of automatic control systems is the **servomechanism** in which the output signal is maintained as closely as possible at a given reference input signal. If the reference signal is constant, the system is called a **regulator**. The hydraulic system is a servomechanism since we require to keep the head H_2 constant; indeed the system is a regulator.

Kids of control

It is not our intention to describe the physical components of these controls; see the Bibliography.

(i) On-off control (bang-bang control)

Here the only control is to turn a device on or off; for example a time-switch on a central heating boiler. To be really effective, the state of the system at which control is to be effected needs to be anticipated.

(ii) Proportional control

The error or deviation between the required output and the actual output is to be measured and an adjustment made proportional to this error. For the hydraulic system this means that the 'extra' input u is given by $u = K_1(\bar{H}_2 - H_2) = -K_1 h_2$ where K_1 is contant. Hence if the head H_2 is less than the required value H_2 an input over and above the steady-state value Q is provided. If H_2 exceeds $\bar{H}_2$ the input is reduced below the value Q . However, whereas it is relatively straightforward to design such a control device, error-proportion is not always a good control strategy. In this example, clearly time lags are involved for the liquid to adjust to the new desired level and for the desired level to be transmitted immediately to the control.

(iii) Rate or Derivative control

In order to provide a means of anticipating the changes required we can institute a control which produces an input proportional to the derivative of the output. To see the advantage of this form of control, suppose that there is a sudden rise in demand which causes the head H_2 to fall quickly. This information will be fed back to the control which increases the input: however, by the time this increased input is making its effect felt, it is possible that the demand has fallen below the steady-state level and the system has begun to overcompensate. It is not hard to imagine that if the demand oscillates about its mean value a situation could occur with the exact opposite of the required correction being applied. The oscillations in the head could grow (the phenomenon of hunting) and instability would then manifest itself. If we take account of the rate at which the head H_2 is changing, it should be feasible to reduce the possibility of oscillations building up.

(iv) Proportional plus Derivative control (PPD)

It is common practice to combine controls (i) and (ii).

(v) Integral control

In this form of control, the error signal is integrated or, in the case of sampled-data systems, it is summed up. A correction is made which is proportional to the integral, or sum. All hydraulic systems employ integral control, sometimes to an undesirable extent. As far as steady-state errors are concerned, integral control is superior to derivative control. Often, derivative controls are better for adjusting transients. Sometimes, both forms of control must be used.

Positive and negative feedback

If the measurement signal is subtracted from the input signal (before amplification) the strategy is called **negative feedback**. See Figure 1.7.

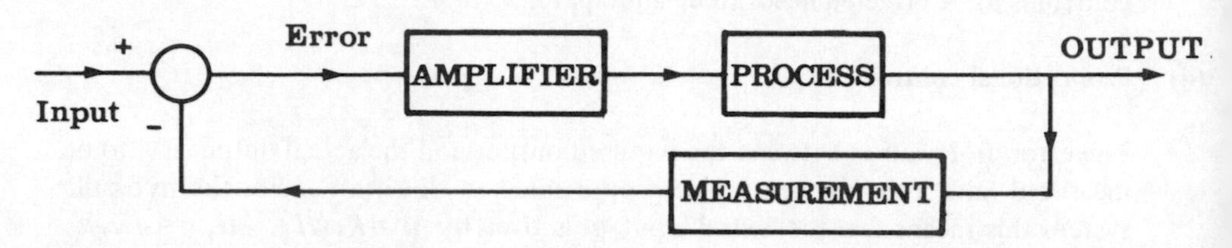

Figure 1.7

In the electro-mechanical example of the last section, it may be desirable to drive the motor at a given angular velocity. If we measure the output angular velocity and the error signal is fed back and the result amplified by changing the armature voltage e_a (proportional to torque) we can make the motor slow down or speed up as required. If, instead of **subtracting** the output from its required value we **add** it to this value then we have **positive feedback**. This will lead to instability, since the error is being amplified.

External disturbances to the system have not yet been taken into account and it should be clear that a positive feedback would merely cause the perturbations to grow, whereas a negative feedback would possibly damp them out.

Examples of control applied to systems of Section 1.2

(i) Hydraulic example

Suppose we adopt an error proportional control: $u = -K_1 h_2$. Then (1.7) and (1.8) become, on substituting $q_2 = K_3 h_2/(2\bar{H}_2^{\,1/2})$ (See Problem on page 13.)

$$\frac{dh_1}{dt} = -\frac{K_2}{2(\bar{H}_1 - \bar{H}_2)^{1/2}A_1}h_1 + \left(-\frac{K_1}{A_1} + \frac{K_2}{2A_1(\bar{H}_1 - \bar{H}_2)^{1/2}}\right)h_2 \tag{1.22}$$

$$\frac{dh_2}{dt} = \frac{K_2}{2A_2(\bar{H}_1 - \bar{H}_2)^{1/2}}h_1 - \left(\frac{K_3}{2A_2\bar{H}_2^{\,1/2}} - \frac{K_2}{2A_2(\bar{H}_1 - \bar{H}_2)^{1/2}}\right)h_2 \tag{1.23}$$

With suitable definition of constants we write these equations as

$$\frac{dh_1}{dt} = \alpha_1 h_1 + \alpha_2 h_2 \tag{1.24}$$

$$\frac{dh_2}{dt} = \alpha_3 h_1 + \alpha_4 h_2 \tag{1.25}$$

If we write $h_1 = a_1 e^{\lambda t}$, $h_2 = a_2 e^{\lambda t}$ we obtain, on cancelling $e^{\lambda t}$ throughout and rearranging,

$$a_1(\alpha_1 - \lambda) + a_2\alpha_2 = 0 \tag{1.26}$$

$$a_1\alpha_3 + a_2(\alpha_4 - \lambda) = 0 \tag{1.27}$$

From work on determinants we know that except for $a_1 = a_2 = 0$ solutions of (1.26) and (1.27) exist if

$$(\alpha_1 - \lambda)(\alpha_4 - \lambda) - \alpha_2\alpha_3 = 0 \tag{1.28}$$

For stability the real part of λ must be negative (why?). Work out for yourselves the requirements on the coefficients α_i and thence the requirements on the original constants $\bar{H}_1$, $\bar{H}_2$, K_1, K_2, A_1, A_2.

You are invited to consider the effect of PPD control on this system.

(ii) Inverted pendulum

Suppose the pendulum is to be controlled by $u = K_1\theta$ where K_1 is constant; this neglects time lags in action. If it is further assumed that the initial values are $x(0) = \dot{x}(0) = \dot{\theta}(0) = 0$ and $\theta(0) = \theta_0$ it can be shown that

$$\theta = \theta_0 \cos \omega t \tag{1.29}$$

where $\omega^2 = 3[K_1 - (m + M)g]/[(m + 4M)l]$.

Note that ω^2 may be negative. You should derive equation (1.29) for yourselves and then investigate the different possibilities for ω and their consequences.

In particular, you should show that a value of K_1 too small results in instability and then you can investigate the resulting forms of $x(t)$. K_1 is often called the **gain** of the control.

Suppose, alternatively, a derivative control $u = K_2 \dot{\theta}$ is imposed. You should be able to derive the equation

$$\ddot{\theta} + 2a\dot{\theta} + \omega\theta = 0 \tag{1.30}$$

under the same initial conditions as before and with $a = 1.5K_1/[(m + 4M)l]$. Again we invite you to find the solution for θ and see that oscillations are damped out. In each case draw suitable graphs of $\theta(t)$.

Controllability and Observability

Before we close this section we look very briefly at two fundamental concepts in control theory.

A system is said to be **controllable** if it is possible to effect a desired change of state by imposing a suitable control. This does not say *how* to choose the control, merely that one exists. In Figure 1.8 is depicted an uncontrollable system.

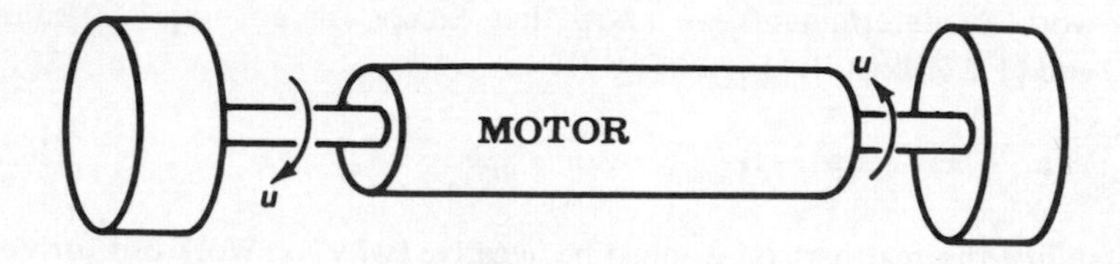

Figure 1.8

Two rotating flywheels are given the same torque u but they are not otherwise connected. It is not possible to position these two wheels independently from the one torque.

In many practical problems it is not possible to measure the state directly, therefore we need to consider the concept of observability. If in the example of Figure 1.8 we replace the motor by a device which measures the difference between the angular speeds of the wheels (which may now rotate independently) then it is not possbile to evaluate the independent positions of the wheels from the measurement. The system is said to be **unobservable.** If the state can be determined from the measurement then the system is **observable.**

Note that both controllability and observability have special interpretations if the equations are linear. See Chapter 2. Further note that these concepts are properties of the system and not the variables used to describe it.

Problem

In the hydraulic example show that $q_2 = K_3 h_2/(2\bar{H}_2^{1/2})$.

1.4 STATE SPACE

In Section 1.1 a somewhat vague reference was made to the **state** of a system. In this section we attempt to present a clearer idea of this concept. Intuitively, we may say that the state of a system is the minimum amount of information necessary to describe the condition of the system at some time t_0 such that knowledge of the inputs for $t > t_0$ and of the dynamical equations allows determination of the condition of the system for $t > t_0$. In the case of the oven example of Section 1.2 we may define the **state variables** of the system as $x_1(t) = T_1 - T_0$, $x_2(t) = T_2 - T_0$ since we are probably more concerned with temperature differences than with actual temperatures. Initially, $x_1(0)$, $x_2(0)$ are known and equations (1.1) and (1.2) allow a determination of the values $x_1(t)$, $x_2(t)$ for any $t > 0$.

In the case of the electro-mechanical example, suitable state variables are $x_1 = \theta$, $x_2 = \dot{\theta}$, $x_3 = i_a$. The equations (1.11) and (1.12) permit a determination of x_1, x_2 and x_3 for any $t > 0$.

The state variables are often written as components of a **state vector** $x(t)$. In the last example,

$$\mathbf{x} = \begin{bmatrix} x_1(t) \\ x_2(t) \\ x_3(t) \end{bmatrix} = \begin{bmatrix} x_1 \\ x_2 \\ x_3 \end{bmatrix}$$

The totality of state vectors constitutes **state space**: each state vector corresponds to a point in this space.

The equations for the oven may now be written in terms of the chosen state variables. Since

$$\dot{T}_0 = 0$$

it follows that

$$\frac{\mathrm{d}}{\mathrm{d}t}(T_1 - T_0) = \dot{T}_1 \qquad \text{and} \qquad \frac{\mathrm{d}}{\mathrm{d}t}(T_2 - T_0) = \dot{T}_2$$

Hence (1.1) becomes

$$C_1 \dot{x}_1 = -A_0 h_0 x_1 + A_2 h_2 (x_2 - x_1) + u \tag{1.31}$$

Similarly, from (1.2) we obtain

$$C_2\dot{x}_2 = A_2h_2(x_1 - x_2) \tag{1.32}$$

It is convenient to express these equations in matrix form as

$$\dot{\mathbf{x}} = \mathbf{A}\mathbf{x} + \mathbf{B}\mathbf{u} \tag{1.33}$$

where

$$\dot{\mathbf{x}} \equiv \begin{bmatrix} \dot{x}_1 \\ \dot{x}_2 \end{bmatrix}, \quad \mathbf{A} = \begin{bmatrix} -(A_0h_0 + A_2h_2)/C_1 & A_2h_2/C_1 \\ A_2h_2/C_2 & -A_2h_2/C_2 \end{bmatrix}$$

$$\mathbf{B} = \begin{bmatrix} 1/C_1 & 0 \\ 0 & 0 \end{bmatrix}, \quad \mathbf{u} = \begin{bmatrix} u \\ 0 \end{bmatrix}$$

A is sometimes referred to as the **system matrix** and **B** as the **distribution matrix** or **input matrix.**

Example 1

Figure 1.9 depicts a system which consists of two flywheels connected by shafts. The input control is the angle of the input shaft.

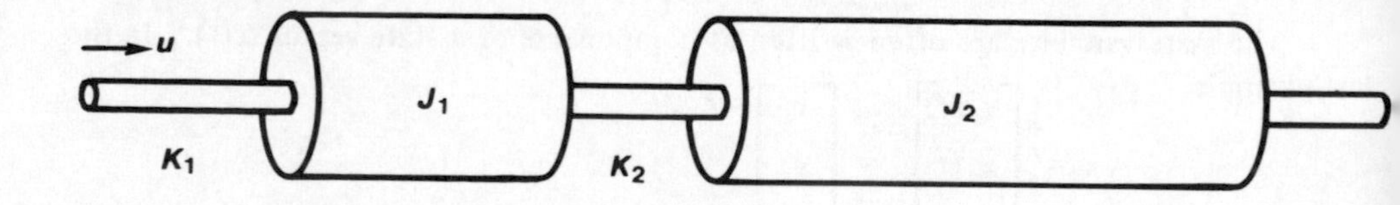

Figure 1.9

The moments of inertia of the wheels are J_1, J_2 kg m² respectively and the stiffness of the input and connecting shafts are K_1, K_2, N m rad^{-1}.

State variables are chosen as follows:

x_1 = angular position of the first wheel, $\quad x_2 = \dot{x}_1$

x_3 = angular position of the second wheel, $\quad x_4 = \dot{x}_3$

The **state equations** which we ask you to verify are

$$\dot{x}_1 = x_2$$

$$\dot{x}_2 = \frac{K_2}{J_1}x_3 - \frac{(K_1 + K_2)}{J_1}x_1 + \frac{K_2}{J_1}u$$

$$\dot{x}_3 = x_4$$

$$\dot{x}_4 = \frac{-K_2}{J_2} x_3 + \frac{K_2}{J_2} x_1$$

or in the form (1.33)

$$\dot{\mathbf{x}} = \begin{bmatrix} 0 & 1 & 0 & 0 \\ \frac{-(K_1 + K_2)}{J_1} & 0 & \frac{K_2}{J_1} & 0 \\ 0 & 0 & 0 & 1 \\ \frac{K_2}{J_2} & 0 & \frac{-K_2}{J_2} & 0 \end{bmatrix} \mathbf{x} + \begin{bmatrix} 0 \\ \frac{K_2}{J_1} \\ 0 \\ 0 \end{bmatrix} \mathbf{u} \tag{1.34}$$

Note that where there is a control component in only one equation, we replace $\mathbf{u}$ by u and $\mathbf{B}$ becomes a column vector.

Example 2 (Electro-mechanical example)

The chosen state variables are $x_1 = \theta$, $x_2 = \dot{\theta}$, $x_3 = i_a$. Then (1.11) and (1.12) become

$$\dot{x}_1 = x_2$$

$$\dot{x}_2 = f_1 x_2 + f_2 x_3$$

$$\dot{x}_3 = f_3 x_2 + f_4 x_3 + Cu$$

or

$$\dot{\mathbf{x}} = \begin{bmatrix} 0 & 1 & 0 \\ 0 & f_1 & f_2 \\ 0 & f_3 & f_4 \end{bmatrix} \mathbf{x} + \begin{bmatrix} 0 \\ 0 \\ C \end{bmatrix} u \tag{1.35}$$

The measurements to be made are i_a the armature current and θ the shaft position.

Note that the choice of state variables is *not* unique. For example, equation (1.9) can be written with i_a as its subject and substitution into (1.10) leads to an equation of the form

$$\dddot{\theta} = a\ddot{\theta} + b\dot{\theta} + de_a \tag{1.36}$$

for suitable constants a, b and d. Check this.

This time we choose state variables $x_1 = \theta$, $x_2 = \dot{\theta}$, $x_3 = \ddot{\theta}$ so that the equations of state may be written as

$$\dot{\mathbf{x}} = \begin{bmatrix} 0 & 1 & 0 \\ 0 & 0 & 1 \\ 0 & b & a \end{bmatrix} \mathbf{x} + \begin{bmatrix} 0 \\ 0 \\ d \end{bmatrix} u \tag{1.37}$$

This time the only measurement to be made is that of θ.

Compare these two choices of state variables. Is there any preference for one choice over the other?

Why state space?

We should first mention that the examples so far considered have been **linear** systems. The concept of state space is not confined to linear systems, but the matrix formulation (1.33) is. Note further that we can study uncontrolled systems by putting $u = 0$.

What then are the advantages of methods based on state variables, the so-called **state space methods**? First, note that the state-space formulations have given rise to a set of first-order differential equations, which is often easier to handle than a single high-order equation. Suppose a system is governed by the equation

$$a_n \frac{\mathrm{d}^n x}{\mathrm{d}t^n} + ... + a_1 \dot{x} + a_0 x = u \tag{1.38}$$

then, if the chosen state variables $x_r = \dfrac{\mathrm{d}^r x}{\mathrm{d}t^r}$, it follows that

$$\dot{x}_1 = x_2, \; \dot{x}_2 = x_3, \; ....., \dot{x}_{n-1} = x_n \tag{1.39}$$

and the set of state equations is completed by

$$\dot{x}_n = -\frac{a_{n-1}}{a_n} x_n - ..... - \frac{a_0}{a_n} x_1 + \frac{1}{a_n} u \tag{1.40}$$

In the case of a discrete system, the governing equation will be a **difference equation.** Consider the governing equation

$$a_n x(i) + ... + a_1 x(i - n + 1) + a_0 x(i - n) = u(i) \tag{1.41}$$

where the variable $x(t)$ is sampled at $t = 0, 1, 2, ...$

Then we choose state variables so that

$$x_1(i + 1) = x_2(i), \; x_2(i + 1) = x_3(i), \; .... \; x_{n-1}(i + 1) = x_n(i) \tag{1.42}$$

and the set is completed by

$$x_n(i+1) = \frac{a_{n-1}}{a_n} x_n(i) + \ldots + \frac{a_0}{a_n} x_1(i) + \frac{1}{a_n} u(i) \tag{1.43}$$

The formulation (1.33) becomes

$$\mathbf{x}(i+1) = \mathbf{A}\mathbf{x}(i) + \mathbf{B}\mathbf{u}(i) \tag{1.44}$$

where $\mathbf{x}$ has n components, $\mathbf{u}$ has m components, $\mathbf{A}$ and $\mathbf{B}$ are $n \times n$ and $n \times m$ matrices respectively.

A second important feature is that multi-input multi-output systems can be handled by state space methods; this is not usually possible by a more classical, transform-orientated approach.

Thirdly, linear systems whose parameters are time-varying can be handled in essentially the same way as time-invariant linear systems.

Fourth, with linear systems, a matrix approach is useful for computing.

Problems

1. Write the equations for the inverted pendulum of Section 1.2 in state-space form. Choose as state variables $x_1 = \theta$, $x_2 = \dot{\theta}$, $x_3 = x$, $x_4 = \dot{x}$.

2. Repeat Problem 1 for the hydraulic example. Choose as state variables $x_1 = h_1$ and $x_2 = h_2$.

3. Refer to Problem 1 of Section 1.2. Show that the choice of state variables $x_1 = \phi(t)$, $x_2 = \dot{\phi}(t)$ together with an output variable $y(t) = \phi(t)$ leads to the state equations

$$\begin{bmatrix} x_1 \\ x_2 \end{bmatrix} = \begin{bmatrix} 0 & 1 \\ 0 & 0 \end{bmatrix} \begin{bmatrix} x_1 \\ x_2 \end{bmatrix} + \begin{bmatrix} 0 \\ \frac{1}{J} \end{bmatrix} \mu(t), \qquad y(t) = (1 \;,\; 0) \begin{bmatrix} x_1 \\ x_2 \end{bmatrix}$$

4. Defining x_1 and x_2 as state variables in Problem 2 of Section 1.2, and denoting the ratio V/F by θ obtain the linearised state equations

$$\begin{bmatrix} \dot{x}_1 \\ \dot{x}_2 \end{bmatrix} = \begin{bmatrix} -\frac{1}{2\theta} & 0 \\ 0 & -\frac{1}{\theta} \end{bmatrix} \begin{bmatrix} x_1 \\ x_2 \end{bmatrix} + \begin{bmatrix} 1 & 1 \\ \frac{c_1 - c_0}{V_0} & \frac{c_2 - c_0}{V_0} \end{bmatrix} \begin{bmatrix} u_1 \\ u_2 \end{bmatrix}$$

Let $y_1(t) = F(t) - F_0$ and $y_2(t) = c(t) - c_0$ and use the appropriate approximations to obtain the linearised output equation

$$\mathbf{y}(t) = \begin{bmatrix} \frac{1}{2\theta} & 0 \\ 0 & 1 \end{bmatrix} \mathbf{x}(t)$$

5. The flight of a rocket in two dimensions ($x - z$ plane) is governed by the equations

$$\ddot{x} = F\cos\theta, \qquad \ddot{z} = F\sin\theta - g$$

where $u_1 = F$ is the thrust/unit mass and $u_2 = \theta$ is the thrust angle. Let F and θ be regarded as input control variables. Choose as state variables $x_1 = x$, $x_2 = \dot{x}$, $x_3 = z$, $x_4 = \dot{z}$ and show that the system can be described by

$$\dot{\mathbf{x}} = \mathbf{A}\mathbf{x} + \mathbf{B}\mathbf{u}, \qquad \mathbf{y} = \mathbf{C}\mathbf{x}$$

where

$$\mathbf{A} = \begin{bmatrix} 0 & 1 & 0 & 0 \\ 0 & 0 & 0 & 0 \\ 0 & 0 & 0 & 1 \\ 0 & 0 & 0 & 0 \end{bmatrix}, \quad \mathbf{B} = \begin{bmatrix} 0 & 0 & 0 & 0 \\ 0 & 1 & 0 & 0 \\ 0 & 0 & 0 & 0 \\ 0 & 0 & 0 & 1 \end{bmatrix},$$

$$\mathbf{u} = \begin{bmatrix} 0 \\ u_1 \cos u_2 \\ 0 \\ u_1 \sin u_2 - g \end{bmatrix}, \quad \mathbf{C} = (0 \quad 0 \quad 1 \quad 0)$$

6. A system governed by the differential equation

$$\dddot{x} + 16\ddot{x} + 18\dot{x} + 4x = 10u$$

has transfer function

$$\frac{10}{s^3 + 16s^2 + 18s + 4}$$

Define the state variables $x_1 = x$, $x_2 = \dot{x}$, $x_3 = \ddot{x}$ and show that the system state equations can be written in the terminology of Problem 5 if

$$\mathbf{A} = \begin{bmatrix} 0 & 1 & 0 \\ 0 & 0 & 1 \\ -4 & -18 & -16 \end{bmatrix}, \quad \mathbf{B} = \begin{bmatrix} 0 & 0 & 0 \\ 0 & 0 & 0 \\ 0 & 0 & 10 \end{bmatrix}, \quad \mathbf{u} = \begin{bmatrix} 0 \\ 0 \\ u \end{bmatrix}, \quad \mathbf{C} = (1, 0, 0)$$

7. The linearised equations for a satellite are

$$\begin{aligned} I\ddot{\theta}_1 + \omega I\dot{\theta}_3 &= T_1 \\ I\ddot{\theta}_2 &= T_2 \\ I\ddot{\theta}_3 - \omega I\dot{\theta}_1 &= T_3 \end{aligned}$$

where θ_1, θ_2, θ_3 are angular displacements related to a fixed set of axes and T_1, T_2, T_3 are applied torques, I is the moment of inertia and ω is the angular frequency of the axes.

Show that with state variables $x_1 = \theta_1$, $x_2 = \dot{\theta}_1$ etc., the state equations are such that

$$\mathbf{A} = \begin{bmatrix} 0 & 1 & 0 & 0 & 0 & 0 \\ 0 & 0 & 0 & 0 & 0 & -\omega \\ 0 & 0 & 0 & 1 & 0 & 0 \\ 0 & 0 & 0 & 0 & 0 & 0 \\ 0 & 0 & 0 & 0 & 0 & 1 \\ 0 & \omega & 0 & 0 & 0 & 0 \end{bmatrix}, \quad \mathbf{B} = \begin{bmatrix} 0 & 0 & 0 & 0 & 0 & 0 \\ 0 & I^{-1} & 0 & 0 & 0 & 0 \\ 0 & 0 & 0 & 0 & 0 & 0 \\ 0 & 0 & 0 & I^{-1} & 0 & 0 \\ 0 & 0 & 0 & 0 & 0 & 0 \\ 0 & 0 & 0 & 0 & 0 & I^{-1} \end{bmatrix}, \quad \mathbf{u} = \begin{bmatrix} 0 \\ T_1 \\ 0 \\ T_2 \\ 0 \\ T_3 \end{bmatrix}.$$

8. The motion of the Apollo XI LEM as it landed on the lunar surface is governed by the equation

$$\ddot{x} = -\frac{k\dot{m}}{m} - g$$

where x is the altitude above the lunar surface, m is the total LEM mass, $\dot{m} \leqslant 0$, and $k > 0$ is the constant velocity of exhaust gases.

Choose suitable state variables and derive the state equations.

9. A system has two inputs and two outputs and is described by the equations

$$\ddot{y}_1 + 3\dot{y}_1 + 2y_1 = u_1 + 2\dddot{u}_2 + 2\dot{u}_2, \qquad \ddot{y}_2 + 4\dot{y}_2 + 3y_2 = \ddot{u}_2 + 3\dot{u}_2 + u_1$$

Choose appropriate state variables and produce the state equations.

10. Find a state variable representation for the system

$$\dddot{y} + e^{-t}\dot{y} + e^{-2t}y = u$$

11. Consider the circuit depicted below: the voltage across R_3 is the output.

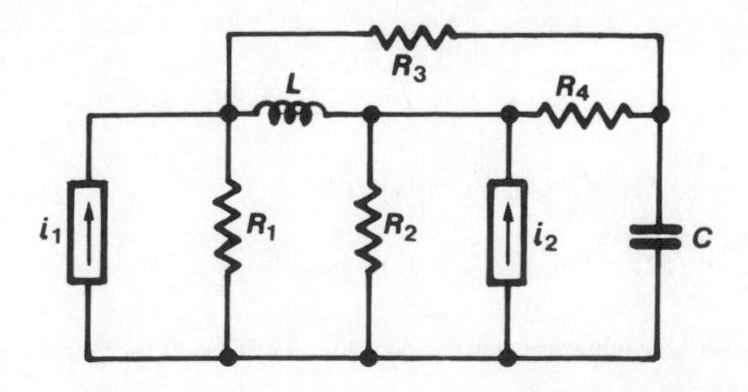

Choose as state variables x_1 = the capacitor voltage and x_2 = the inductor current, $u_1 = i_1$ and $u_2 = i_2$. Derive the state equations

$$\dot{\mathbf{x}} = \begin{bmatrix} \dfrac{-(R_1 + R_2 + R_3 + R_4)}{C(R_1 + R_3)(R_2 + R_4)} & \dfrac{R_2 R_3 - R_1 R_4}{C(R_2 + R_4)(R_1 + R_3)} \\ \dfrac{R_1 R_4 - R_2 R_3}{L(R_1 + R_3)(R_2 + R_4)} & \dfrac{-1}{L}\left(\dfrac{R_1 R_3}{R_1 + R_3} + \dfrac{R_2 R_4}{R_2 + R_4}\right) \end{bmatrix} \mathbf{x}$$

$$+ \begin{bmatrix} \dfrac{R_1}{C(R_1 + R_3)} & \dfrac{R_2}{C(R_2 + R_4)} \\ \dfrac{R_1 R_3}{L(R_1 + R_3)} & \dfrac{-R_2 R_4}{L(R_2 + R_4)} \end{bmatrix} \mathbf{u}$$

$$\mathbf{y} = \left[\dfrac{-R_3}{R_1 + R_3}, \dfrac{R_1 R_3}{R_1 + R_3}\right] \mathbf{x} + \left[\dfrac{R_1 R_3}{R_1 + R_3}, 0\right] \mathbf{u}$$

12. A rocket moving vertically above the Earth is governed by the equation

$$m(t)\ddot{h} = T(t) - f(\dot{h}) - \frac{m(t)k^2 g_0}{(k+h)^2}$$

where $h(t)$ is the height above the Earth's surface, $m(t)$ is the instantaneous mass of the rocket, $f(\dot{h})$ is the drag force, $T(t) = C\dot{m}$ is the thrust force (C is constant), k is the radius of the Earth and g_0 is the acceleration due to gravity at the Earth's surface. Take $\dot{x}_1 = h$, $x_2 = \dot{h}$ and $x_3 = m = u$. Find the (non-linear) state equations.

13. A satellite rotates under three mutually orthogonal gas jets. The components of angular velocity in the body-fixed Cartesian axes are $(\omega_1, \omega_2, \omega_3)$ and the input control torques are (T_1, T_2, T_3). The moments of inertia about the axes are (J_1, J_2, J_3). Derive the equations of motion

$$T_1 = J_1\dot{\omega}_1 + (J_3 - J_2)\omega_2\omega_3$$

$$T_2 = J_2\dot{\omega}_2 + (J_1 - J_3)\omega_1\omega_3$$

$$T_3 = J_3\dot{\omega}_3 + (J_2 - J_1)\omega_1\omega_2$$

and show that the state variable equations where $x_i = \omega_i$ are linear in the control variables, but not in the state variables.

14. Show that the equation

$$x(k+2) + 3x(k+1) + 2x(k) = 5u(k+1) + 3u(k)$$

can be written as

$$\begin{bmatrix} x_1(k+1) \\ x_2(k+1) \end{bmatrix} = \begin{bmatrix} 0 & 1 \\ -2 & -3 \end{bmatrix} \begin{bmatrix} x_1(k) \\ x_2(k) \end{bmatrix} + \begin{bmatrix} 5 \\ -12 \end{bmatrix} u(k)$$

1.5 REVIEW EXAMPLE

Background

The personal rapid-transit (PRT) system under automatical control is argued to be a viable means of easing urban congestion. The number of passengers per vehicle is small and the vehicles are envisaged as small, electrically propelled and operating with short headways† on a network of single-line guideways. What is required is a system which will keep a reasonable spacing between vehicles without causing the passengers discomfort. Obviously the system must be economical and capable of allowing easy manoeuvering at interchanges.

Moving-cell control

The philosophy to be employed is that of the moving-cell. The idea is that hypothetical 'cells' of equal length move at fixed time intervals along the guideways. The length of each cell should be as small as possible to allow high capacity on the guideways but should be sufficiently great so that when each vehicle is at the centre of a cell there is a safe separation. The length of a cell will depend on guideway velocity, on the value of emergency deceleration and on vehicle length. The idea, then, is that a vehicle is controlled to move within its assigned cell which in turn progresses along the guideway.

The model

The aim is a system which will keep headway and velocity errors small and yet not cause passenger discomfort.

The guideway is assumed to be straight and horizontal.

Let a vehicle move with a velocity $v(t)$ along the guideway so that its displacement from a fixed point is $s(t)$. Then

$$\dot{s}(t) = v(t) \tag{1.45}$$

The acceleration of the vehicle is

$$a(t) = \dot{v}(t) = F(t) - F_D(t) \tag{1.46}$$

† The headway is the distance between consecutive vehicles.

where $F(t)$ is the propulsive force/unit mass applied and $F_D(t)$ is the drag force/unit mass.

The propulsive force is assumed to be given by the equation

$$\dot{F}(t) = -\frac{1}{\tau} F(t) + G(t) \tag{1.47}$$

where τ is the time constant for the propulsion system and $G(t)$ is the input to the propulsion system. Solve (1.47) and see why we call the model of $F(t)$ a first-order lag. Decide what this means in physical terms. The drag force depends on the velocities of the vehicle $v(t)$ and of the wind, $v_w(t)$ and can be approximated by

$$F_D = c_D(v + v_w)^2 = c_0 + c_1 v(t) + c_2 [v(t)]^2 \tag{1.48}$$

the constants c_0, c_1 and c_2 will be determined for a particular vehicle.

The rate of change of acceleration or jerk is given by

$$\dot{F}(t) - c_1 a(t) - 2c_2 v(t) \,.\, a(t) = \dot{F}(t) - a(t)[c_1 + 2c_2 v(t)] \tag{1.49}$$

It has been shown from simulation studies that if the velocity of the vehicle is reasonably close to the desired guideway velocity V the term $[c_1 + 2c_2 v(t)]$ can be safely taken as constant and equal to $[c_1 + 2c_2 V]$. Let the vehicle be assigned to a cell (previously empty) whose centre is displaced a distance $S(t)$ from the fixed point so that

$$\dot{S}(t) = V \tag{1.50}$$

Choice of state variables

It should be reasonably clear that two state variables should be headway errors $x_1(t) = s(t) - S(t)$ and velocity error $x_2(t) = v(t) - V$. We shall choose a control u so that the state variables are sent to zero values. It should be clear that we want x_1 and x_2 to be zero. The third state variable is chosen as the acceleration error,i.e. $x_3(t) = a(t)$ since $\dot{V} = 0$. The acceleration must be zero if the position and velocity errors are to stay zero. Further, if the acceleration error is to be zero we must require that propulsive force is equal to any disturbing forces. If we assume that the disturbing forces are constant then it follows that the propulsive force must tend towards a constant value; hence the derivative of the propulsive force must approach zero. We choose the fourth state variable as this derivative, viz.

$$x_4(t) = \dot{F}(t) \tag{1.51}$$

The control variable is selected as $u(t) = \dot{G}(t)$, the derivative of the propulsion input.

State equations

It should be obvious that

$$\dot{x}_1(t) = x_2(t) \tag{1.52}$$

$$\dot{x}_2(t) = x_3(t) \tag{1.53}$$

Now from (1.49) and (1.51)

$$\dot{x}_3(t) = \dot{a}(t) = x_4(t) - x_3(t)[c_1 + 2c_2 V] \tag{1.54}$$

Also, differentiating (1.47) we obtain

$$\dot{x}_4(t) = \ddot{F}(t) = -\frac{1}{T}\dot{F}(t) + \dot{G}(t)$$

$$\text{i.e.}\quad \dot{x}_4(t) = -\frac{1}{T}x_4(t) + u(t) \tag{1.55}$$

In matrix form the state equations are

$$\dot{\mathbf{x}} = \begin{bmatrix} 0 & 1 & 0 & 0 \\ 0 & 0 & 1 & 0 \\ 0 & 0 & -d & 1 \\ 0 & 0 & 0 & -\frac{1}{T} \end{bmatrix} \mathbf{x} + \begin{bmatrix} 0 \\ 0 \\ 0 \\ 1 \end{bmatrix} u \tag{1.56}$$

where $d = c_1 + 2c_2 V$

Modifications

In practice a sampled-data control would probably be employed and the discrete analogue of (1.56) would be in the form

$$\mathbf{x}[(k+1)T] = \phi(T)\mathbf{x}(kT) + D(T)u(kT) \tag{1.57}$$

where $\mathbf{x}(kT)$ gives the state of the system at the kth sampling point of time, T is the sampling period, $\phi(T)$ and $D(T)$ are the constant transition and driving matrices. (See Chapter 2.)

The control problem would be one of selecting that sequence of inputs $u(kT)$ which minimises some criterion of performance. This is taken up again in Chapter 4.

In practice, the vehicle is likely to be subjected to stochastic disturbances such as wind gusts or even from the vehicle's propulsion system itself.

Further, it may be difficult and costly to measure x_3 and x_4. Also, any measured quantity will be subject to random errors. Clearly an extended model would need to be provided to cater for these difficulties.

Chapter Two

Linear Systems

2.1 INTRODUCTION

In this chapter we concentrate on linear systems, that is, those systems which obey the principle of superposition of solutions. We deduce the forms of solution and develop methods of performing the necessary calculations. These techniques have been put deliberately on a formal footing to enable you to acquire further facility with matrix algebra. The concepts of controllability and observability are considered in more detail. Finally, we take a brief look at discrete systems.

2.2 LAPLACE TRANSFORM SOLUTION OF STATE SPACE EQUATIONS

Consider again the thermal system described in Section 1.2. Suppose that the constants involved are so chosen that the state space equations $\dot{\mathbf{x}} = \mathbf{Ax} + \mathbf{Bu}$ become

$$\dot{\mathbf{x}} = \begin{bmatrix} -3 & 1 \\ 2 & -2 \end{bmatrix} \mathbf{x} + \begin{bmatrix} 1 & 0 \\ 0 & 0 \end{bmatrix} \mathbf{u}$$

or

$$\dot{\mathbf{x}} = \begin{bmatrix} -3 & 1 \\ 2 & -2 \end{bmatrix} \mathbf{x} + \begin{bmatrix} 1 \\ 0 \end{bmatrix} u \tag{2.1}$$

Further suppose that there is a thermocouple which measures the excess temperature of the oven walls over that of the outside air and that we switch the heating coil on at $t = 0$ with a constant input of 2 (in whatever units are appropriate). Initially, all temperatures are assumed to be T_0. We wish to find the temperature inside the oven for $t > 0$.

We attempt the solution via Laplace transforms. We assume you are familiar with this method so an outline only of the solution is provided.

The component equations of (2.1) are transformed into

$$\left.\begin{aligned} (s+3)X_1(s) - X_2(s) &= x_1(0) + \frac{2}{s} \\ -2X_1(s) + (s+2)X_2(s) &= x_2(0) \end{aligned}\right\} \tag{2.2}$$

where $X_1(s)$, $X_2(s)$ are the transforms of $x_1(t)$, $x_2(t)$ respectively and the initial values of the state variables are taken as $x_1(0)$, $x_2(0)$ for the moment, although this particular problem gives them zero values.

Solving (2.2) for the transforms and inverting the transforms produces

$$x_1(t) = \left(\frac{2}{3}e^{-4t} + \frac{1}{3}e^{-t}\right)x_1(0) + \left(-\frac{1}{3}e^{-4t} + \frac{1}{3}e^{-t}\right)x_2(0) + 1 - \frac{1}{3}e^{-4t} - \frac{2}{3}e^{-t}$$

$$x_2(t) = \left(-\frac{2}{3}e^{-4t} + \frac{2}{3}e^{-t}\right)x_1(0) + \left(\frac{1}{3}e^{-4t} + \frac{2}{3}e^{-t}\right)x_2(0) + 1 + \frac{1}{3}e^{-4t} - \frac{4}{3}e^{-t}$$

or,

$$\begin{bmatrix} x_1(t) \\ x_2(t) \end{bmatrix} = \begin{bmatrix} \frac{2}{3}e^{-4t} + \frac{1}{3}e^{-t} & -\frac{1}{3}e^{-4t} + \frac{1}{3}e^{-t} \\ -\frac{2}{3}e^{-4t} + \frac{2}{3}e^{-t} & \frac{1}{3}e^{-4t} + \frac{2}{3}e^{-t} \end{bmatrix} \begin{bmatrix} x_1(0) \\ x_2(0) \end{bmatrix}$$

$$+ \begin{bmatrix} 1 - \frac{1}{3}e^{-4t} - \frac{2}{3}e^{-t} \\ 1 + \frac{1}{3}e^{-4t} - \frac{4}{3}e^{-t} \end{bmatrix} \tag{2.3}$$

Note that the equations (2.3) are satisfied when $t = 0$ and also that as $t \to \infty$ the transient parts leave the solution independent of the initial conditions.

In the case we considered, $x_1(0) = 0 = x_2(0)$ as you can verify from the definition of the state variables. Since $x_2 = T_2 - T_0$ it follows that the required solution is

$$T_2 = T_0 + 1 + \frac{1}{3}e^{-4t} - \frac{4}{3}e^{-t} \tag{2.4}$$

This simple-minded approach has served to give us a solution which we may use as a basis for checking answers obtained by more general methods. To see how tedious the Laplace transform method may become, and therefore why we look for different techniques consider the following example.

A voltmeter

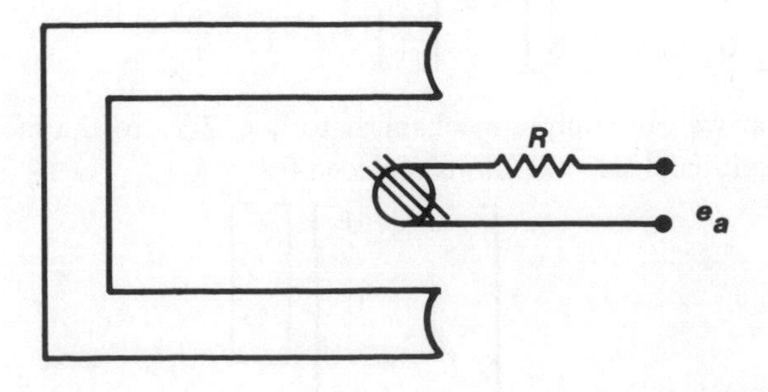

Figure 2.1

Figure 2.1 shows a meter which aims to measure an applied voltage e_a. Let i, L, R be the armature current, inductance and resistance, respectively; k_t is a torque constant; k_g is a constant relating angular velocity and the back e.m.f. and J is the moment of inertia of the rotating part of the meter.

In effect this is the system of Figure 1.3 except that the force $B\dot{\theta}$ due to viscous effects is replaced by a force $K_s\theta$ where K_s is a spring constant. The system equations are

$$\left.\begin{aligned} J\ddot{\theta} &= -K_s\,\theta + K_t\,i_a \\ L\,\frac{\mathrm{d}i_a}{\mathrm{d}t} + Ri_a + e_b &= e_a \\ e_b &= K_g\,\dot{\theta} \end{aligned}\right\} \tag{2.5}$$

The input is the voltage e_a and the output is the angle θ which can be measured directly.

We choose state variables $x_1 = \theta$, $x_2 = \dot{\theta}$, $x_3 = i_a$ and e_a is denoted by u. We leave you to show that the state equations are

$$\dot{\mathbf{x}} = \begin{bmatrix} 0 & 1 & 0 \\ -\dfrac{K_s}{J} & 0 & \dfrac{K_t}{J} \\ 0 & -\dfrac{K_g}{L} & -\dfrac{R}{L} \end{bmatrix} \mathbf{x} + \begin{bmatrix} 0 \\ 0 \\ \dfrac{1}{L} \end{bmatrix} u \tag{2.6}$$

with output given by

$$\mathbf{y} = (1, 0, 0)\mathbf{x} \tag{2.7}$$

For ease of working, simple values of the system parameters are selected. The state equations are

$$\dot{\mathbf{x}} = \begin{bmatrix} 0 & 1 & 0 \\ -1 & 0 & 2 \\ 0 & -5 & -6 \end{bmatrix} \mathbf{x} + \begin{bmatrix} 0 \\ 0 \\ 1 \end{bmatrix} u \tag{2.8}$$

Note that we can apply a mechanical torque $T(t)$ to the meter so that the second term on the right-hand side of (2.6) is replaced by

$$\begin{bmatrix} 0 & 0 \\ -\dfrac{1}{J} & 0 \\ 0 & \dfrac{1}{L} \end{bmatrix} \begin{bmatrix} T \\ e_a \end{bmatrix}$$

Check this out for yourself. This is an example of *multi-input*.

Applying Laplace transforms to (2.8) we have, with the obvious notation

$$\begin{bmatrix} s & -1 & 0 \\ 1 & s & -2 \\ 0 & 5 & s+6 \end{bmatrix} \begin{bmatrix} X_1(s) \\ X_2(s) \\ X_3(s) \end{bmatrix} = \begin{bmatrix} x_1(0) \\ x_2(0) \\ x_3(0) \end{bmatrix} + \begin{bmatrix} 0 \\ 0 \\ U(s) \end{bmatrix} \tag{2.9}$$

This can be cast in matrix form as

$$(s\mathbf{I} - \mathbf{A})\mathbf{X}(s) = \mathbf{x}(0) + \mathbf{U}(s)$$

For the solution we need to invert the matrix $(s\mathbf{I} - \mathbf{A})$. This can be done by Gauss-Jordan elimination to obtain the result

$$(s\mathbf{I} - \mathbf{A})^{-1} = \frac{1}{(s^3 + 6s^2 + 11s + 6)} \begin{bmatrix} (s^2 + 6s + 10) & (s+6) & 2 \\ -(s+6) & s(s+6) & 2s \\ 5 & -5s & (s^2+1) \end{bmatrix}$$

The natural response of the system is found by inverting the transformed equation $\mathbf{X}(s) = (s\mathbf{I} - \mathbf{A})^{-1}\,\mathbf{x}(0)$.

The result as should be verified, is $\mathbf{x}(t) =$

$$\begin{bmatrix} \frac{5}{2}e^{-t} - 2e^{-2t} + \frac{1}{2}e^{-3t} & \frac{5}{2}e^{-t} - 4e^{-2t} + \frac{3}{2}e^{-3t} & e^{-t} - 2e^{-2t} + e^{-3t} \\ -\frac{5}{2}e^{-t} + 4e^{-2t} - \frac{3}{2}e^{-3t} & -\frac{5}{2}e^{-t} + 8e^{-2t} - \frac{9}{2}e^{-3t} & -e^{-t} + 4e^{-2t} - 3e^{-3t} \\ \frac{5}{2}e^{-t} - 5e^{-2t} + \frac{5}{2}e^{-3t} & \frac{5}{2}e^{-t} - 10e^{-2t} + \frac{15}{2}e^{-3t} & e^{-t} - 5e^{-2t} + 5e^{-3t} \end{bmatrix} \tag{2.10}$$

We look for more general (and hopefully less tedious) methods of solving the state equations which can be applied to large-scale systems.

2.3 BACKGROUND THEORY

In this section we lay down a firm theoretical foundation from which to launch our methods of solution. In the next section, the special case of time-invariance will be studied. We begin by stating without proof a theorem which sets up the conditions for the **existence** and **uniqueness** of the solution of a system of differential equations

$$\dot{\mathbf{x}} = \mathbf{A}(t)\mathbf{x} + \mathbf{b}(t) \tag{2.11}$$

Notice the slightly different form of this system as opposed to (1.33). $\mathbf{A}(t)$ and $\mathbf{b}(t)$ are matrices whose elements may be time-dependent. In the case where the elements are not dependent on time we shall write $\mathbf{A}$ and $\mathbf{b}$.

Theorem 2.1

If $\mathbf{A}(t)$ and $\mathbf{b}(t)$ are continuous† in the **interval** $a \leqslant t \leqslant c$ and t_0 is such that $a \leqslant t_0 \leqslant c$ and if $\mathbf{x}_0$ is a given vector then the system (2.11) has a unique solution $\mathbf{x}(t)$ which satisfies $\mathbf{x}(t_0) = \mathbf{x}_0$ and which exists for all t such that $a \leqslant t \leqslant c$.

It follows that $\mathbf{x}(t)$ can be defined for all *finite* values of t, positive and negative, **provided** $\mathbf{A}(t)$ and $\mathbf{b}(t)$ are continuous for *all* values of t. The proof of the theorem can be found in Sánchez (1968).

Fundamental Matrix

Now consider the **homogeneous** system

$$\dot{\mathbf{x}} = \mathbf{A}(t)\mathbf{x} \tag{2.12}$$

and let $\mathbf{x}$ be an $n \times 1$ matrix. First note that $\mathbf{x}(t_0) = \mathbf{0}$ implies that $\mathbf{x}(t) = \mathbf{0}$ for all time; since zero is *a* solution , it must be *the* solution.

Assume that $\mathbf{A}(t)$ is continuous on an interval $a \leqslant t \leqslant c$; then it is straightforward to show that if $\mathbf{x}_1(t)$ and $\mathbf{x}_2(t)$ are any two solutions of (2.12), $\mathbf{x}(t) = \alpha_1 \mathbf{x}_1(t) + \alpha_2 \mathbf{x}_2(t)$ is also a solution, where α_1 and α_2 are any two scalar constants. Hence the solutions of (2.12) form an n-dimensional vector space (over the field of complex numbers).

From work on vector spaces you should now expect to study linear independence, and this is now done, but first three definitions are given.

A matrix whose columns are solutions of (2.12) is called a **solution matrix** of the system.

A set of n solutions of (2.12) is called a **fundamental set of solutions** if it is linearly independent.

A matrix whose columns are a fundamental set of solutions of (2.12) is a **fundamental matrix** of the system and is denoted $\Phi(t)$. ††

† A matrix is continuous if its elements are continuous functions.

†† Some authors write $\Phi(t, t_0)$; we shall reconcile this difference later.

Example

The system (2.12) where $\mathbf{A}(t) = \begin{bmatrix} \frac{2}{t} & \frac{-2}{t^2} \\ 1 & 0 \end{bmatrix}$ for $t \neq 0$ has among its set of solutions $\mathbf{x}_1(t) = \begin{bmatrix} 1 \\ t \end{bmatrix}$ and $\mathbf{x}_2(t) = \begin{bmatrix} 2t \\ t^2 \end{bmatrix}$, as you can verify.

Now $\alpha_1 \mathbf{x}_1(t) + \alpha_2 \mathbf{x}_2(t) = \mathbf{0}$ if and only if $\alpha_1 + 2t\alpha_2 = 0$ and $t\alpha_1 + t^2\alpha_2 = 0$, i.e. if and only if $\alpha_1 = 0 = \alpha_2$. Hence $\{\mathbf{x}_1(t), \mathbf{x}_2(t)\}$ is a fundamental set. A fundamental matrix for this system is $\Phi(t) = \begin{bmatrix} 1 & 2t \\ t & t^2 \end{bmatrix}$ as also are $\begin{bmatrix} 2t & 1 \\ t^2 & t \end{bmatrix}$, $\begin{bmatrix} 4t & 3 \\ 2t^2 & 3t \end{bmatrix}$ and $\begin{bmatrix} 1+2t & 2t \\ t+t^2 & t^2 \end{bmatrix}$.

Try to find some more fundamental matrices. (Remember that in this example the interval of definition must exclude $t = 0$.)

Note that $\alpha\Phi(t)$ is also a fundamental matrix for any scalar constant α. Note also that if $\mathbf{C}$ is a constant column vector then $\mathbf{x}(t) = \Phi(t)\mathbf{C}$ is *the* solution of (2.12). Since each column of $\Phi(t)$ is a solution $\mathbf{x}_j(t)$ of (2.12) it further follows that $\dot{\mathbf{x}}_j(t) = \mathbf{A}(t)\mathbf{x}_j(t)$ and hence:

Theorem 2.2

The matrix $\Phi(t)$ is a solution of the equation

$$\dot{\Phi} = \mathbf{A}(t)\Phi(t) \tag{2.13}$$

Verify (2.13) for $\Phi(t) = \begin{bmatrix} 1 & 2t \\ t & t^2 \end{bmatrix}$ and then prove the following result.

Theorem 2.3

If $\Phi(t)$ is a fundamental matrix of (2.12) then so is $\Phi(t)\mathbf{C}$ where $\mathbf{C}$ is a non-singular matrix whose elements are constant. Indeed every fundamental matrix of (2.12) is of this form.

(Hint: post-multiply (2.13) by $\mathbf{C}$ and show that the columns of $\Phi\mathbf{C}$ are linearly independent.)

For example, the fundamental matrix $\begin{bmatrix} 1+2t & 2t \\ t+t^2 & t^2 \end{bmatrix}$ can be written as the product

$\begin{bmatrix} 1 & 2t \\ t & t^2 \end{bmatrix}\begin{bmatrix} 1 & 0 \\ 1 & 1 \end{bmatrix}$. Now we have an infinite number of fundamental matrices for the system (2.12) and we seek one which satisfies the condition that $\Phi(t_0) = \mathbf{I}$, the unit $n \times n$ matrix for reasons of simplicity, as will be seen later. The following result demonstrates that such a possibility exists.

Theorem 2.4

A fundamental matrix may be selected with

$$\Phi(t_0) = \mathbf{I} \tag{2.14}$$

For proof suppose that we have a fundamental matrix $\Phi(t)$ for which $\Phi(t_0) \neq \mathbf{I}$. If we choose a constant matrix $\mathbf{C}$ so that $\mathbf{C} = [\Phi(t_0)]^{-1}$ it follows from Theorem 2.3 that $\Psi(t) = \Phi(t)\mathbf{C} = \Phi(t)[\Phi(t_0)]^{-1}$ is a fundamental matrix, and $\Psi(t_0) = \Phi(t_0)\mathbf{C} = \mathbf{I}$. (We have assumed that $\Phi^{-1}(t_0)$ exists. Since the columns of $\Phi(t)$ are linearly independent its determinant is non-zero and the inverse exists.)

Note that the fundamental matrix

$$\Phi(t) = \begin{bmatrix} -1 + \dfrac{2t}{t_0} & \dfrac{2}{t_0} & -\dfrac{2t}{t_0^2} \\ -t + \dfrac{t^2}{t_0} & \dfrac{2t}{t_0} & -\dfrac{t^2}{t_0^2} \end{bmatrix}$$

satisfies (2.14).

Theorem 2.5

The solution of the system $\dot{\mathbf{x}} = \mathbf{A}(t)\mathbf{x}$ which satisfies the condition $\mathbf{x}(t_0) = \mathbf{x}_0$ may be written $\mathbf{x}(t) = \Phi(t) \,.\, [\Phi(t_0)]^{-1}\mathbf{x}_0$. We know that $\mathbf{x}(t) = \Phi(t)\mathbf{C}$ is a solution of the system and since $\mathbf{x}(t_0) = \Phi(t_0)\mathbf{C}$, $\mathbf{C} = [\Phi(t_0)]^{-1}$, $\mathbf{x}(t_0) = [\Phi(t_0)]^{-1}\mathbf{x}_0$. Note that if we choose $\Phi(t_0) = \mathbf{I}$, the solution may be written as

$$\mathbf{x}(t) = \Phi(t)\mathbf{x}_0 \tag{2.15}$$

State Transition Matrix

We now reconcile our notation for a fundamental matrix with the notation $\Phi(t, t_0)$. Some authors write the solution of the equation $\dot{\mathbf{x}} = \mathbf{A}(t)\mathbf{x}(t)$, with the condition

$$\mathbf{x}(t_0) = \mathbf{x}_0, \quad \text{as} \quad \mathbf{x}(t) = \Phi(t, t_0)\mathbf{x}_0 \tag{2.16}$$

and term the matrix $\Phi(t, t_0)$ the **state transition matrix**. The notation emphasises that the initial state $\mathbf{x}_0$ at time t_0 has undergone a transition to a new state $\mathbf{x}(t)$ at time t and that the matrix Φ maps the initial vector to the new vector in state space.

Differentiating (2.16) we find that $\quad \dot{\mathbf{x}}(t) = \dfrac{\partial \Phi}{\partial t}\mathbf{x}_0$

and since $\dot{\mathbf{x}} = \mathbf{A}(t)\mathbf{x}(t) = \mathbf{A}(t) \,.\, \Phi(t, t_0)\mathbf{x}_0$ it follows that

$$\frac{\partial \Phi}{\partial t} = \mathbf{A}(t)\Phi(t, t_0) \tag{2.17}$$

By putting $t = t_0$ in (2.16) we obtain

$$\Phi(t_0, t_0) = \mathbf{I} \tag{2.18}$$

When $\Phi(t_0, t_0) = \mathbf{I}$ it is usual to write the transition matrix as $\Phi(t)$.

Note that the jth column of $\Phi(t, t_0)$ represents the response of the system with the initial condition $x_i(0) = \delta_{ij}(\delta_{ij} = 1$ if $i = j$, and $= 0$ otherwise). The following properties of Φ are important

(i) $\Phi(t_2, t_0) = \Phi(t_2, t_1) \,.\, \Phi(t_1, t_0)$ (2.19)

(ii) $\Phi(t_0, t_1) = [\Phi(t_1, t_0)]^{-1}$ (2.20)

(iii) $\Phi(t_1, t_0) = \Phi(t_1)[\Phi(t_0)]^{-1}$ (2.21)

(iv) $\det \Phi(t_1, t_0) = \exp \left\{ \int_{t_0}^{t_1} [\operatorname{tr} \mathbf{A}(\tau)] \, d\tau \right\}$ (2.22)

where $\operatorname{tr} \mathbf{A}(\tau)$ is the notation for the *trace* of the matrix $\mathbf{A}(\tau)$, i.e. the sum of its diagonal elements.

Property (ii) follows from (i) by putting $t = t_0$ and using (2.18). To examine property (iv) we make a useful definition.

If $\mathbf{x}_1(t), \mathbf{x}_2(t), \ldots\ldots \mathbf{x}_n(t)$ are solutions of (2.12) with components given by $\mathbf{x}_i(t) = [\mathbf{x}_{1i}(t), \mathbf{x}_{2i}(t), \ldots\ldots \mathbf{x}_{ni}(t)]^{\mathrm{T}}$ then the scalar function

$$W(t) = \begin{vmatrix} x_{11} & x_{12} \cdots\cdots & x_{1n} \\ x_{21} & x_{22} \cdots\cdots & x_{2n} \\ \vdots & & \\ x_{n1} & x_{n2} \cdots\cdots & x_{nn} \end{vmatrix} \tag{2.23}$$

is called the **Wronskian of**

$$\mathbf{x}_1(t), \mathbf{x}_2(t), \ldots\ldots, \mathbf{x}_n(t)$$

Theorem 2.6

If $a < t_0 < b$ then, with the above definition,

$$W(t) = W(t_0) \exp \left\{ \int_{t_0}^{t} \mathrm{tr}\, \mathbf{A}(\tau) \mathrm{d}\tau \right\} \quad \text{for} \quad a < t < b$$

The proof hinges on the result that $\dot{W} = [\mathrm{tr}\, \mathbf{A}(t)] W$. This we leave for you to pursue.

From the result that the exponential of a matrix is never zero it follows that either

$$W(t) \equiv 0 \quad \text{for} \quad a < t < b$$

or

$$W(t) \quad \text{never vanishes in} \quad a < t < b$$

Theorem 2.7

A necessary and sufficient condition for a solution matrix to be a fundamental matrix is that (2.23) is never zero for $a < t < b$.

The determinant of a solution matrix is simply the Wronskian (2.23). If this is non-zero then the columns of the solution matrix are linearly independent and hence the solution matrix is a fundamental matrix. To see whether a solution matrix is a fundamental matrix we need only test its determinant (at any value of t) to check for a non-zero value.

Determination of the form of $\Phi(t)$

First refer to Appendix A for results on functions of a matrix. Using the definition of the exponential of a matrix, we have the following result.

Theorem 2.8

The fundamental matrix of the system (2.12) can be written

$$\Phi(t) = \exp \left(\int_{t_0}^{t} \mathbf{A}(\tau) \mathrm{d}\tau \right) \tag{2.24}$$

provided that the matrices $\mathbf{A}(t)$ and $\mathbf{B}(t) = \int_{t_0}^{t} \mathbf{A}(\tau) \mathrm{d}\tau$ commute.

For the proof first note that $\frac{d\mathbf{B}}{dt} = \mathbf{A}$. Via the series for $\exp(\mathbf{B})$ it can be shown that[†], provided $\mathbf{AB} = \mathbf{BA}$, $\frac{d}{dt}[\exp(\mathbf{B})] = \mathbf{A}(\mathbf{I} + \mathbf{B} + \frac{\mathbf{B}^2}{2!} + ...) = \mathbf{A}\exp(\mathbf{B})$

Hence $\exp(\mathbf{B})$ satisfies (2.13) and is therefore a solution matrix. Since $|\exp(\mathbf{B})| \neq 0$ the columns of $\exp(\mathbf{B})$ must be linearly independent and it is therefore a fundamental matrix. Putting $t = t_0$ in (2.24) produces the result $\Phi(t_0) = \exp(\mathbf{0}) = \mathbf{I}$. Hence the solution of (2.12) may be written

$$\mathbf{x}(t) = \exp\left(\int_{t_0}^{t} \mathbf{A}(\tau)d\tau\right)\mathbf{x}_0 \tag{2.25}$$

However, (2.25) is not easy to employ. Fortunately we often deal with models where the elements of $\mathbf{A}(t)$ vary slowly with time so that a reasonable approximation is to take the elements of $\mathbf{A}$ as constants. This is dealt with in the next two sections.

Nonhomogeneous equations

Consider the nonhomogeneous system

$$\dot{\mathbf{x}} = \mathbf{A}(t)\mathbf{x} + \mathbf{b}(t) \tag{2.11}$$

where $\mathbf{A}(t)$ and $\mathbf{b}(t)$ are continuous on some interval. ($\mathbf{b}(t)$ is sometimes referred to as a forcing term.)

Theorem 2.9

Any two solutions of (2.11) differ at most by a solution of the associated homogeneous system (2.12).

This theorem gives an analogue of the complementary function/particular integral method of solving differential equations. You are left to outline the proof.

To obtain a solution of (2.11) the method of variation of parameters is used on the solution of (2.12), viz. $\mathbf{x} = \Phi(t)\mathbf{c}$ where $\mathbf{c}$ is a constant vector. We look for a solution of (2.11) as $\mathbf{x}(t) = \Phi(t)\mathbf{c}(t)$ where $\mathbf{c}(t_0) = \mathbf{x}_0$. (Why?)

Substitution into (2.11), and using (2.13), produces

$$\dot{\mathbf{x}} = \dot{\Phi}\mathbf{c}(t) + \Phi(t)\dot{\mathbf{c}} = \mathbf{A}(t)\Phi(t)\mathbf{c}(t) + \Phi(t)\dot{\mathbf{c}} = \mathbf{A}(t)\Phi(t)\mathbf{c}(t) + \mathbf{b}(t)$$

[†] Verify this for yourself; note that, for example,

$$\frac{d}{dt}(\mathbf{B}.\mathbf{B}) = \mathbf{AB} + \mathbf{BA} = 2\mathbf{AB} \text{ only if } \mathbf{AB} = \mathbf{BA}.$$

Hence $\Phi(t)\dot{\mathbf{c}} = \mathbf{b}(t)$ or $\dot{\mathbf{c}} = [\Phi(t)]^{-1}\mathbf{b}(t)$. Therefore

$$\mathbf{c}(t) = \mathbf{c}(t_0) + \int_{t_0}^{t} [\Phi(\tau)]^{-1}\,\mathbf{b}(\tau)\mathrm{d}\tau$$

Using the initial condition on $\mathbf{c}(t)$ we may write *the* solution of (2.11), guaranteed unique by Theorem 2.1, as

$$\mathbf{x}(t) = \Phi(t)\mathbf{x}_0 + \Phi(t)\int_{t_0}^{t} [\Phi(\tau)]^{-1}\,\mathbf{b}(\tau)\mathrm{d}\tau \tag{2.26}$$

Problems

1. Prove properties (i), (ii) and (iii) for $\Phi(t, t_0)$.

2. A system is governed by the equation $\dot{\mathbf{x}}(t) = \begin{bmatrix} 0 & 1 \\ 0 & t \end{bmatrix}\mathbf{x}.$

   Show that the transition matrix is

$$\begin{bmatrix} 1 & \displaystyle\int_{t_0}^{t} f(t_0, u)\mathrm{d}u \\ 0 & f(t_0, u) \end{bmatrix}$$

   where $f(t_0, u) = \exp\left[\tfrac{1}{2}(u^2 - t_0^2)\right]$.

3. A fundamental matrix for a particular system is given by

$$\Phi(t, t_0) = \begin{bmatrix} e^{(t-t_0)} & e^{2(t-t_0)} \\ 0 & 1 + e^{(t-t_0)} \end{bmatrix}$$

   Find the corresponding matrix $\mathbf{A}(t)$.

4. Find the transition matrix for the system $\dot{\mathbf{x}} = \begin{bmatrix} \sin t & \sin t \\ \sin t & \sin t \end{bmatrix}\mathbf{x}$

   and verify properties (i) to (iv).

5. Repeat Problem 4 for the system $\dot{\mathbf{x}} = \begin{bmatrix} 4 & e^{-t} \\ -e^{-t} & 4 \end{bmatrix}\mathbf{x}$

6. For the system $\dot{x}_1 = x_2$, $\dot{x}_2 = x_1$ show that a fundamental matrix is

$$\begin{bmatrix} \cos t & \sin t \\ -\sin t & \cos t \end{bmatrix}$$

and hence that the Wronskian is identically equal to 1.

7. Show that

$$\begin{bmatrix} e^{t} & e^{5t} \\ -e^{t} & 3e^{5t} \end{bmatrix}$$

is a fundamental matrix for the system $\dot{x}_1 = 2x_1 + x_2$, $\dot{x}_2 = 3x_1 + 4x_2$; deduce that the Wronskian is $4e^{6t}$. (Hint: $W(0) = \det \Phi(0)$).

2.4 TIME-INVARIANT EQUATIONS

The special case where the coefficients of $\mathbf{A}(t)$ are in fact time-independent is now considered. We shall see that the fundamental matrix depends only on the length of the time interval, not on its actual end-points. It follows from (2.24) that for this special case

$$\Phi(t) = \exp(\mathbf{A}[t - t_0]) \tag{2.27}$$

When $t_0 = 0$, as will now be assumed to be the case,

$$\Phi(t) = \exp(\mathbf{A}t) \tag{2.28}$$

and then $\Phi(0) = \mathbf{I}$.

The solution of the system

$$\dot{\mathbf{x}} = \mathbf{A}\mathbf{x} \tag{2.29}$$

with $\mathbf{x}(0) = \mathbf{x}_0$ can be written

$$\mathbf{x}(t) = e^{\mathbf{A}t}\,\mathbf{x}_0 \tag{2.30}$$

Note that

$$\Phi(t - \tau) = \Phi(t)\,.\,[\Phi(\tau)]^{-1} \tag{2.31}$$

for any constant τ (cf. property (iii): (2.21)). This follows since $\Phi(t - \tau)$ satisfies $\dot{\Phi} = \mathbf{A}\Phi$ and at $t = \tau$ both sides of (2.31) reduce to $\mathbf{I}$. Since the right-hand-side is in the form of a constant matrix multiplying $\Phi(t)$ it is a solution of (2.29) as is the left-hand-side of (2.31). But the solution of (2.29) is unique and hence (2.31) is established.

It then follows that the solution of

$$\dot{\mathbf{x}} = \mathbf{A}\mathbf{x}(t) + \mathbf{b}(t) \tag{2.32}$$

which satisfies $\mathbf{x}(0) = \mathbf{x}_0$ can be written (cf. (2.26))

$$\mathbf{x}(t) = e^{\mathbf{A}t}\,\mathbf{x}_0 + \int_0^t e^{\mathbf{A}(t-\tau)}\,\mathbf{b}(\tau)d\tau \tag{2.33}$$

Computation of $e^{\mathbf{A}t}$

The thermal system of Section 1.2 had system matrix $\mathbf{A} = \begin{bmatrix} -3 & 1 \\ 2 & -2 \end{bmatrix}$. To find $e^{\mathbf{A}t}$ we employ the series identity

$$e^{\mathbf{A}t} = \mathbf{I} + \mathbf{A}t + \frac{\mathbf{A}^2 t^2}{2!} + \frac{\mathbf{A}^3 t^3}{3!} + \ldots\ldots \tag{2.34}$$

Now $\mathbf{A}^2 = \begin{bmatrix} 11 & -5 \\ -10 & 6 \end{bmatrix}$, $\mathbf{A}^3 = \begin{bmatrix} -43 & 21 \\ 42 & -22 \end{bmatrix}$ etc. so that (2.34) becomes

$$e^{\mathbf{A}t} = \begin{bmatrix} 1 & 0 \\ 0 & 1 \end{bmatrix} + \begin{bmatrix} -3t & t \\ 2t & -2t \end{bmatrix} + \begin{bmatrix} \frac{11}{2}t^2 & -\frac{5}{2}t^2 \\ -5t^2 & 3t^2 \end{bmatrix} + \begin{bmatrix} -\frac{43t^3}{6} & \frac{7}{2}t^3 \\ 7t^3 & -\frac{11}{3}t^3 \end{bmatrix} + \ldots$$

$$= \begin{bmatrix} 1 - 3t + \frac{11}{2}t^2 - \frac{43}{6}t^3 + \ldots.. & t - \frac{5}{2}t^2 + \frac{7}{2}t^3 + \ldots.. \\ 2t - 5t^2 + 7t^3 + \ldots.. & 1 - 2t + 3t^2 - \frac{11}{3}t^3 + \ldots.. \end{bmatrix}$$

Comparison with (2.3) shows that this last matrix should in fact be

$$\begin{bmatrix} \frac{2}{3}e^{-4t} + \frac{1}{3}e^{-t} & -\frac{1}{3}e^{-4t} + \frac{1}{3}e^{-t} \\ -\frac{2}{3}e^{-4t} + \frac{2}{3}e^{-t} & \frac{1}{3}e^{-4t} + \frac{2}{3}e^{-t} \end{bmatrix}$$

and you can check by expanding these exponentials as far as the terms in t^3 that agreement is achieved. However, it is a very tedious process to use the series (2.34) even for a 2×2 matrix. Sometimes, **A** has a form which permits a short-cut, as the next example shows.

Example

Find $e^{\mathbf{A}t}$ where $\mathbf{A} = \begin{bmatrix} 4 & 1 \\ 0 & 4 \end{bmatrix}$. Now $\mathbf{A} = \begin{bmatrix} 4 & 0 \\ 0 & 4 \end{bmatrix} + \begin{bmatrix} 0 & 1 \\ 0 & 0 \end{bmatrix}$ so that

$$e^{\mathbf{A}t} = \exp\left\{\begin{bmatrix} 4 & 0 \\ 0 & 4 \end{bmatrix} t\right\} . \exp\left\{\begin{bmatrix} 0 & 1 \\ 0 & 0 \end{bmatrix} t\right\}$$

since the two parts of the right-hand-side commute.

However, $\begin{bmatrix} 0 & 1 \\ 0 & 0 \end{bmatrix}^2 = \begin{bmatrix} 0 & 0 \\ 0 & 0 \end{bmatrix}$ and therefore

$$\exp\left\{\begin{bmatrix} 0 & 1 \\ 0 & 0 \end{bmatrix} t\right\} = \mathbf{I} + \begin{bmatrix} 0 & 1 \\ 0 & 0 \end{bmatrix} t$$

Further $\exp\{4\mathbf{I}t\} = \begin{bmatrix} e^{4t} & 0 \\ 0 & e^{4t} \end{bmatrix}$ as you can verify. Hence $e^{\mathbf{A}t} = e^{4t} \begin{bmatrix} 1 & t \\ 0 & 1 \end{bmatrix}$.

This is clearly a specialised technique applicable to only very few instances, but worth employing if possible.

(A) Direct series computation

We could use the series (2.34) computing as many terms as necessary until the *remainder* was less than a prescribed value. For *remainder* we might read *matrix norm* as defined in Appendix A. This method is, however, generally too time consuming, even for a computer.

(B) Use of Sylvester's Interpolation Formula

In a manner similar to Lagrange's interpolation formula we have a formula which relies on a function of a matrix being specified at its eigenvalues. Three cases are considered; these depend on the nature of **A**.

(i) Eigenvalues distinct

We assume that a function $f(\mathbf{A})$ of an $n \times n$ matrix **A** is expressible as a convergent (infinite) series. Then we can write

$$f(\mathbf{A}) = \sum_{k=1}^{n} \mathbf{Z}_k \, f(\lambda_k) \tag{2.35}$$

where λ_k are the eigenvalues of A and

$$\mathbf{Z}_k = \prod_{\substack{j=1 \\ j \neq k}}^{n} \frac{(\mathbf{A} - \lambda_j \mathbf{I})}{(\lambda_k - \lambda_j)} \equiv \frac{(\mathbf{A} - \lambda_1 \mathbf{I}) \,.....\, (\mathbf{A} - \lambda_{k-1}\mathbf{I}).(\mathbf{A} - \lambda_{k+1}\mathbf{I}) \,.....\, (\mathbf{A} - \lambda_n \mathbf{I})}{(\lambda_k - \lambda_1) \,.....\, (\lambda_k - \lambda_{k-1}).(\lambda_k - \lambda_{k+1}) \,.....\, (\lambda_k - \lambda_n)} \tag{2.36}$$

Formulae (2.35) and (2.36) comprise **Sylvester's Interpolation Formula.** A verification is given in Froberg (1965).

Note that $\mathbf{Z}_k$ are independent of the form of the function f. If two functions have the same n values at λ_k then they have the same matrix function $f(\mathbf{A})$.

Remembering that a matrix satisfies a polynomial equation of least degree, m, called the **minimum polynomial**, we can effectively replace all terms of degree $\geqslant m$ in the infinite series expansion $f(\mathbf{A})$ by terms involving powers of **A** no greater than $(m - 1)$, (this needs proving). Hence any convergent series $f(\mathbf{A})$ can be written as a polynomial of degree $(m - 1)$ or less. The series (2.35) is of this type, since distinct eigenvalues for **A** implies that the minimum polynomial and characteristic polynomial are identical, and the degree of the terms of (2.35) is $(n - 1)$. Further the series is unique.

Example

The oven matrix $\mathbf{A} = \begin{bmatrix} -3 & 1 \\ 2 & -2 \end{bmatrix}$ has eigenvalues $\lambda_1 = -1$, $\lambda_2 = -4$. We seek $f(\mathbf{A}) = e^{\mathbf{A}t}$. Hence $f(\lambda) = e^{\lambda t}$.

Therefore

$$\mathbf{Z}_1 = \frac{\mathbf{A} - (-4)\mathbf{I}}{-1 + 4} = \frac{\mathbf{A} + 4\mathbf{I}}{3}$$

and

$$\mathbf{Z}_2 = \frac{\mathbf{A} - (-1)\mathbf{I}}{(-4 - (-1))} = \frac{\mathbf{A} + \mathbf{I}}{-3}$$

From Sylvester's formula

$$f(\mathbf{A}) = e^{\mathbf{A}t} = e^{-t} \,.\, \frac{1}{3}(\mathbf{A} + 4\mathbf{I}) + \frac{e^{-4t}}{-3}(\mathbf{A} + \mathbf{I})$$

$$= \frac{1}{3} e^{-t} \begin{bmatrix} 1 & 1 \\ 2 & 2 \end{bmatrix} - \frac{1}{3} e^{-4t} \begin{bmatrix} -2 & 1 \\ 2 & -1 \end{bmatrix}$$

$$= \begin{bmatrix} \frac{1}{3} e^{-t} + \frac{2}{3} e^{-4t} & \frac{1}{3} e^{-t} - \frac{1}{3} e^{-4t} \\ \frac{2}{3} e^{-t} - \frac{2}{3} e^{-4t} & \frac{2}{3} e^{-t} + \frac{1}{3} e^{-4t} \end{bmatrix}$$

, as before.

Example

For the voltmeter problem $\mathbf{A} = \begin{bmatrix} 0 & 1 & 0 \\ -1 & 0 & 2 \\ 0 & -5 & -6 \end{bmatrix}$. The eigenvalues of **A** are $\lambda = -1$, $\lambda = -2$, $\lambda = -3$. Then

$$f(\mathbf{A}) = e^{\mathbf{A}t}$$

$$= e^{-t}\,\frac{(\mathbf{A}+2\mathbf{I})(\mathbf{A}+3\mathbf{I})}{(-1+2)(-1+3)} + e^{-2t}\,\frac{(\mathbf{A}+\mathbf{I})(\mathbf{A}+3\mathbf{I})}{(-2+1)(-2+3)} + e^{-3t}\,\frac{(\mathbf{A}+\mathbf{I})(\mathbf{A}+2\mathbf{I})}{(-3+1)(-3+2)}$$

We leave you to show that this reduces to the form encountered earlier. Note that the computation can become tedious even for a 3 x 3 matrix. An alternative procedure is to note that $f(\mathbf{A})$ is the solution of the determinantal equation:

$$\begin{vmatrix} 1 & 1 & & 1 & \mathbf{I} \\ \lambda_1 & \lambda_2 & & \lambda_n & \mathbf{A} \\ \lambda_1^2 & \lambda_2^2 & \cdots\cdots & \lambda_n^2 & \mathbf{A}^2 \\ \vdots & \vdots & & \vdots & \vdots \\ \lambda_1^{n-1} & \lambda_2^{n-1} & & \lambda_n^{n-1} & \mathbf{A}^{n-1} \\ f(\lambda_1) & f(\lambda_2) & & f(\lambda_n) & f(\mathbf{A}) \end{vmatrix} = 0 \qquad (2.37)$$

We can solve (2.37) with the oven matrix to obtain

$$\begin{vmatrix} 1 & 1 & \mathbf{I} \\ -1 & -4 & \mathbf{A} \\ e^{-t} & e^{-4t} & e^{\mathbf{A}t} \end{vmatrix} = 0, \text{ i.e. } e^{\mathbf{A}t} = \frac{\mathbf{A}}{3}(e^{-t} - e^{-4t}) + \frac{\mathbf{I}}{3}(4e^{-t} - e^{-4t})$$

which readily gives the required result.

For the meter matrix we solve

$$\begin{vmatrix} 1 & 1 & 1 & \mathbf{I} \\ -1 & -2 & -3 & \mathbf{A} \\ 1 & 4 & 9 & \mathbf{A}^2 \\ e^{-t} & e^{-2t} & e^{-3t} & e^{\mathbf{A}t} \end{vmatrix} = 0$$

and then (eventually)

$$e^{\mathbf{A}t} = \frac{1}{2}\mathbf{A}^2(e^{-t} - 2e^{-2t} - e^{-3t}) + \frac{1}{2}\mathbf{A}(5e^{-t} - 8e^{-2t} + 3e^{-3t}) + \mathbf{I}(3e^{-t} - 3e^{-2t} + e^{-3t})$$

To reduce, or perhaps redirect, the computing effort, note that the oven example can be dealt with by first solving the equations

$$a_0 - a_1 = e^{-t}$$

$$a_0 - 4a_1 = e^{-4t}$$

for a_0 and a_1 and then writing $e^{\mathbf{A}t} = a_0\mathbf{I} + a_1\mathbf{A}$.

This is a special case of solving (2.37) for $f(\mathbf{A})$ by putting $f(\mathbf{A}) = a_0\mathbf{I} + a_1\mathbf{A} + \ldots + a_{n-1}\mathbf{A}^{n-1}$ and solving for a_k, $k = 0, 1, \ldots, n-1$ the equations

$$\left.\begin{array}{l} a_0 + a_1\lambda_1 + \ldots\ldots + a_{n-1}\lambda_1^{n-1} = e^{\lambda_1 t} \\ \vdots \qquad\qquad\qquad\qquad \vdots \\ a_0 + a_1\lambda_n + \ldots\ldots + a_{n-1}\lambda_n^{n-1} = e^{\lambda_n t} \end{array}\right\} \qquad (2.38)$$

This is a consequence of the Cayley-Hamilton theorem: that a matrix satisfies its own characteristic equation.

For the meter matrix we need to solve

$$\begin{aligned} a_0 - a_1 + a_2 &= e^{-t} \\ a_0 - 2a_1 + 4a_2 &= e^{-2t} \\ a_0 - 3a_1 + 9a_2 &= e^{-3t} \end{aligned}$$

and compute $e^{\mathbf{A}t} = a_0\mathbf{I} + a_1\mathbf{A} + a_2\mathbf{A}^2$.

Check that you get the required result via this method.

(ii) Eigenvalues not distinct, minimum polynomial has only simple roots

Suppose the minimum polynomial is of the form $\prod_{i=1}^{m}(\mathbf{A} - \lambda_i\mathbf{I})$ where $m < n$, i.e. the matrix $\mathbf{A}$ has repeated eigenvalues, m distinct, but the minimum polynomial has only m distinct factors. In effect, Sylvester's interpolation formula applies with m replacing n in (2.35) and (2.36).

Example

Consider $\mathbf{A} = \begin{bmatrix} 1 & 0 & 0 \\ 1 & 1 & 1 \\ -1 & 0 & 0 \end{bmatrix}$. The eigenvalues can be shown to be $\lambda_1 = 0$, $\lambda_2 = 1$

$\lambda_3 = 1$. The characteristic equation is therefore

$$\lambda(\lambda - 1)^2 = 0$$

i.e. $\lambda^3 - 2\lambda^2 + \lambda = 0$ and by the Cayley-Hamilton theorem we know that $\mathbf{A}^3 - 2\mathbf{A}^2 + \mathbf{A} = \mathbf{0}$. Note that $\mathbf{A}^2 = \mathbf{A}$ i.e. $\mathbf{A}^2 - \mathbf{A} = \mathbf{0}$ and therefore we have a simple case with which to deal.

$$e^{\mathbf{A}t} = \mathbf{I} + \mathbf{A}t + \mathbf{A}^2 \frac{t^2}{2!} + \mathbf{A}^3 \frac{t^3}{3!} + \ldots\ldots\ldots = \mathbf{I} + \mathbf{A}t + \frac{\mathbf{A}t^2}{2!} + \mathbf{A}\frac{t^3}{3!} + \ldots\ldots\ldots\ldots$$

since $\mathbf{A}^2 = \mathbf{A}$, $\mathbf{A}^3 = \mathbf{A}\,.\,\mathbf{A}^2 = \mathbf{A}\,.\,\mathbf{A} = \mathbf{A}$ etc. But

$$\mathbf{I} = \begin{bmatrix} 1 & 0 & 0 \\ 0 & 1 & 0 \\ 0 & 0 & 1 \end{bmatrix} = \begin{bmatrix} 0 & 0 & 0 \\ -1 & 0 & -1 \\ 1 & 0 & 1 \end{bmatrix} + \begin{bmatrix} 1 & 0 & 0 \\ 1 & 1 & 1 \\ -1 & 0 & 0 \end{bmatrix} = \begin{bmatrix} 0 & 0 & 0 \\ -1 & 0 & -1 \\ 1 & 0 & 1 \end{bmatrix} + \mathbf{A}$$

Then

$$e^{\mathbf{A}t} = \begin{bmatrix} 0 & 0 & 0 \\ -1 & 0 & -1 \\ 1 & 0 & 1 \end{bmatrix} + \mathbf{A}e^t \qquad \text{(why?)}$$

$$= \begin{bmatrix} 0 & 0 & 0 \\ -1 & 0 & -1 \\ 1 & 0 & 1 \end{bmatrix} + \begin{bmatrix} e^t & 0 & 0 \\ e^t & e^t & e^t \\ -e^t & 0 & 0 \end{bmatrix} = \begin{bmatrix} e^t & 0 & 0 \\ -1 + e^t & e^t & -1 + e^t \\ 1 - e^t & 0 & 1 \end{bmatrix}$$

Check this result via the method (i) applied to $\mathbf{A}^2 - \mathbf{A}$ and, when you have read method (iii), apply it to the polynomial $\mathbf{A}^3 - 2\mathbf{A}^2 + \mathbf{A} = \mathbf{0}$. What do you now get for $e^{\mathbf{A}t}$? How do you reconcile the results?

(iii) Minimum polynomial has repeated roots

The matrix $\mathbf{A} = \begin{bmatrix} 1 & 1 & 0 \\ 0 & 1 & 0 \\ 0 & 0 & 2 \end{bmatrix}$ has eigenvalues $\lambda_1 = 1$, $\lambda_2 = 1$, $\lambda_3 = 2$. We leave you to show that, whereas $\mathbf{A}^3 - 4\mathbf{A}^2 + 5\mathbf{A} - 2\mathbf{I} = \mathbf{0}$, no equation of lower degree is satisfied by $\mathbf{A}$; hence the characteristic and minimum polynomials are identical with repeated root $\lambda = 1$. We need to look for a variant on the Sylvester Interpolation Formula.

Consider an $n \times n$ matrix $\mathbf{A}$.

Suppose that of the n eigenvalues λ_1, ..., λ_n, the first m are identical i.e. $\lambda_1 = \lambda_2 = ... = \lambda_m$. Then to replace the first m terms of the right-hand-side of (2.35) we put

$$\mathbf{Z}_{11}f(\lambda_1) + \mathbf{Z}_{12}f'(\lambda_1) + ... + \mathbf{Z}_{1m}f^{(m-1)}(\lambda_1) \tag{2.39}$$

where $f^{(r)}(\lambda_1)$ is the rth derivative of $f(\lambda)$ with respect to λ evaluated at $\lambda = \lambda_1$. The $\mathbf{Z}_{1r}$ will be evaluated by an artifice to be explained later.

For the present example we write,

$$f(\mathbf{A}) = \mathbf{Z}_{11}f(\lambda_1) + \mathbf{Z}_{12}f'(\lambda_1) + \mathbf{Z}_3 f(\lambda_3) \tag{2.40}$$

It is known that

$$\mathbf{Z}_3 = \frac{(\mathbf{A}-\mathbf{I})(\mathbf{A}-\mathbf{I})}{(2-1)(2-1)} = \mathbf{A}^2 - 2\mathbf{A} + \mathbf{I} = \begin{bmatrix} 0 & 0 & 0 \\ 0 & 0 & 0 \\ 0 & 0 & 1 \end{bmatrix}$$

Now (2.40) is valid for any function f for which $f(\mathbf{A})$ is convergent. Therefore we choose special forms for f. First, $f(\lambda) = 1$ so that $f(\lambda_1) = f(\lambda_3) = 1$, $f'(\lambda_1) = 0$, $f(\mathbf{A}) = \mathbf{I}$. Then $\mathbf{I} = \mathbf{Z}_{11} \,.\, 1 + 0 + \mathbf{Z}_3 \,.\, 1$. Hence

$$\mathbf{Z}_{11} = \begin{bmatrix} 1 & 0 & 0 \\ 0 & 1 & 0 \\ 0 & 0 & 0 \end{bmatrix}$$

Next, $f(\lambda) = \lambda - 1$ so that $f(\lambda_1) = 0$, $f(\lambda_3) = 1$, $f'(\lambda_1) = 1$, $f(\mathbf{A}) = \mathbf{A} - \mathbf{I}$. Then $\mathbf{A} - \mathbf{I} = \mathbf{Z}_{11} \,.\, 0 + \mathbf{Z}_{12} \,.\, 1 + \mathbf{Z}_3 \,.\, 1$. Hence

$$\mathbf{Z}_{12} = \begin{bmatrix} 0 & 1 & 0 \\ 0 & 0 & 0 \\ 0 & 0 & 0 \end{bmatrix}$$

Now note that if $f(\lambda) = e^{\lambda t}$, $f'(\lambda) = te^{\lambda t}$. Finally,

$$e^{\mathbf{A}t} = \begin{bmatrix} 1 & 0 & 0 \\ 0 & 1 & 0 \\ 0 & 0 & 0 \end{bmatrix} e^t + \begin{bmatrix} 0 & 1 & 0 \\ 0 & 0 & 0 \\ 0 & 0 & 0 \end{bmatrix} te^t + \begin{bmatrix} 0 & 0 & 0 \\ 0 & 0 & 0 \\ 0 & 0 & 1 \end{bmatrix} e^{2t}$$

i.e.

$$e^{\mathbf{A}t} = \begin{bmatrix} e^t & te^t & 0 \\ 0 & e^t & 0 \\ 0 & 0 & e^{2t} \end{bmatrix} \tag{2.41}$$

The alternative approach, parallel to that for case (i) is to solve the system

$$\left.\begin{aligned} a_0 + a_1\lambda_1 + a_2\lambda_2 &= f(\lambda_1) \\ 0 + a_1 + 2a_2\lambda_1 &= f'(\lambda_1) \\ a_0 + a_1\lambda_3 + a_2\lambda_3^2 &= f(\lambda_3) \end{aligned}\right\} \tag{2.42}$$

and write $$f(\mathbf{A}) = a_0\mathbf{I} + a_1\mathbf{A} + a_2\mathbf{A}^2.$$

In our example, this means solving the equations

$$\left.\begin{aligned} a_0 + a_1 + a_2 &= e^t \\ 0 + a_1 + 2a_2 &= te^t \\ a_0 + 2a_1 + 4a_2 &= e^{2t} \end{aligned}\right\}$$

We leave you to show that this yields the result (2.41).

(C) Leverrier's Algorithm

This is an algorithm for determining $(s\mathbf{I} - \mathbf{A})^{-1}$ as required in the Laplace Transform approach. The rules are:

$$(s\mathbf{I} - \mathbf{A})^{-1} = \frac{[s^{n-1}\mathbf{I} + s^{n-2}\mathbf{B}_2 + \ldots\ldots + s\,\mathbf{B}_{n-1} + \mathbf{B}_n]}{k(s)}$$

where $k(s)$ = characteristic polynomial of $\mathbf{A}$ in $s = s^n + k_1 s^{n-1} + \ldots\ldots + k_{n-1}s + k_n$ and the scalars k_j and matrices $\mathbf{B}_j$ are given by

$$\left.\begin{aligned} \mathbf{B}_1 &= \mathbf{I};\ k_1 = -\text{tr}\ \mathbf{AB}_1 = -\text{tr}\mathbf{A} \\ \mathbf{B}_2 &= \mathbf{AB}_1 + k_1\mathbf{I};\ k_2 = -\frac{1}{2}\text{tr}\ \mathbf{AB}_2 \\ \mathbf{B}_3 &= \mathbf{AB}_2 + k_2\mathbf{I};\ k_3 = -\frac{1}{3}\text{tr}\ \mathbf{AB}_3 \\ \mathbf{B}_n &= \mathbf{AB}_{n-1} + k_{n-1}\mathbf{I};\ k_n = -\frac{1}{n}\text{tr}\ \mathbf{AB}_n \end{aligned}\right\} \tag{2.43}$$

A useful check is $\mathbf{AB}_n + k_n\mathbf{I} = \mathbf{0}$. We apply this algorithm to the meter matrix

$$\mathbf{A} = \begin{bmatrix} 0 & 1 & 0 \\ -1 & 0 & 2 \\ 0 & -5 & -6 \end{bmatrix}$$

Then $k_1 = -\text{tr}\mathbf{A} = 6$.

$$\mathbf{B}_2 = \mathbf{A}\,.\,\mathbf{I} + 6\mathbf{I} = \begin{bmatrix} 6 & 1 & 0 \\ -1 & 6 & 2 \\ 0 & -5 & 0 \end{bmatrix}$$

$$\mathbf{AB}_2 = \begin{bmatrix} 0 & 1 & 0 \\ -1 & 0 & 2 \\ 0 & -5 & -6 \end{bmatrix} \begin{bmatrix} 6 & 1 & 0 \\ -1 & 6 & 2 \\ 0 & -5 & 0 \end{bmatrix} = \begin{bmatrix} -1 & 6 & 2 \\ -6 & -11 & 0 \\ 5 & 0 & -10 \end{bmatrix};$$

$$k_2 = -\frac{1}{2}(-22) = 11$$

$$\mathbf{B}_3 = \mathbf{AB}_2 + 11\mathbf{I} = \begin{bmatrix} 10 & 6 & 2 \\ -6 & 0 & 0 \\ 5 & 0 & 1 \end{bmatrix}$$

$$\mathbf{AB}_3 = \begin{bmatrix} 0 & 1 & 0 \\ -1 & 0 & 2 \\ 0 & -5 & -6 \end{bmatrix} \begin{bmatrix} 10 & 6 & 2 \\ -6 & 0 & 0 \\ 5 & 0 & 1 \end{bmatrix} = \begin{bmatrix} -6 & 0 & 0 \\ 0 & -6 & 0 \\ 0 & 0 & -6 \end{bmatrix};$$

$$k_3 = -\frac{1}{3}(-18) = 6$$

Check: $\mathbf{AB}_3 + 6\mathbf{I} = \mathbf{0}$. Then

$$(s\mathbf{I} - \mathbf{A})^{-1} = \frac{s^2 \begin{bmatrix} 1 & 0 & 0 \\ 0 & 1 & 0 \\ 0 & 0 & 1 \end{bmatrix} + s \begin{bmatrix} 6 & 1 & 0 \\ -1 & 6 & 2 \\ 0 & -5 & 0 \end{bmatrix} + \begin{bmatrix} 10 & 6 & 2 \\ -6 & 0 & 0 \\ 5 & 0 & 1 \end{bmatrix}}{s^3 + 6s^2 + 11s + 6}$$

$$= \frac{1}{(s^3 + 6s^2 + 11s + 6)} \begin{bmatrix} (s^2 + 6s + 10) & (s + 6) & 2 \\ -(s + 6) & (s^2 + 6s) & 2s \\ 5 & -5s & s^2 + 1 \end{bmatrix}$$

which agrees with the result on page 27.

In the next section we adopt a different approach to the solution of the state equations: that of reducing the equations to a canonical (or standard) form.

Problems

1. Evaluate the transition matrix $e^{\mathbf{A}t}$ for each of the following systems using as many of the methods of this section as you can. The systems are of the form $\dot{\mathbf{x}} = \mathbf{A}\mathbf{x}$ where $\mathbf{A}$ is given below.

(a) $\begin{bmatrix} 0 & 2 \\ 2 & -3 \end{bmatrix}$ (b) $\begin{bmatrix} 0 & 1 & 1 \\ 0 & 0 & 1 \\ 0 & 0 & 0 \end{bmatrix}$ (c) $\begin{bmatrix} -3 & 1 & 0 \\ 1 & -3 & 0 \\ 0 & 0 & -3 \end{bmatrix}$

(d) $\begin{bmatrix} -1 & -2 & -3 \\ 0 & 1 & -1 \\ -1 & -1 & 2 \end{bmatrix}$

2. Find $(s\mathbf{I} - \mathbf{A})^{-1}$ where $\mathbf{A}$ is given by

(a) $\begin{bmatrix} 5 & \sqrt{2} \\ \sqrt{2} & 4 \end{bmatrix}$ (b) $\begin{bmatrix} -1 & 0 & 0 \\ 0 & -4 & 4 \\ 0 & -1 & 0 \end{bmatrix}$

3. A possible algorithm to determine $e^{\mathbf{A}t}$ is as follows:

Select an interval h, compute $(\mathbf{I} + h\mathbf{A})$ and $(\mathbf{I} + \frac{1}{2}h\mathbf{A})$; note that $\Phi(h) \doteq \mathbf{I} + h\mathbf{A}$. Hence $[\Phi(h/2)]^2 \doteq \Phi(h)$. Compare $(\mathbf{I} + \frac{1}{2}h\mathbf{A})^2$ with $(\mathbf{I} + h\mathbf{A})$. If the difference exceeds the selected error, compute $(\mathbf{I} + \frac{1}{4}h\mathbf{A})^2$ and compare with $(\mathbf{I} + \frac{1}{2}h\mathbf{A})$. Continue until the difference between successive functions is small enough.

Then, if $e^{\mathbf{A}t_0}$ is known, $e^{\mathbf{A}(t_0+h)}$ can be calculated, then $e^{\mathbf{A}(t_0+2h)}$, etc. Choosing suitable values for h, try this algorithm on the matrices (a) and (b) of Problem 1 and compare with the algebraically determined results.

4. If $\mathbf{A} = \begin{bmatrix} 0 & 1 \\ -1 & 0 \end{bmatrix}$ show that $\mathbf{A}^2 = -\mathbf{I}$, $\mathbf{A}^3 = -\mathbf{A}$, $\mathbf{A}^4 = \mathbf{I}$ and hence that

$$e^{\mathbf{A}t} = \begin{bmatrix} \cos t & \sin t \\ -\sin t & \cos t \end{bmatrix}.$$ Deduce that if $\mathbf{A} = \begin{bmatrix} 2 & 1 \\ -1 & 2 \end{bmatrix}$,

$$e^{\mathbf{A}t} = e^{2t}\begin{bmatrix} \cos t & \sin t \\ -\sin t & \cos t \end{bmatrix}.$$

5. Show that $e^{(\mathbf{A}+\mathbf{B})} = e^{\mathbf{A}} . e^{\mathbf{B}}$ if and only if $\mathbf{AB} = \mathbf{BA}$.

6. Find $e^{\mathbf{A}t}$ for the examples of Section 1.2.

7. Consider the matrix

$$\mathbf{A} = \begin{bmatrix} 4 & 1 & 0 & 0 & 0 \\ & 4 & 1 & 0 & 0 \\ & & 4 & 0 & 0 \\ & \bigcirc & & 4 & 0 \\ & & & & 4 \end{bmatrix}$$

Show that $(\mathbf{A} - 4\mathbf{I})^3 = \mathbf{0}$, but $(\mathbf{A} - 4\mathbf{I})^2 \neq \mathbf{0}$. Hence find $e^{(\mathbf{A}-4\mathbf{I})t}$ and $e^{\mathbf{A}t}$.

2.5 REDUCTION TO CANONICAL FORMS

Consider the system

$$\begin{bmatrix} \dot{x}_1 \\ \dot{x}_2 \\ \dot{x}_3 \\ \dot{x}_4 \end{bmatrix} = \begin{bmatrix} 2 & 0 & 0 & 0 \\ 0 & 3 & 0 & 0 \\ 0 & 0 & 1 & 0 \\ 0 & 0 & 0 & 2 \end{bmatrix} \begin{bmatrix} x_1 \\ x_2 \\ x_3 \\ x_4 \end{bmatrix} \tag{2.44}$$

Since the matrix $\mathbf{A}$ is diagonal, the system equations decouple into 4 equations each in one variable. These are straightforward to solve.

Example

To find the general solution of the system $\dot{\mathbf{x}} = \begin{bmatrix} 0 & 1 \\ -4 & -5 \end{bmatrix} \mathbf{x}$ we first find the eigenvalues and eigenvectors of $\mathbf{A} = \begin{bmatrix} 0 & 1 \\ -4 & -5 \end{bmatrix}$. These are $\lambda = -1, -4$ with eigenvectors $\begin{bmatrix} 1 \\ -1 \end{bmatrix}$ and $\begin{bmatrix} 1 \\ -4 \end{bmatrix}$ respectively. The matrix $\mathbf{T} = \begin{bmatrix} 1 & 1 \\ -1 & -4 \end{bmatrix}$ has inverse $\mathbf{T}^{-1} = \begin{bmatrix} 4/3 & 1/3 \\ -1/3 & -1/3 \end{bmatrix}$. The rows of $\mathbf{T}^{-1}$ are (4/3, 1/3) and (–1/3, –1/3). The solution of the system may be written in the **spectral form**

$$\mathbf{x} = \left[\frac{4}{3}x_1(0) + \frac{1}{3}x_2(0)\right] e^{-t} \begin{bmatrix} 1 \\ -1 \end{bmatrix} + \left[-\frac{1}{3}x_1(0) - \frac{1}{3}x_2(0)\right] e^{-4t} \begin{bmatrix} 1 \\ -4 \end{bmatrix}$$

You can verify by substitution that this is the general solution.

The matrix $\mathbf{T}^{-1}\mathbf{A}\mathbf{T} = \begin{bmatrix} -1 & 0 \\ 0 & -4 \end{bmatrix}$. Find $\mathbf{z} = \mathbf{T}^{-1}\mathbf{x}$ and note that

$\dot{\mathbf{z}} = \mathbf{T}^{-1}\dot{\mathbf{x}} = \mathbf{T}^{-1}\mathbf{ATz}$. Solve this equation for $\mathbf{z}$ and recover $\mathbf{x} = \mathbf{Tz}$. Verify therefore that you get the quoted solution for $\mathbf{x}$.

Now look at the system

$$\dot{\mathbf{x}} = \begin{bmatrix} \lambda_1 & 1 & 0 \\ 0 & \lambda_1 & 1 \\ 0 & 0 & \lambda_1 \end{bmatrix} \mathbf{x} \tag{2.45}$$

If we solve the last scalar equation viz. $\dot{x}_3 = \lambda_1 x_3(t)$ we obtain $x_3(t) = e^{\lambda_1 t} x_3(0)$. The second equation is $\dot{x}_2 = \lambda_1 x_2(t) + x_3(t)$ and hence $x_2(t) = e^{\lambda_1 t} x_2(0) + te^{\lambda_1 t} x_3(0)$. Substituting this into the first scalar equation leads to the result

$$x_1(t) = e^{\lambda_1 t} x_1(0) + te^{\lambda_1 t} x_2(0) + \frac{t^2}{2!} e^{\lambda_1 t} x_3(0)$$

Whilst not as easy to solve as the system (2.44), this system is still relatively simple.

The matrix in (2.45) is a **Jordan matrix**. This is a matrix with main diagonal elements (some possibly zero), 1's or zeros on the diagonal above the leading diagonal and zeros elsewhere. A diagonal matrix is a special kind of Jordan matrix.

Here are some more examples of a Jordan matrix.

$$\begin{bmatrix} \lambda_1 & 1 \\ 0 & \lambda_1 \end{bmatrix} \begin{bmatrix} \lambda_1 & 1 & 0 & 0 \\ 0 & \lambda_1 & 0 & 0 \\ 0 & 0 & \lambda_1 & 0 \\ 0 & 0 & 0 & \lambda_2 \end{bmatrix} \begin{bmatrix} \lambda_1 & 1 & 0 & 0 & 0 \\ 0 & \lambda_1 & 1 & 0 & 0 \\ 0 & 0 & \lambda_1 & 0 & 0 \\ 0 & 0 & 0 & \lambda_2 & 1 \\ 0 & 0 & 0 & 0 & \lambda_2 \end{bmatrix}$$

Each matrix can be thought of as being compounded from **Jordan blocks** (which are really themselves Jordan matrices) and from blocks of zero elements. They can be symbolised as

$$\begin{bmatrix} \mathbf{J}_1 & & & \\ & \mathbf{J}_2 & & \\ & & \mathbf{J}_3 & \\ & & & \mathbf{J}_4 \end{bmatrix} \quad \text{etc.}$$

The Jordan blocks appear along the leading diagonal.

It is clear that if the matrix $\mathbf{A}$ is of Jordan form (or, better still, is diagonal), then the system equations are easy to solve. Under certain conditions, a matrix $\mathbf{A}$ can be transformed to a diagonal matrix $\mathbf{D}$ via $\mathbf{D} = \mathbf{T}^{-1}\mathbf{AT}$ where $\mathbf{T}$ is a matrix whose columns are the eigenvectors (assumed distinct) of $\mathbf{A}$. If this is not possible then the hope is that a matrix can be transformed to Jordan form. In the next section we consider a special transformation when the matrix has been derived from a single nth order differential equation; for this section we concentrate on a general approach.

Transformation to canonical form is useful for modelling: it allows an easier analysis of controllability and observability and it is more straightforward in the determination of a solution for $\mathbf{x}(t)$. However, any physical significance of the parameters of $\mathbf{A}$ is lost; this might not be important if a model is the main requirement.

The theorem which gives the necessary theoretical basis for the techniques we develop is quoted without proof. The definition of similarity eases the statement of the theorem.

Two matrices $\mathbf{A}$ and $\mathbf{B}$ are said to be **similar** if there exists a non-singular matrix $\mathbf{T}$ such that $\mathbf{B} = \mathbf{T}^{-1}\mathbf{AT}$. The transformation from $\mathbf{A}$ to $\mathbf{B}$ is called a **similarity transformation.** Notice that $\mathbf{A} = \mathbf{TBT}^{-1}$ and if we write $\mathbf{U} = \mathbf{T}^{-1}$ then $\mathbf{A} = \mathbf{U}^{-1}\mathbf{BU}$ where $\mathbf{U}$ is non-singular.

The theorem states that an n x n matrix $\mathbf{A}$ is similar to a matrix $\mathbf{J}$ which possesses the n eigenvalues of $\mathbf{A}$ along its leading diagonal, zeros and/or 1's on the diagonal immediately above the leading diagonal, and zeros elsewhere.

We have not stated whether the eigenvalues are distinct: if they are, then the matrix $\mathbf{J}$ is diagonal with elements the eigenvalues of $\mathbf{A}$. If an eigenvalue of $\mathbf{A}$, λ_1 say, has multiplicity 3 then there will be three λ_1's along the leading diagonal. The presence of a 1 or zero depends on whether or not the multiple eigenvalue has linearly independent eigenvectors.

Note that an eigenvalue λ_1 of multiplicity > 1 may have more than one Jordan block associated with it: this depends on the number of linearly independent eigenvectors associated with λ_1.

Determining the transformation to Jordan form

First a few results on diagonalising a matrix are collected.

(i) Start from $\dot{\mathbf{x}} = \mathbf{Ax}(t)$ and look for a solution in the form $\mathbf{x}(t) = e^{\lambda t}\mathbf{c}$ where $\mathbf{c}$ is a constant vector to be determined and λ is a scalar constant to be determined. Substitution into the system equation produces $\lambda e^{\lambda t}\mathbf{c} = \mathbf{A}e^{\lambda t}\mathbf{c}$ or $(\mathbf{A} - \lambda\mathbf{I})\mathbf{c} = \mathbf{0}$. Hence λ is an eigenvalue of $\mathbf{A}$ and $\mathbf{c}$ is its corresponding eigenvector.

(ii) If eigenvalues $\lambda_1, \lambda_2, \ldots\ldots \lambda_m$ of an n x n matrix are distinct the eigenvectors $\mathbf{c}_1, \mathbf{c}_2, \ldots\ldots, \mathbf{c}_m$ which correspond are linearly independent.

(iii) Any set of characteristic solutions of $\dot{\mathbf{x}} = \mathbf{Ax}(t)$ which correspond to distinct eigenvalues is a linearly independent set. Indeed if the matrix $\mathbf{A}$ has n distinct eigenvalues the corresponding set of characteristic solutions is a fundamental set of solutions of the system.

(iv) A fundamental matrix can be constructed as

$$\Phi(t) = (e^{\lambda_1 t}\,\mathbf{c}_1,\; e^{\lambda_2 t}\,\mathbf{c}_2, \dots\dots\dots, e^{\lambda_n t}\,\mathbf{c}_n)$$

where λ_r and $\mathbf{c}_r$ may be complex.

Example

Find a fundamental matrix if $\mathbf{A} = \begin{bmatrix} 2 & 4 \\ -4 & 2 \end{bmatrix}$. It can be readily shown that the eigenvalues of $\mathbf{A}$ are $\lambda_1 = 2 + 4i$ and $\lambda_2 = 2 - 4i$. Corresponding eigenvectors are $\mathbf{c}_1 = c_1 \begin{bmatrix} 1 \\ i \end{bmatrix}$ and $\mathbf{c}_2 = c_2 \begin{bmatrix} i \\ 1 \end{bmatrix}$ where c_1 and c_2 are scalar constants.

Then, by choosing $c_1 = 1 = c_2$ it follows that $\mathbf{x}_1(t) = e^{(2+4i)t} \begin{bmatrix} 1 \\ i \end{bmatrix}$ and $\mathbf{x}_2(t) = e^{(2-4i)t} \begin{bmatrix} i \\ 1 \end{bmatrix}$ are solutions. Then $\Phi(t) = \begin{bmatrix} e^{(2+4i)t} & i\,e^{(2-4i)t} \\ i\,e^{(2+4i)t} & e^{(2-4i)t} \end{bmatrix}$

Note that the real and imaginary parts of $\mathbf{x}_1(t)$ and $\mathbf{x}_2(t)$ are each solutions.

Postmultiplying Φ by the constant matrix $\frac{1}{2}\begin{bmatrix} 1 & -i \\ -i & 1 \end{bmatrix} = [\Phi(0)]^{-1}$ as Theorem 2.3 allows, the real part of Φ, viz. $\begin{bmatrix} e^{2t}\cos 4t & e^{2t}\sin 4t \\ -e^{2t}\sin 4t & e^{2t}\cos 4t \end{bmatrix}$ is obtained.

Generalised eigenvectors

We now examine the situation when there are not n distinct eigenvectors. First a definition.

Suppose λ is an eigenvalue of $\mathbf{A}$ and for some integer $p > 1$ there is a vector $\mathbf{v}$ such that $(\mathbf{A} - \lambda\mathbf{I})^p\,\mathbf{v} = \mathbf{0}$ but $(\mathbf{A} - \lambda\mathbf{I})^{p-1}\,\mathbf{v} \neq \mathbf{0}$. Then $\mathbf{v}$ is referred to as a **generalised eigenvector of index** p corresponding to λ.

Example 1

Consider the matrix $\mathbf{A} = \begin{bmatrix} 1 & -1 & 1 \\ 0 & -1 & 4 \\ 0 & -1 & 3 \end{bmatrix}$. The characteristic polynomial equation is $(\lambda - 1)^3 = 0$ and hence eigenvalues are $\lambda_1 = 1$, $\lambda_2 = 1$, $\lambda_3 = 1$. To find the eigenvectors, we solve $(\mathbf{A} - \lambda\mathbf{I})\mathbf{v} = \mathbf{0}$, i.e. $(\mathbf{A} - \mathbf{I})\mathbf{v} = \mathbf{0}$, for $\mathbf{v}$.

Then $\begin{bmatrix} 0 & -1 & 1 \\ 0 & -2 & 4 \\ 0 & -1 & 2 \end{bmatrix} \begin{bmatrix} u_1 \\ u_2 \\ u_3 \end{bmatrix} = \begin{bmatrix} 0 \\ 0 \\ 0 \end{bmatrix}$ and this leads to $u_3 = 0$, $u_2 = 0$ with u_1 arbitrary.

A typical eigenvector is $\mathbf{v}_1 = \begin{bmatrix} \alpha \\ 0 \\ 0 \end{bmatrix}$. To find the two other (generalised) eigenvectors, consider $(\mathbf{A} - \lambda\mathbf{I})^2\ \mathbf{v} = \mathbf{0}$.

Suppose we define $\mathbf{v}_2$ by the equation $\mathbf{A}\mathbf{v}_2 = \lambda\mathbf{v}_2 + \mathbf{v}_1$. Then $(\mathbf{A} - \lambda\mathbf{I})\mathbf{v}_2 = \mathbf{v}_1$ and $(\mathbf{A} - \lambda\mathbf{I})^2\ \mathbf{v}_2 = (\mathbf{A} - \lambda\mathbf{I})\mathbf{v}_1 = \mathbf{0}$. Therefore we solve $(\mathbf{A} - \mathbf{I})\mathbf{v}_2 = \mathbf{v}_1$ that is,

$\begin{bmatrix} 0 & -1 & 1 \\ 0 & -2 & 4 \\ 0 & -1 & 2 \end{bmatrix} \mathbf{v}_2 = \begin{bmatrix} \alpha \\ 0 \\ 0 \end{bmatrix}$. It follows that $\mathbf{v}_2 = \begin{bmatrix} \beta \\ -2\alpha \\ -\alpha \end{bmatrix}$ where β is arbitrary.

(We could have solved $(\mathbf{A} - \lambda\mathbf{I})^2\,\mathbf{v} = \mathbf{0}$ directly. The matrix $(\mathbf{A} - \lambda\mathbf{I})^2 = \begin{bmatrix} 0 & 1 & -2 \\ 0 & 0 & 0 \\ 0 & 0 & 0 \end{bmatrix}$ which leads to the form for $\mathbf{v}_2$ above. Note that $\mathbf{v}_1$ also satisfies $(\mathbf{A} - \lambda\mathbf{I})^2\,\mathbf{v} = 0$.) Continuing in like vein, we solve the equation

$(\mathbf{A} - \lambda\mathbf{I})\mathbf{v}_3 = \mathbf{v}_2$ i.e. $\begin{bmatrix} 0 & -1 & 1 \\ 0 & -2 & 4 \\ 0 & -1 & 2 \end{bmatrix} \mathbf{v}_3 = \begin{bmatrix} \beta \\ -2\alpha \\ -\alpha \end{bmatrix}$

to obtain $\mathbf{v}_3 = \begin{bmatrix} \gamma \\ -\alpha - 2\beta \\ -\alpha - \beta \end{bmatrix}$ where γ is arbitrary.

(Note that $(\mathbf{A} - \lambda\mathbf{I})^3 = \mathbf{0}$ and so the equation $(\mathbf{A} - \lambda\mathbf{I})^3\,\mathbf{v} = \mathbf{0}$ is satisfied by any vector belonging to E_n, the space of vectors with three real coordinates.)

$\mathbf{v}_1$, $\mathbf{v}_2$ and $\mathbf{v}_3$ as chosen are in fact linearly independent vectors. This means that if a matrix $\mathbf{T}$ is defined by $\mathbf{T} = (\mathbf{v}_1, \mathbf{v}_2, \mathbf{v}_3)$, i.e.

$$\mathbf{T} = \begin{bmatrix} \alpha & \beta & \gamma \\ 0 & -2\alpha & -\alpha - 2\beta \\ 0 & -\alpha & -\alpha - \beta \end{bmatrix}$$

then $\mathbf{T}^{-1}$ will exist.

To be specific we select $\alpha = 1$, $\beta = 0$, $\gamma = 0$ so that $\mathbf{T} = \begin{bmatrix} 1 & 0 & 0 \\ 0 & -2 & -1 \\ 0 & -1 & -1 \end{bmatrix}$ and

$$\mathbf{T}^{-1} = \begin{bmatrix} 1 & 0 & 0 \\ 0 & 1 & -1 \\ 0 & -1 & 2 \end{bmatrix}.$$

You can verify for yourself that the required matrix is $\mathbf{B} = \mathbf{T}^{-1}\mathbf{A}\mathbf{T} = \begin{bmatrix} 1 & 1 & 0 \\ 0 & 1 & 1 \\ 0 & 0 & 1 \end{bmatrix}$.

The presence of two 1's immediately above the leading diagonal is a consequence of two generalised eigenvectors.

Example 2

Consider $\mathbf{A} = \begin{bmatrix} 2 & 4 & -2 \\ 0 & 2 & 0 \\ 0 & 0 & 2 \end{bmatrix}$. The eigenvalues are $\lambda_1 = 2$, $\lambda_2 = 2$, $\lambda_3 = 2$.

First we solve $(\mathbf{A} - 2\mathbf{I})\mathbf{v} = \mathbf{0}$. You can check that the component equations require two linearly independent eigenvectors to be found and the only restriction is that $u_3 = 2u_2$. Suitable choices† are $\begin{bmatrix} \alpha \\ 0 \\ 0 \end{bmatrix}$ and $\begin{bmatrix} 0 \\ \beta \\ 2\beta \end{bmatrix}$ where α and β are arbitrary.

Corresponding to these two linearly independent eigenvectors, there will be two Jordan blocks. We could therefore reduce $\mathbf{A}$ to one of two forms:

$$\begin{bmatrix} 2 & 1 & 0 \\ 0 & 2 & 0 \\ 0 & 0 & 2 \end{bmatrix} \quad \text{or} \quad \begin{bmatrix} 2 & 0 & 0 \\ 0 & 2 & 1 \\ 0 & 0 & 2 \end{bmatrix}$$

Which form we obtain depends on the order of the eigenvectors and generalised eigenvector.

The strategy adopted is to choose a particular vector from the set of linear combinations of the two given eigenvectors and use it to generate a generalised eigenvector of index 2; then choose a second vector from the set of linear combinations

† It is always useful to try and copy multiples of the unit vectors. In this case $\begin{bmatrix} 0 \\ 0 \\ 1 \end{bmatrix}$ is useless. Why?

of the eigenvectors so that we obtain a linearly independent set.

Consider the vector $\mathbf{v}_1 = \begin{bmatrix} \alpha \\ \beta \\ 2\beta \end{bmatrix}$ which is the general linear combination of the eigenvectors. Then solve $\begin{bmatrix} 0 & 4 & -2 \\ 0 & 0 & 0 \\ 0 & 0 & 0 \end{bmatrix} \mathbf{v}_2 = \begin{bmatrix} \alpha \\ \beta \\ 2\beta \end{bmatrix}$. Hence $\beta = 0$; now choose $\mathbf{v}_1 = \begin{bmatrix} \alpha \\ 0 \\ 0 \end{bmatrix}$. It then follows that $\mathbf{v}_2 = \begin{bmatrix} \gamma \\ \delta \\ 2\delta - \alpha \end{bmatrix}$ where γ and δ are arbitrary. Then we look for a third linearly independent vector. Suppose we take $\alpha = 1$, $\gamma = 0$, $\delta = 0$ so that $\mathbf{v}_1 = \begin{bmatrix} 1 \\ 0 \\ 0 \end{bmatrix}$ and $\mathbf{v}_2 = \begin{bmatrix} 0 \\ 0 \\ -1 \end{bmatrix}$; we want a value of β so that $\mathbf{v}_3 = \begin{bmatrix} 1 \\ \beta \\ 2\beta \end{bmatrix}$, one of the set of linear combinations with $\alpha = 1$, is independent from $\mathbf{v}_1$ and $\mathbf{v}_2$. A suitable choice is $\beta = 1$ giving $\mathbf{v}_3 = \begin{bmatrix} 1 \\ 1 \\ 2 \end{bmatrix}$. Then $\mathbf{T} = \begin{bmatrix} 1 & 0 & 1 \\ 0 & 0 & 1 \\ 0 & -1 & 2 \end{bmatrix}$. Verify that $\mathbf{T}^{-1}\mathbf{AT} = \begin{bmatrix} 2 & 1 & 0 \\ 0 & 2 & 0 \\ 0 & 0 & 2 \end{bmatrix}$. Note that one 1 above the leading diagonal implies one generalised eigenvector.

Formalisation of the method

Let us now endeavour to find a formal background for these ideas. We quote the following theorem without proof.

If $\lambda_1, \lambda_2, \ldots\ldots, \lambda_k$ are the distinct eigenvalues of an $n \times n$ matrix **A** multiplicities $m_1, m_2, \ldots\ldots, m_k$ respectively then the subspace $\mathbf{X}_j$ of generalised eigenvectors corresponding to λ_j has the following properties:

(i) the dimension of $\mathbf{X}_j$ is m_j

(ii) the vector $\mathbf{Ax}_i \in \mathbf{X}_j$ if $\mathbf{x}_i \in \mathbf{X}_j$

(iii) $(\mathbf{A} - \lambda_j)^{m_j}\, \mathbf{x} = \mathbf{0}$ for any $\mathbf{x} \in \mathbf{X}_j$

It follows that to each eigenvalue λ_j of multiplicity m_j there is a subspace $\mathbf{X}_j$ of dimension m_j which is spanned by the eigenvector of $(\mathbf{A} - \lambda\mathbf{I})^{m_j}$. (Note that if a vector $\mathbf{x}$ is an eigenvector of a matrix $\mathbf{B}$ so that $(\mathbf{B} - \lambda\mathbf{I})\mathbf{x} = \mathbf{0}$ then

$$(\mathbf{B} - \lambda\mathbf{I})^2\mathbf{x} = (\mathbf{B} - \lambda\mathbf{I})[(\mathbf{B} - \lambda\mathbf{I})\mathbf{x}] = (\mathbf{B} - \lambda\mathbf{I})\mathbf{0} = \mathbf{0}$$

and therefore $\mathbf{x}$ is an eigenvector of $(\mathbf{B} - \lambda\mathbf{I})^2$ and indeed, of $(\mathbf{B} - \lambda\mathbf{I})^p$ for $p \geqslant 1$.) The space E_n is the direct sum of $X_1, X_2, \ldots\ldots, X_k$†.

Now consider the system $\dot{\mathbf{x}} = \mathbf{A}\mathbf{x}$. Suppose we make the linear transformation $\mathbf{y} = \mathbf{T}\mathbf{x}$ where $\mathbf{T}^{-1}$ exists. Then $\dot{\mathbf{y}} = \mathbf{T}\dot{\mathbf{x}} = \mathbf{T}\mathbf{A}\mathbf{x} = \mathbf{T}\mathbf{A}\mathbf{T}^{-1}\mathbf{y}$ or $\dot{\mathbf{y}} = \mathbf{B}\mathbf{y}$ where $\mathbf{B} = \mathbf{T}\mathbf{A}\mathbf{T}^{-1}$. This transformation has not destroyed any system properties, since similar matrices have the same characteristic polynomial. This follows because

$$|\mathbf{B} - \lambda\mathbf{I}| = |\mathbf{T}\mathbf{A}\mathbf{T}^{-1} - \lambda\mathbf{I}| = |\mathbf{T}(\mathbf{A} - \lambda\mathbf{I})\mathbf{T}^{-1}|$$

$$= |\mathbf{T}| \cdot |\mathbf{A} - \lambda\mathbf{I}| \cdot |\mathbf{T}^{-1}| = |\mathbf{A} - \lambda\mathbf{I}|$$

Therefore two similar matrices have the same eigenvalues and the same number of distinct eigenvectors.

Fundamental matrices

(i) When the matrix $\mathbf{A}$ has n linearly independent eigenvectors $\mathbf{v}_i$ corresponding to eigenvalues λ_i (not necessarily distinct), it is similar to the diagonal matrix

$$\mathbf{B} = \begin{bmatrix} \lambda_1 & & & \\ & \lambda_2 & & \mathbf{0} \\ & & \ddots & \\ \mathbf{0} & & & \lambda_n \end{bmatrix}$$

whose transformation matrix $\mathbf{T} = (\mathbf{v}_1, \mathbf{v}_2, \ldots\ldots, \mathbf{v}_n)$.

Now $\Phi(t) = e^{\mathbf{B}t} = \mathbf{I} + \mathbf{B}t + \dfrac{\mathbf{B}^2 t^2}{2!} + \ldots\ldots\ldots\ldots$ and

$$\mathbf{B}^k = \begin{bmatrix} \lambda_1^k & & & \\ & \lambda_2^k & & \mathbf{0} \\ & & \ddots & \\ \mathbf{0} & & & \lambda_n^k \end{bmatrix}$$

† If $X_1 \cap X_2 = \{0\}$ and $Y = X_1 \cup X_2$, Y is called the **direct sum** of X_1 and X_2.

Therefore

$$\Phi(t) = \begin{bmatrix} e^{\lambda_1 t} & & & \\ & e^{\lambda_2 t} & & \\ & & \ddots & \\ & & & e^{\lambda_n t} \end{bmatrix}$$

(ii) In the case where there are not n linearly independent eigenvectors we have already established the result of reduction to a Jordan matrix. Let the distinct eigenvalues be $\lambda_0, \lambda_1, \ldots\ldots, \lambda_r$. Let

$$\mathbf{J} = \begin{bmatrix} \mathbf{J}_0 & & & \\ & \mathbf{J}_1 & & \\ & & \ddots & \\ & & & \mathbf{J}_p \end{bmatrix}$$

where $\mathbf{J}_0$ is the diagonal matrix

$$\begin{bmatrix} \lambda_0 & & \\ & \ddots & \\ & & \lambda_r \end{bmatrix}$$

and

$$\mathbf{J}_i = \begin{bmatrix} \lambda_{r+i} & 1 & & & \\ & \lambda_{r+i} & 1 & & \\ & & \ddots & \ddots & \\ & & & \lambda_{r+i} & 1 \\ & & & & \lambda_{r+i} \end{bmatrix}, \quad i = 1, 2, \ldots\ldots, p$$

It can be shown that

$$e^{\mathbf{J}t} = \begin{bmatrix} e^{\mathbf{J}_0 t} & & & \\ & e^{\mathbf{J}_1 t} & & \\ & & \ddots & \\ & & & e^{\mathbf{J}_p t} \end{bmatrix}$$

It is left to you to show that

$$e^{\mathbf{J}_0 t} = \begin{bmatrix} e^{\lambda_1 t} & & & \bigcirc \\ & e^{\lambda_2 t} & & \\ & & \ddots & \\ \bigcirc & & & e^{\lambda_r t} \end{bmatrix}$$

What about $e^{\mathbf{J}_i t}$? We claim that if $\mathbf{J}_i = \lambda_{r+i}\mathbf{I} + \mathbf{M}$ then $e^{\mathbf{J}_i t} = e^{\lambda_{r+i} t} e^{\mathbf{M}t}$ since $\lambda_{r+i}\mathbf{I}$ and $\mathbf{M}$ commute. (Write down the form of $\mathbf{M}$ and check the commutative relationship.) Clearly, the size of $\mathbf{M}$ is the size of $\mathbf{J}_i$. Further,

$$e^{\mathbf{M}t} = \mathbf{I} + \mathbf{M}t + \mathbf{M}^2 \frac{t^2}{2!} + \dots..$$

Hence it is possible, in theory, to find the fundamental matrix via the eigenvalues of $\mathbf{A}$.

Problems

1. Reduce the following matrices to Jordan form:

(a) $\begin{bmatrix} 8 & -8 & -2 \\ 4 & -3 & -2 \\ 3 & -4 & 1 \end{bmatrix}$ (b) $\begin{bmatrix} -1 & 2 & -1 \\ 0 & -1 & 0 \\ 0 & 0 & -1 \end{bmatrix}$ (c) $\begin{bmatrix} -1 & 1 & -1 \\ 0 & 1 & -4 \\ 0 & 1 & -3 \end{bmatrix}$

(d) $\begin{bmatrix} -1 & 0 & 0 \\ 0 & -1 & 0 \\ 0 & 0 & -1 \end{bmatrix}$ (e) $\begin{bmatrix} 1 & 0.6 & -0.8 \\ 0 & 1 & 0 \\ 0 & 0 & 1 \end{bmatrix}$ (f) $\begin{bmatrix} 0 & 1 & 0 \\ -1 & 2 & 0 \\ 1 & 0 & 1 \end{bmatrix}$

Identify the Jordan blocks in each case.

2. Find the eigenvectors and generalised eigenvectors of the matrices below. Reduce (a) and (b) to Jordan form.

(a) $\begin{bmatrix} 1 & 2 & 3 \\ 0 & 1 & 4 \\ 0 & 0 & 1 \end{bmatrix}$ (b) $\begin{bmatrix} 4 & 2 & 1 \\ 0 & 6 & 1 \\ 0 & -4 & 2 \end{bmatrix}$ (c) $\begin{bmatrix} 2 & 0 & 1 & 0 \\ 0 & 0 & 0 & 1 \\ 0 & 0 & 0 & 0 \\ 0 & 0 & 0 & 0 \end{bmatrix}$

3. Show that the matrix $\mathbf{A} = \begin{bmatrix} 1 & 1 \\ 2 & 0 \end{bmatrix}$ can be reduced to Jordan form via the transformation

$\mathbf{J} = \mathbf{P}^{-1}\mathbf{A}\mathbf{P}$ where

$\mathbf{P} = \begin{bmatrix} 1 & 1 \\ -2 & 1 \end{bmatrix}.$

4. A system is described by the equation

$$\dot{\mathbf{x}} = \begin{bmatrix} -2 & -2 & 0 \\ 0 & 0 & 1 \\ 0 & -3 & -4 \end{bmatrix} \mathbf{x} \quad \text{with} \quad \mathbf{x}(0) = \begin{bmatrix} 10 \\ 5 \\ 2 \end{bmatrix}$$

Show that the transformation $\mathbf{x} = \mathbf{Mq}$ where $\mathbf{M}$ is the modal matrix (i.e. its columns are eigenvectors of $\mathbf{A}$) produces an uncoupled system and solve this. Hence find a solution for $\mathbf{x}(t)$.

5. If the system of Problem 4 is subjected to a control

$$\begin{bmatrix} 1 & 0 \\ 0 & 1 \\ 1 & 1 \end{bmatrix} \begin{bmatrix} t \\ 1 \end{bmatrix}$$

show that the solution of the state equations is

$$\mathbf{x} = \begin{bmatrix} -14e^{-t} + 31.75e^{-2t} + (58/9)e^{-3t} + t/6 - 47/36 \\ 7e^{-t} - (29/9)e^{-3t} + t/3 + 11/9 \\ -7e^{-t} + (29/3)e^{-3t} - 2/3 \end{bmatrix}$$

2.6 STATE SPACE AND CLASSICAL APPROACH COMPARED

Consider the equation

$$\dddot{z} + a_1 \ddot{z} + a_2 \dot{z} + a_3 z = u \qquad (2.46)$$

By choosing state variables $x_1 = z$, $x_2 = \dot{z}$, $x_3 = \ddot{z}$ we may write (2.46) in the state space form as

$$\dot{\mathbf{x}} = \begin{bmatrix} 0 & 1 & 0 \\ 0 & 0 & 1 \\ -a_3 & -a_2 & -a_1 \end{bmatrix} \mathbf{x} + \begin{bmatrix} 0 \\ 0 \\ 1 \end{bmatrix} u \qquad (2.47)$$

The first matrix on the r.h.s. of (2.47) is said to be in **companion form**†. It can be diagonalised or reduced to Jordan form by application of a particular form of the transformation matrix $\mathbf{T}$.

† Other companion forms for the same problem would have the a_i in the first or last column or in the first row, with modification of the positions of the 1's.

Example 1

Consider $\mathbf{A} = \begin{bmatrix} 0 & 1 & 0 \\ 0 & 0 & 1 \\ -6 & -11 & -6 \end{bmatrix}$ as a companion form and let $\mathbf{x} = \mathbf{V}\mathbf{y}$, where

$\mathbf{V}$ is the **Vandermonde Matrix** $\begin{bmatrix} 1 & 1 & 1 \\ \lambda_1 & \lambda_2 & \lambda_3 \\ \lambda_1^2 & \lambda_2^2 & \lambda_3^2 \end{bmatrix}$

in which λ_1, λ_2, λ_3 are the eigenvalues of $\mathbf{A}$.

The eigenvalues of $\mathbf{A}$ are $\lambda_1 = -1$, $\lambda_2 = -2$, $\lambda_3 = -3$ so that

$$\mathbf{V} = \begin{bmatrix} 1 & 1 & 1 \\ -1 & -2 & -3 \\ 1 & 4 & 9 \end{bmatrix}$$

From the equation $\dot{\mathbf{x}} = \mathbf{A}\mathbf{x}$, it follows that $\dot{\mathbf{y}} = \mathbf{V}^{-1}\mathbf{A}\mathbf{V}\mathbf{y}$. We leave you to show that

$$\dot{\mathbf{y}} = \begin{bmatrix} -3 & 0 & 0 \\ 0 & -2 & 0 \\ 0 & 0 & -1 \end{bmatrix} \mathbf{y}$$

In equation (2.47) if $u = 0$, $\mathbf{y} = \begin{bmatrix} e^{-3t} & 0 & 0 \\ 0 & e^{-2t} & 0 \\ 0 & 0 & e^{-t} \end{bmatrix} \mathbf{y}(0)$

and hence $\mathbf{x} = \begin{bmatrix} 1 & 1 & 1 \\ -1 & -2 & -3 \\ 1 & 4 & 9 \end{bmatrix} \begin{bmatrix} e^{-3t} & 0 & 0 \\ 0 & e^{-2t} & 0 \\ 0 & 0 & e^{-t} \end{bmatrix} \begin{bmatrix} 1 & 1 & 1 \\ -1 & -2 & -3 \\ 1 & 4 & 9 \end{bmatrix}^{-1} \mathbf{x}(0)$

(Note that the eigenvectors of $\mathbf{A}$ are the columns of the Vandermonde matrix.)

Example 2

Consider $\mathbf{A} = \begin{bmatrix} 0 & 1 & 0 \\ 0 & 0 & 1 \\ -2 & -5 & -4 \end{bmatrix}$. The eigenvalues of $\mathbf{A}$ are $\lambda_1 = -2$, $\lambda_2 = -1$,

$\lambda_3 = -1$. This time a modified Vandermonde matrix

$$\mathbf{V} = \begin{bmatrix} 1 & 1 & 0 \\ \lambda_1 & \lambda_2 & 1 \\ \lambda_1^2 & \lambda_2^2 & 2\lambda_2 \end{bmatrix}$$

is required. Notice that the third column is the derivative of the second†. In this

instance $\mathbf{V} = \begin{bmatrix} 1 & 1 & 0 \\ -2 & -1 & 1 \\ 4 & 1 & -2 \end{bmatrix}$.

This leads to a Jordan form $\mathbf{J} = \begin{bmatrix} -2 & 0 & 0 \\ 0 & -1 & 1 \\ 0 & 0 & -1 \end{bmatrix}$

for $\mathbf{V}^{-1}\mathbf{AV}$.

We might ask under what conditions a linear state space system $\dot{\mathbf{x}} = \mathbf{Ax} + \mathbf{Bu}$ can be put into the form equivalent to (2.47). We quote the following result (Barnett, 1975). The system $\dot{\mathbf{x}} = \mathbf{Ax} + \mathbf{Bu}$ where $\mathbf{A}$ is $n \times n$ can be transformed via $\mathbf{y} = \mathbf{Tx}$ for a non-singular matrix $\mathbf{T}$ into the form analagous to (2.47) provided that

$$\text{the rank of the matrix } (\mathbf{B},\ \mathbf{AB},\ \mathbf{A}^2\mathbf{B}, \ldots\ldots, \mathbf{A}^{n-1}\mathbf{B}) \text{ is } n \tag{2.48}$$

This is the **Kalman criterion for controllability**, referred to in the next section.

The required form for $\mathbf{T}$ is $\begin{bmatrix} \mathbf{t} \\ \mathbf{tA} \\ \cdot \\ \cdot \\ \cdot \\ \mathbf{tA}^{n-1} \end{bmatrix}$

where $\mathbf{t}$ is a row vector with n components chosen so that $\mathbf{T}^{-1}$ exists. The necessary conditions are

$$\mathbf{tB} = 0,\ \mathbf{tAB} = 0, \ldots\ldots, \mathbf{tA}^{n-2}\mathbf{B} = 0,\ \mathbf{tA}^{n-1}\mathbf{B} = 1 \tag{2.49}$$

† With a triple eigenvalue there would be

$$\begin{bmatrix} 1 & 0 & 0 \\ \lambda & 1 & 0 \\ \lambda^2 & 2\lambda & 2 \end{bmatrix}$$

Example 3

The thermal system studied earlier was

$$\dot{\mathbf{x}} = \begin{bmatrix} -3 & 1 \\ 2 & -2 \end{bmatrix} \mathbf{x} + \begin{bmatrix} 1 \\ 0 \end{bmatrix} u \qquad (2.1)$$

Let $\mathbf{t} = (t_1, t_2)$ so that $\mathbf{tB} = t_1$ and $\mathbf{tAB} = -3t_1 + 2t_2$. Hence from (2.49) $t_1 = 0$, $t_2 = ½$. Then

$$\mathbf{T} = \begin{bmatrix} 0 & ½ \\ 1 & -1 \end{bmatrix}, \quad \mathbf{T}^{-1} = \begin{bmatrix} 2 & 1 \\ 2 & 0 \end{bmatrix}, \quad \mathbf{TAT}^{-1} = \begin{bmatrix} 0 & -1 \\ -11 & 3 \end{bmatrix}$$

Note that the matrix $(\mathbf{B}, \mathbf{AB}) = \begin{bmatrix} 1 & -3 \\ 0 & 2 \end{bmatrix}$ and has full rank 2.

2.7 CONTROLLABILITY AND OBSERVABILITY FOR TIME-INVARIANT SYSTEMS

Consider the system given by

$$\dot{\mathbf{x}} = \begin{bmatrix} -1 & 0 \\ 0 & -3 \end{bmatrix} \mathbf{x}(t) + \begin{bmatrix} 0 \\ 1 \end{bmatrix} u(t)$$

The equivalent scalar equations are

$$\dot{x}_1(t) = -x_1(t)$$

$$\dot{x}_2(t) = -3x_2(t) + u(t)$$

from which it is clear that the state $x_1(t)$ cannot be influenced by the control $u(t)$. Now consider the output of a system given by

$$\mathbf{y} = \begin{bmatrix} 0 & 2 \\ 0 & 3 \end{bmatrix} \mathbf{x}(t)$$

or

$$y_1(t) = 2x_2(t)$$

$$y_2(t) = 3x_2(t)$$

from which we see that the variable $x_1(t)$ is not capable of observation. Sometimes an unstable mode is not observable and this leads to erroneous use of techniques such as those of the transfer function in analysing the system. See page 66.

We now consider the concepts of state controllability† and observability of a system.

Example 1

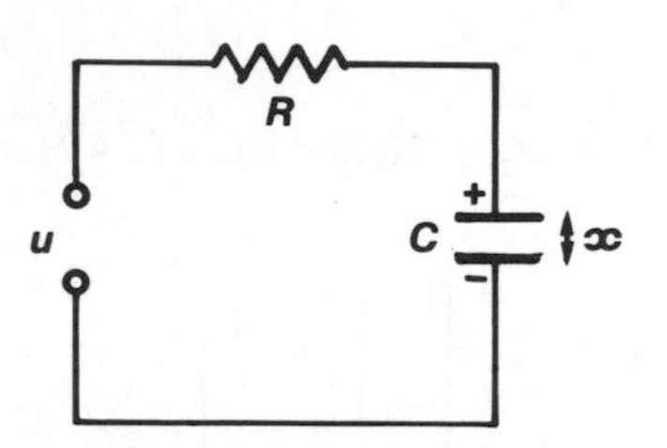

Figure 2.2

The system depicted in Figure 2.2 is governed by the equation

$$\dot{\mathbf{x}} = -\frac{1}{RC}\mathbf{x} + \frac{1}{RC}u$$

It is controllable.

The system depicted in Figure 2.3 is governed by the equation

$$\dot{\mathbf{x}} = \begin{bmatrix} -\dfrac{1}{R_1C_1} & 0 \\ 0 & \dfrac{-1}{R_2C_2} \end{bmatrix}\mathbf{x} + \begin{bmatrix} \dfrac{1}{R_1C_1} \\ 0 \end{bmatrix}u$$

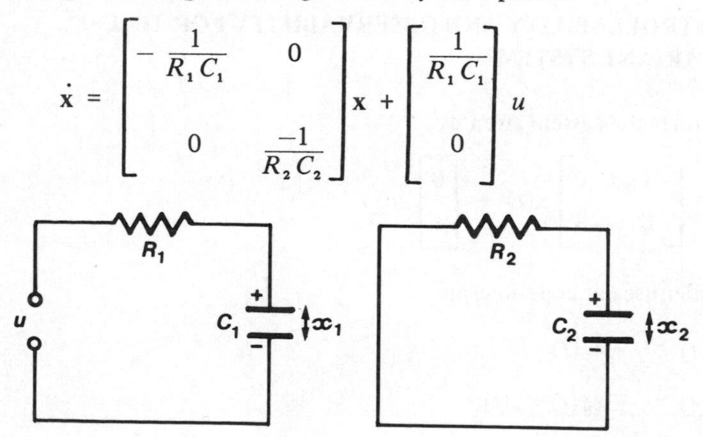

Figure 2.3

It is uncontrollable. (Why?) A formal definition of controllability follows.

A system is said to be *controllable* if it is possible to find a control vector $\mathbf{u}(t)$ which, in a specified finite time t_f, will transfer the system from an arbitrary state $\mathbf{x}_0$ to an arbitrary state $\mathbf{x}_f$.

Example 2

The system depicted in Figure 2.4 is observable.

† The concept of output controllability also exists.

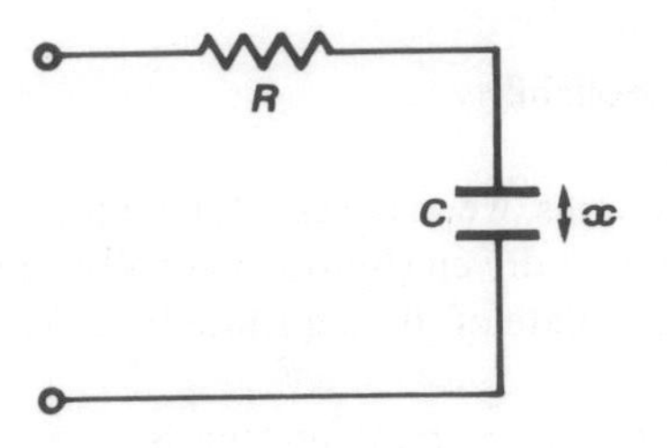

Figure 2.4

However, the system depicted in Figure 2.5 is unobservable.

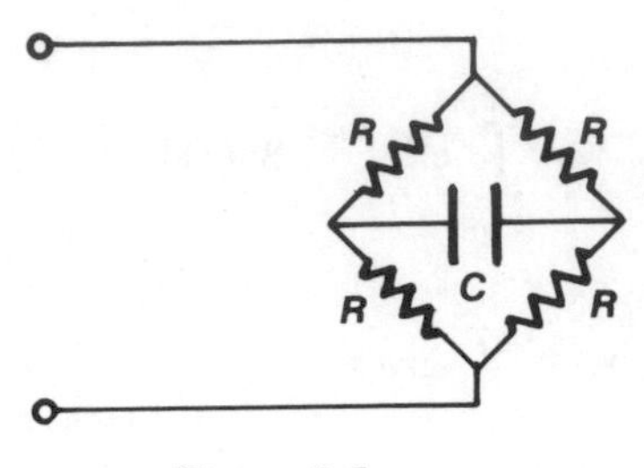

Figure 2.5

A system is said to be **observable** within the period $t_0 < t < t_f$ if the initial state $\mathbf{x}_0$ can be determined from the output $\mathbf{y}(t)$ observed over the same period.

Gilbert's criterion for controllability

Suppose we have a system with distinct eigenvalues which has been diagonalised to

$$\begin{bmatrix} \dot{\zeta}_1 \\ \dot{\zeta}_2 \\ \vdots \\ \dot{\zeta}_n \end{bmatrix} = \begin{bmatrix} \lambda_1 & & & \bigcirc \\ & \lambda_2 & & \\ & & \ddots & \\ \bigcirc & & & \lambda_n \end{bmatrix} \begin{bmatrix} \zeta_1 \\ \zeta_2 \\ \vdots \\ \zeta_n \end{bmatrix} + \begin{bmatrix} \beta_{11} & \beta_{12} & \cdots & \beta_{1n} \\ \beta_{21} & \beta_{22} & \cdots & \beta_{2n} \\ \vdots & \vdots & & \vdots \\ \beta_{n1} & \beta_{n2} & \cdots & \beta_{nn} \end{bmatrix} \begin{bmatrix} u_1 \\ u_2 \\ \vdots \\ u_n \end{bmatrix} \qquad (2.50)$$

Since the state variables are decoupled from each other, a necessary condition for controllability is that the matrix $[\beta_{ij}]$ must have no zero rows, otherwise the controls will not be able to influence all the state variables. Actually, controllability implies rather more than 'influence', since we require to effect a transfer between two arbitrary states in a specified finite time. (In fact, we can control a finite number of states by one control.) This interpretation in terms of the modes λ_i, is physically meaningful, but does necessitate reduction to canonical form. A criterion which relates more directly to the definition of controllability is now developed.

Kalman's criterion for controllability

For time-invariant systems, we may state the requirement for controllability as the requirement that the system be driven (by the control) from an arbitrary initial state $\mathbf{x}(0)$ at time $t = 0$ to a final state of $\mathbf{0}$ in a finite time T. (Verify this.)

The state solution for time-invariant systems is

$$\mathbf{x}(t) = e^{\mathbf{A}t}\,\mathbf{x}(0) + \int_0^t e^{\mathbf{A}(t-\tau)}\,\mathbf{B}\mathbf{u}(\tau)d\tau \tag{2.33}$$

At time T, therefore,

$$\mathbf{0} = e^{\mathbf{A}T}\,\mathbf{x}(0) + \int_0^T e^{\mathbf{A}(T-\tau)}\,\mathbf{B}\mathbf{u}(\tau)d\tau \tag{2.51}$$

Multiplication throughout by $e^{-\mathbf{A}T}$ gives

$$\mathbf{0} = \mathbf{x}(0) + \int_0^T e^{-\mathbf{A}\tau}\,\mathbf{B}\mathbf{u}(\tau)d\tau$$

Writing $e^{-\mathbf{A}\tau}$ as an infinite series and using the fact that the matrix $\mathbf{A}$ satisfies a minimur polynomial of degree p, we can then write

$$e^{-\mathbf{A}\tau} = \sum_{r=0}^{p-1} a_r(\tau)\mathbf{A}^r$$

where $a_r(\tau)$ are coefficients dependent on τ. Then

$$\mathbf{x}(0) = -\sum_{r=0}^{p-1} \mathbf{A}^r\mathbf{B} \int_0^T a_r(\tau)\mathbf{u}(\tau)d\tau$$

i.e.

$$\mathbf{x}(0) = -\sum_{r=0}^{p-1} \alpha_r\,\mathbf{A}^r\,\mathbf{B} \tag{2.52}$$

where α_r are constants.

Consider now the $(n \times pn)$ matrix $\mathbf{Q} = [\mathbf{B}, \mathbf{AB}, \mathbf{A}^2\mathbf{B}, \ldots, \mathbf{A}^{p-1}\mathbf{B}]$. (2.53)

The right-hand-side of (2.52) is a linear combination of the columns of **Q**. Since **x**(0) is arbitrary, *any* vector in state space must be such a linear combination. For the system to be controllable, then, the columns of **Q** must span the state space. A necessary condition for the system to be controllable is that

$$\text{rank of } \mathbf{Q} = n$$

(Note that if $p < n$ the matrices $\mathbf{A}^r\mathbf{B}$ where $r = p, \ldots, (n-1)$ are comprised of columns which are linearly dependent on the columns of **Q** and therefore a necessary condition for the system to be controllable is that the rank of the matrix $\mathbf{P} = [\mathbf{B}, \mathbf{AB}, \mathbf{A}^2\mathbf{B}, \ldots, \mathbf{A}^{n-1}\mathbf{B}]$ is n. This avoids the need to find the degree of the minimum polynomial. Conversely, if the rank of **P** is less than n, say r, then the set of all initial states which can be transferred to the origin in finite time is r-dimensional and therefore the system cannot be completely state controllable.)

Example 1

For the thermal system, $\mathbf{A} = \begin{bmatrix} -3 & 1 \\ 2 & -2 \end{bmatrix}$, $\mathbf{B} = \begin{bmatrix} 1 \\ 0 \end{bmatrix}$. Here $n = 2$ and $\mathbf{AB} = \begin{bmatrix} -3 \\ 2 \end{bmatrix}$. The matrix $\mathbf{P} \equiv [\mathbf{B}, \mathbf{AB}] = \begin{bmatrix} 1 & -3 \\ 0 & 2 \end{bmatrix}$ and clearly has rank 2. The system is then state controllable, i.e. we can control (in the sense defined at the start of this section) both interior and wall temperatures by the heating coil.

Example 2

For the voltmeter, $\mathbf{A} = \begin{bmatrix} 0 & 1 & 0 \\ -1 & 0 & 2 \\ 0 & -5 & -6 \end{bmatrix}$, $\mathbf{B} = \begin{bmatrix} 0 \\ 0 \\ 1 \end{bmatrix}$ and $n = 3$. Therefore

$\mathbf{AB} = \begin{bmatrix} 0 \\ 2 \\ -6 \end{bmatrix}$ and $\mathbf{A}^2\mathbf{B} = \begin{bmatrix} 2 \\ -12 \\ 26 \end{bmatrix}$. The matrix $\mathbf{P} = \begin{bmatrix} 0 & 0 & 2 \\ 0 & 2 & -12 \\ 1 & -6 & 26 \end{bmatrix}$ and has rank 3.

Again the system is controllable.

Example 3

The bridge system of Figure 2.6 gives rise to the equations

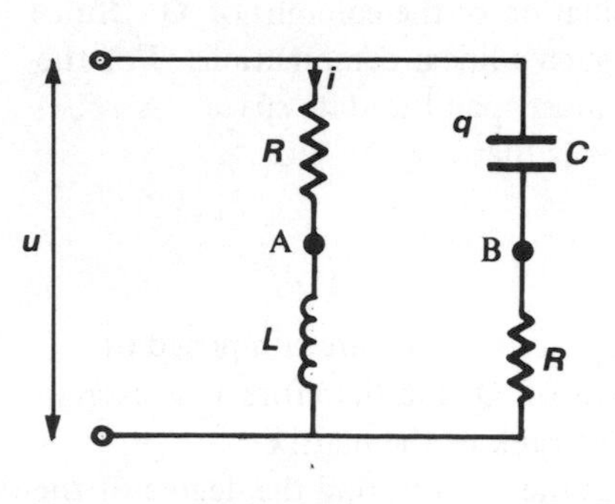

$$\dot{\mathbf{x}} = \begin{bmatrix} -\frac{1}{RC} & 0 \\ 0 & -\frac{R}{L} \end{bmatrix} \mathbf{x} + \begin{bmatrix} \frac{1}{R} \\ \frac{1}{L} \end{bmatrix} u$$

where the state variables are

$$x_1 = q, \quad x_2 = i$$

Figure 2.6

The system is controllable if $-\frac{1}{L^2} + \frac{1}{LCR^2} > 0$ but uncontrollable if $R^2 = L/C$, for then the bridge is balanced in the sense that an ammeter connected across AB indicates zero current. (Why is this?)

Note the result that a system can be transferred into the canonical form if and only if it is controllable.

Kalman's criterion does not say *how* to choose a control vector to perform the change of state. Barnett (1975) provides a theorem which gives a formula for the determination of a suitable control.

Criterion for Observability

Again consider the system $\dot{\mathbf{x}} = \mathbf{Ax} + \mathbf{Bu}$ with $\mathbf{y} = \mathbf{Cx}$ where $\mathbf{C}$ is an $m \times n$ matrix. We have defined a system to be observable if knowledge of the output over a finite interval $t_0 < t < t_1$ allows us to determine the initial state $\mathbf{x}(t_0)$.

Without loss of generality take t_0 to be zero. (Why?) We have

$$\mathbf{y} = \mathbf{C}e^{\mathbf{A}t}\,\mathbf{x}(0) + \mathbf{C} \int_0^t e^{\mathbf{A}(t-\tau)}\,\mathbf{Bu}(\tau)d\tau \tag{2.26}$$

where $\mathbf{A}$, $\mathbf{B}$ and $\mathbf{C}$ are known, as is $\mathbf{u}(t)$, so that the second term on the right-hand side is a completely known quantity. We require to know whether knowledge of $\mathbf{C}e^{\mathbf{A}t}\,\mathbf{x}(0)$ for $0 < t < t_1$ allows us to determine $\mathbf{x}(0)$ uniquely.

A parallel argument to that used to establish the controllability criterion is employed. This leads to the equation (cf(2.52))

$$\mathbf{C}e^{\mathbf{A}t}\,\mathbf{x}(0) = \sum_{r=0}^{p-1} b_r(t)\mathbf{C}\mathbf{A}^r\,\mathbf{x}(0)$$

However, the full proof of the observability criterion is beyond our scope and we merely state it here.

A necessary and sufficient condition for the system to be observable is that the matrix

$$\mathbf{R} = \begin{bmatrix} \mathbf{C} \\ \mathbf{CA} \\ \vdots \\ \mathbf{CA}^{n-1} \end{bmatrix} \qquad \text{has rank } n.$$

Example

For the meter we had $\mathbf{C} = [1,\ 0,\ 0]$. Now $\mathbf{CA} = [0,\ 1,\ 0]$ and $\mathbf{CA}^2 = [-1,\ 0,\ 2]$.

Therefore $\mathbf{R} = \begin{bmatrix} 1 & 0 & 0 \\ 0 & 1 & 0 \\ -1 & 0 & 2 \end{bmatrix}$ and has rank 3. The system is observable.

Comments

The similarity between the criteria for observability and controllability is emphasised in the following result.

The system $\dot{\mathbf{x}} = \mathbf{Ax}(t) + \mathbf{Bu}(t)$, $\mathbf{y}(t) = \mathbf{Cx}(t)$ is controllable if and only if the dual system $\dot{\mathbf{x}} = -\mathbf{A}^{\mathrm{T}}(t)\mathbf{x}(t) + \mathbf{C}^{\mathrm{T}}\mathbf{u}(t)$, $\mathbf{y}(t) = \mathbf{B}^{\mathrm{T}}\mathbf{x}(t)$ is observable and vice versa. (If the matrices have complex elements the transpose operation is replaced by conjugate transpose.)

The Gilbert criterion for observability is: If there are distinct eigenvalues and the transformation to diagonal form is accomplished via $\mathbf{x} = \mathbf{Ty}$ then the system is observable, if the matrix $\mathbf{CT}$ contains no zero columns.

It is worthy of note that a general linear time invariant system can be split into four mutually exclusive parts: controllable and observable, controllable but unobservable, uncontrollable but observable, uncontrollable and unobservable. Of course, some of these parts may be absent in particular systems.

The **transfer function** model represents only the controllable and observable part of a system.

Example

The system depicted in the transfer function diagram of Figure 2.7 has transfer function $G(s) = \frac{s-1}{s+5}$. In the transfer function approach we would conclude that the system is stable.

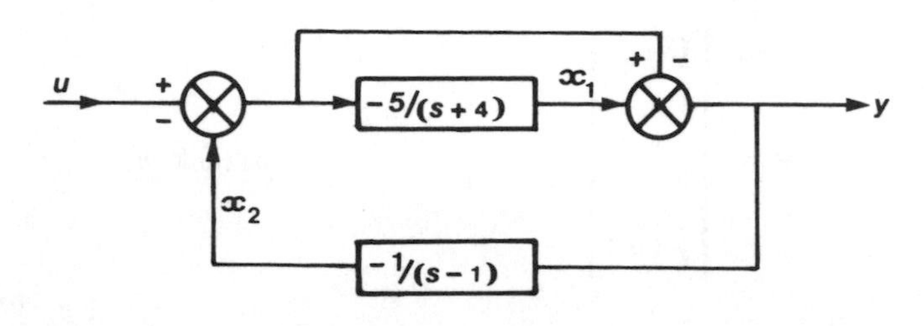

Figure 2.7

However, the state model is

$$\dot{\mathbf{x}} = \begin{bmatrix} -4 & 5 \\ 1 & 0 \end{bmatrix} \mathbf{x} + \begin{bmatrix} -5 \\ 1 \end{bmatrix} u$$

with $\mathbf{y} = (1, \ -1)\mathbf{x} + u$.

The eigenvalues of $\mathbf{A}$ are -5 and 1 and this implies instability (see Chapter 3). The control $u(t)$ influences the dynamics of the unobservable, controllable part of the system and this happens to be unstable. Its fluctuations under the action of the control do not appear at the output since the state model is unobservable.

Finally, we mention the concept of **reconstructibility**, which some authors prefer to observability. A system is said to be reconstructible if knowledge of the output variable for $t < t_1$ implies that there is a time $t_0 < t_1$ at which the system can be completely determined; if $\mathbf{x}(t_0)$ is known then $\mathbf{x}(t)$ can be determined.

The system is reconstructed from past output records, whereas **observability** implies that the system state at t_0 can be gleaned from *future* observations. For time invariant systems the one property implies the other.

Problems

1. Is the system $\dot{\mathbf{x}} = \begin{bmatrix} 1 & 2 & -1 \\ 0 & 1 & 0 \\ 1 & -4 & 3 \end{bmatrix} \mathbf{x} + \begin{bmatrix} 0 \\ 0 \\ 1 \end{bmatrix} u$ with $\mathbf{y} = (1, \ -1, \ 1)\mathbf{x}$

(i) controllable? (ii) observable?

2. For which of the three vectors **b** below is the system controllable?

$$\dot{\mathbf{x}} = \begin{bmatrix} 1 & 1 & 0 \\ 0 & 1 & 0 \\ 0 & 0 & 1 \end{bmatrix} \mathbf{x} + \begin{bmatrix} 0 \\ 1 \\ 1 \end{bmatrix} u_1 + \mathbf{b} u_2$$

(i) $\begin{bmatrix} 0 \\ 0 \\ 0 \end{bmatrix}$ (ii) $\begin{bmatrix} 0 \\ 0 \\ 1 \end{bmatrix}$ (iii) $\begin{bmatrix} 1 \\ 0 \\ 0 \end{bmatrix}$

3. Show that the system

$$\dot{\mathbf{x}} = \begin{bmatrix} -0.75 & -0.25 \\ -0.5 & -0.5 \end{bmatrix} \mathbf{x} + \begin{bmatrix} 1 \\ 1 \end{bmatrix} u, \quad \mathbf{y} = (4,\ 2)\mathbf{x}$$

is neither controllable nor observable.

4. Show that the system $\dot{\mathbf{x}} = \begin{bmatrix} 2 & -5 \\ -4 & 0 \end{bmatrix} \mathbf{x} + \begin{bmatrix} 1 \\ -1 \end{bmatrix} u$ with $\mathbf{y} = (1,\ 1)\mathbf{x}$ is controllable and observable.

5. Show that the system $\dot{\mathbf{x}} = \mathbf{A}\mathbf{x} + \mathbf{b}u$ with $\mathbf{A} = \begin{bmatrix} -3 & 1 \\ -2 & 1.5 \end{bmatrix}$ is controllable if $\mathbf{b} = \begin{bmatrix} 0 \\ 1 \end{bmatrix}$ but not controllable if $\mathbf{b} = \begin{bmatrix} 1 \\ 4 \end{bmatrix}$.

6. Show that the system

$$\dot{\mathbf{x}} = \begin{bmatrix} -1 & -2 & -2 \\ 0 & -1 & 1 \\ 1 & 0 & -1 \end{bmatrix} \mathbf{x} + \begin{bmatrix} 2 \\ 0 \\ 1 \end{bmatrix} u; \qquad \mathbf{y} = (1,\ 1,\ 0)\mathbf{x}$$

is controllable and observable.

7. The system $\dot{\mathbf{x}} = \begin{bmatrix} 2 & 0 & 0 \\ 0 & 2 & 0 \\ 0 & 3 & 1 \end{bmatrix} \mathbf{x}$, $\mathbf{y} = (1,\ 1,\ 1)\mathbf{x}$ is not observable, but if

$$\mathbf{y} = \begin{bmatrix} 1 & 1 & 1 \\ 1 & 2 & 3 \end{bmatrix} \mathbf{x}$$ then it is observable. Verify these statements.

2.8 LINEAR FEEDBACK

Given the time invariant system $\dot{\mathbf{x}} = \mathbf{Ax} + \mathbf{Bu}$, assume linear feedback, i.e. each control variable is a linear combination of the state variables so that $\mathbf{u} = \mathbf{Gx}$ where $\mathbf{G}$ is a **gain matrix**. Then we obtain the **closed-loop** system

$$\dot{\mathbf{x}} = (\mathbf{A} + \mathbf{BG})\mathbf{x} \tag{2.54}$$

Two important results will be stated. We first define Λ to be a set of n arbitrary complex numbers of which the non-real ones occur in conjugate pairs. $\mathbf{A}$ and $\mathbf{B}$ are assumed to contain real elements.

(i) If the system $\dot{\mathbf{x}} = \mathbf{Ax} + \mathbf{Bu}$ is controllable, a matrix $\mathbf{G}$ can be found such that the characteristic roots of $(\mathbf{A} + \mathbf{BG})$ are the set Λ. Since the solution of (2.54) can be shown to depend on the characteristic roots of $(\mathbf{A} + \mathbf{BG})$ it follows that the choice of characteristic roots will influence the behaviour of the closed-loop system.

(ii) Similarly, if the system $\dot{\mathbf{x}} = \mathbf{Ax}$, $\mathbf{y} = \mathbf{Cx}$ is observable then there is a real column vector $\mathbf{d}$ such that the characteristic roots of $(\mathbf{A} + \mathbf{dC})$ are the set Λ.

2.9 DISCRETE SYSTEMS

This section contains a brief mention of discrete systems. First consider the uncontrolled system

$$\mathbf{x}(k + 1) = \mathbf{Ax}(k) \tag{2.55}$$

where $\mathbf{x}(k_0) = \mathbf{x}_0$.

The **notation** $\mathbf{x}(k)$ is used for $\mathbf{x}(kT)$. It is clear that $\mathbf{x}(k_0 + 1) = \mathbf{Ax}(k_0)$, $\mathbf{x}(k_0 + 2) = \mathbf{A}^2\mathbf{x}(k_0)$, etc. and therefore the solution of (2.55) may be written as

$$\mathbf{x}(k) = \mathbf{A}^{(k-k_0)}\mathbf{x}_0 \tag{2.56}$$

The matrix $\mathbf{A}^k$ may be found by methods modified from those which were used to find $e^{\mathbf{A}t}$. For example remember that Sylvester's formula in the case of distinct eigenvalues

λ_k was

$$f(\mathbf{A}) = \sum_{k=1}^{n} \mathbf{Z}_k f(\lambda_k) \tag{2.35}$$

where

$$\mathbf{Z}_k = \prod_{\substack{j=1 \\ j \neq k}}^{n} \frac{(\mathbf{A} - \lambda_j \mathbf{I})}{(\lambda_k - \lambda_j)} \tag{2.36}$$

Put $f(\mathbf{A}) = \mathbf{A}^k$. For the oven example, $\mathbf{A} = \begin{bmatrix} -3 & 1 \\ 2 & -2 \end{bmatrix}$ with eigenvalues $\lambda_1 = -1$, $\lambda_2 = -4$.

Then $\mathbf{Z}_1 = \frac{1}{3}(\mathbf{A} + 4\mathbf{I})$, $\mathbf{Z}_2 = -\frac{1}{3}(\mathbf{A} + \mathbf{I})$. Hence

$$\mathbf{A}^k = \frac{1}{3}(-1)^k(\mathbf{A} + 4\mathbf{I}) - \frac{1}{3}(-4)^k(\mathbf{A} + \mathbf{I})$$

The state transition matrix is defined by

$$\Phi(k, k_0) = \mathbf{A}^{k-k_0}$$

so that the solution (2.56) may be written

$$\mathbf{x}(k) = \Phi(k, k_0)\mathbf{x}(k_0) \tag{2.57}$$

The following properties may be quoted for Φ

(i) $\Phi(k + 1, k_0) = \mathbf{A}\Phi(k, k_0)$

(ii) $\Phi(k, k) = \mathbf{I}$

(iii) $\Phi(k_0, k) = [\Phi(k, k_0)]^{-1}$, if $\mathbf{A}^{-1}$ exists

(iv) $\Phi(k, k_0) = \Phi(k, k_1)\Phi(k_1, k_0)$ for $k_0 \leqslant k_1 \leqslant k$

(Compare these with the properties of Φ for the continuous case.)

For the system $\mathbf{x}(k + 1) = \mathbf{A}\mathbf{x}(k) + \mathbf{B}\mathbf{u}(k)$ it can be shown (the task is left to you) that the solution is

$$\mathbf{x}(k) = \Phi(k, k_0)\mathbf{x}(k_0) + \Phi(k, k_0) \sum_{r=k_0}^{k-1} \Phi(k_0, r + 1)\mathbf{B}\mathbf{u}(r) \tag{2.58}$$

(Again, compare the continuous case.)

For time varying matrices the system may be written

$$\mathbf{x}(k+1) = \mathbf{A}(kT)\mathbf{x}(k) + \mathbf{B}(kT)\mathbf{u}(k)$$

The solution is

$$\mathbf{x}(k) = \Phi(k, k_0)\mathbf{x}(k_0) + \Phi(k, k_0) \sum_{r=k_0}^{k-1} \Phi(k_0, r+1)\mathbf{B}(rT)\mathbf{u}(r) \qquad (2.59)$$

where $\Phi(k, k_0) = \prod_{r=k_0}^{k-1} \mathbf{A}(rT)$.

Note that if $\mathbf{A}$ is diagonal with elements $\lambda_1, \lambda_2, \ldots\ldots, \lambda_n$, then $\Phi(k, 0)$ is diagonal with elements $\lambda_1^k, \lambda_2^k, \ldots\ldots, \lambda_n^k$.

Discretization of continuous case

Consider $\dot{\mathbf{x}} = \mathbf{A}\mathbf{x} + \mathbf{B}\mathbf{u}$ with solution

$$\mathbf{x}(t) = e^{\mathbf{A}t}\,\mathbf{x}(0) + e^{\mathbf{A}t} \int_0^t e^{-\mathbf{A}\tau}\,\mathbf{B}\mathbf{u}(\tau)d\tau$$

and assume that we make sampling measurements of $\mathbf{x}$ at intervals of T seconds. Further assume that components of $\mathbf{u}(t)$ are constant during these intervals, i.e. $\mathbf{u}(t) = \mathbf{u}(kT)$ for the kth period of sampling. Then

$$\mathbf{x}([k+1]T) = e^{\mathbf{A}T}\,\mathbf{x}(kT) + e^{\mathbf{A}T} \int_0^T e^{-\mathbf{A}\tau}\,\mathbf{B}\mathbf{u}(kT)d\tau$$

$$= e^{\mathbf{A}T}\,\mathbf{x}(kT) + \int_0^T e^{\mathbf{A}(T-\tau)}\,\mathbf{B}\mathbf{u}(kT)d\tau$$

If we define $\mathbf{G}(T) = e^{\mathbf{A}T}$,

$$\mathbf{F}(T) = e^{-\mathbf{A}T}\left(\int_0^T e^{-\mathbf{A}\tau}\,d\tau\right)\mathbf{B}$$

and

$$\mathbf{H}(T) = \mathbf{G}(T)\mathbf{F}(T) = \left(\int_0^T e^{-\mathbf{A}\tau}\, d\tau \right)\mathbf{B}$$

Then

$$\mathbf{x}([k+1]T) = \mathbf{G}(T)\mathbf{x}(kT) + \mathbf{H}(T)\mathbf{u}(kT) \tag{2.60}$$

as you can verify.

Example

Given the system $\dot{\mathbf{x}} = \begin{bmatrix} -3 & 1 \\ 2 & -2 \end{bmatrix} \mathbf{x} + \begin{bmatrix} 1 \\ 0 \end{bmatrix} u$ we find the equivalent discrete system. Now

$$\mathbf{G}(T) = e^{\mathbf{A}T} = \begin{bmatrix} \frac{2}{3}e^{-4T} + \frac{1}{3}e^{-T} & -\frac{1}{3}e^{-4T} + \frac{1}{3}e^{-T} \\ -\frac{2}{3}e^{-4T} + \frac{2}{3}e^{-T} & \frac{1}{3}e^{-4T} + \frac{2}{3}e^{-T} \end{bmatrix}$$

Also

$$\mathbf{H}(T) = \left(\int_0^T e^{-\mathbf{A}\tau}\, d\tau \right)\mathbf{B}$$

$$= \begin{bmatrix} -\frac{1}{6}e^{-4T} - \frac{1}{3}e^{-T} & +\frac{1}{12}e^{-4T} - \frac{1}{3}e^{-T} \\ +\frac{1}{6}e^{-4T} - \frac{2}{3}e^{-T} & -\frac{1}{12}e^{-4T} - \frac{2}{3}e^{-T} \end{bmatrix} \begin{bmatrix} 1 \\ 0 \end{bmatrix}$$

$$= \begin{bmatrix} -\frac{1}{6}e^{-4T} - \frac{1}{3}e^{-T} \\ \frac{1}{6}e^{-4T} - \frac{2}{3}e^{-T} \end{bmatrix}$$

Therefore $\mathbf{x}([k+1]T) = \mathbf{G}(T) + \mathbf{H}(T)\mathbf{u}(kT)$ with the above matrices. Note that $\mathbf{G}(t)$

and $\mathbf{H}(t)$ may be defined via Laplace transforms as

$$\mathbf{G}(t) = \mathcal{L}^{-1}[(s\mathbf{I} - \mathbf{A})^{-1}] \tag{2.61}$$

$$\mathbf{H}(t) = \left\{\mathcal{L}^{-1}\left[\frac{1}{s}(s\mathbf{I} - \mathbf{A})^{-1}\right]\right\}\mathbf{B} \tag{2.62}$$

Try to derive these results and then apply them to the above example.

Problems

1. Show that the solution of the discrete system

$$\mathbf{x}(k+1) = \begin{bmatrix} 5/12 & 1/12 \\ 1/12 & 5/12 \end{bmatrix} \mathbf{x}(k) \quad \text{with} \quad \mathbf{x}(0) = \begin{bmatrix} 2 \\ 1 \end{bmatrix}$$

is

$$\mathbf{x}(k) = 1.5(0.5)^k \begin{bmatrix} 1 \\ 1 \end{bmatrix} - 0.5(1/3)^k \begin{bmatrix} -1 \\ 1 \end{bmatrix}.$$

2. The discrete system

$$\mathbf{x}(k+1) = \begin{bmatrix} -0.75 & -0.25 \\ -0.25 & -0.75 \end{bmatrix} \mathbf{x}(k) + \begin{bmatrix} 1 \\ 0 \end{bmatrix} u(k)$$

where $\mathbf{x}(0) = \begin{bmatrix} 0 \\ 0 \end{bmatrix}$ and $u(k) \equiv 1$ has output equation $y(k) = (1,\ 0)\mathbf{x}(k) + 4u(k)$.

Show that $y(k) = 4 + (1/12)[7 - 3(-1)^k - 4(0.5)^k]$.

Chapter Three

Stability of Systems

3.1 INTRODUCTION

This chapter is concerned with the stability of a system. To fix some ideas consider a small bead rolling on the surface of Figure 3.1. If placed at any of the points A, B, C, D to F and G, the bead would remain there. These points are **equilibrium points** for the system.

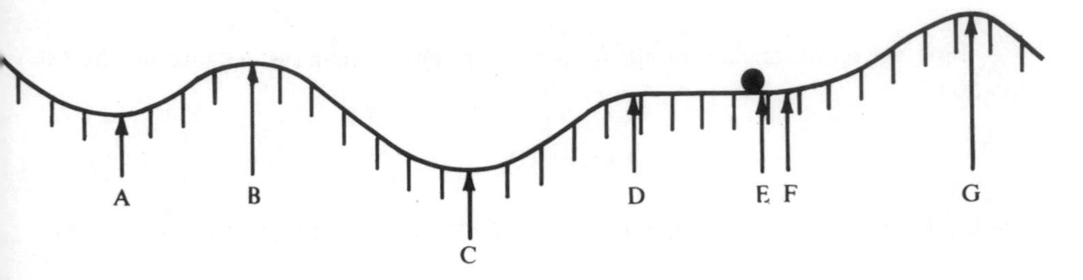

Figure 3.1

After an infinitesimal displacement of the bead away from A or C the bead will return to these points: they are points of **stable equilibrium**. However, an infinitesimal displacement from B or G will lead to the bead moving yet further from these points of **unstable equilibrium**. When the bead is moved slightly from E it will remain in its new position: E is said to be a point of **neutral equilibrium**. These points are points of **ocal stability**; if the bead were displaced sufficiently far from A or C it would not eturn to these points and therefore stability depends on the size of the perturbation.

If we seek to apply these ideas to dynamical systems, the analogy to be drawn is hat of the surface of Figure 3.1 changing with time. A point such as A could be stable or some of the time and unstable at other times. For a dynamical system, the point $\mathbf{x}_c$ s an equilibrium point if $\mathbf{x}(t_0) = \mathbf{x}_c$ and $\mathbf{x}(t) = \mathbf{x}_c$ for $t > t_0$ when no disturbances or nputs to the system occur. Note that the origin of state space is always an equilibrium oint.

In this chapter only isolated equilibrium points will be considered. For a linear ystem, the origin is the only isolated equilibrium point. In a non-linear system, each solated equilibrium point can be removed to the origin by a change of variable: $' = \mathbf{x} - \mathbf{x}_c$. The questions to be answered by stability tests are

i) if the system is displaced from equilibrium will it return to that state, remain near to it, or diverge from it?

(ii) how is the situation affected by inputs or disturbances?

The chapter begins by considering a simple dynamical system which gives rise to a linear second-order differential equation with constant coefficients. For this system a pictorial representation, called a phase plane portrait is developed. Then the general features of such portraits are discussed. Some definitions of stability are given. The Routh-Hurwitz criterion is studied as is the direct method of Liapunov. Some methods are given for constructing Liapunov functions, which give an indication of the stability of a system. Finally, the effects of controls on the stability of a system are discussed.

3.2 PHASE PLANE PORTRAITS

First, we recall standard results for a mass-spring-damping system governed by the equation

$$\ddot{x} + 2k\dot{x} + n^2 x = 0 \tag{3.1}$$

Case (a) k = 0 This is the case of no damping which has the simple harmonic solution

$$x = A \cos nt + B \sin nt \tag{3.2}$$

where A and B are constants. The graph of x against t is shown in Figure 3.2(a) for the case $x(0) = 0$, $\dot{x}(0) > 0$.

Case (b) 0 < k < n This is the case of light damping or underdamping. If we write $k^2 - n^2 = -p^2$ the solution is

$$x = e^{-kt}(A \cos pt + B \sin pt) \tag{3.3}$$

Refer to Figure 3.2(b) for the problem with $x(0) = 0$, $\dot{x}(0) > 0$.

Case (c) k = n Critical damping. The solution is

$$x = (A + Bt)\, e^{-kt} \tag{3.4}$$

See Figure 3.2(c); again, $x(0) = 0$, $\dot{x}(0) > 0$.

Case (d) k > n Overdamping. The solution is

$$x = A\, e^{(-k+q)t} + B\, e^{(-k-q)t} \tag{3.5}$$

where $k^2 - n^2 = q^2$. The graph resembles Figure 3.2(c) if $x(0) = 0$, $\dot{x}(0) > 0$.

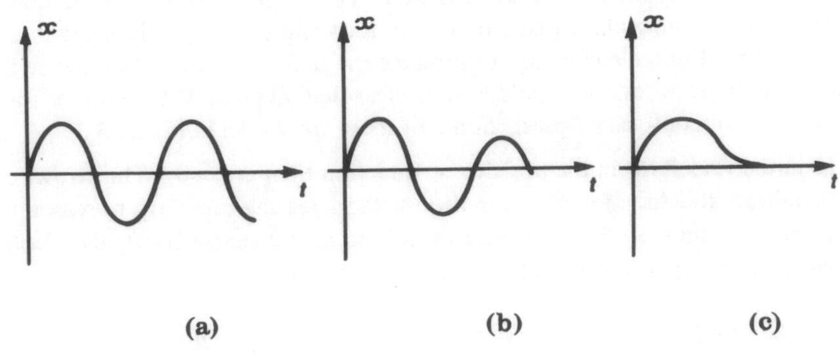

Figure 3.2

If (3.1) is written in terms of state variables $x_1 = x$ and $x_2 = \dot{x}$, then $\dot{x}_1 = x_2$ and $\dot{x}_2 = -2kx_2 - n^2 x_1$ so that $\dot{\mathbf{x}} = \mathbf{A}\mathbf{x}$ where

$$\mathbf{A} = \begin{bmatrix} 0 & 1 \\ -n^2 & -2k \end{bmatrix} \qquad (3.6)$$

The phase variables for the system, viz. x and $\dot{x}$, are also the state variables. A **phase trajectory** of the system is a plot of $\dot{x}$ **against** x, starting from a particular initial condition. The **phase plane** is the $x_1 - x_2$ plane. A **portrait** is a family of typical trajectories. The phase plane portraits for the various cases are now examined.

Case (a) Put $n = 2$ for simplicity. Then we have $\dot{x}_1 = x_2$, $\dot{x}_2 = -4x_1$ so that

$$\frac{dx_2}{dx_1} = \frac{\dot{x}_2}{\dot{x}_1} = \frac{-4x_1}{x_2}$$

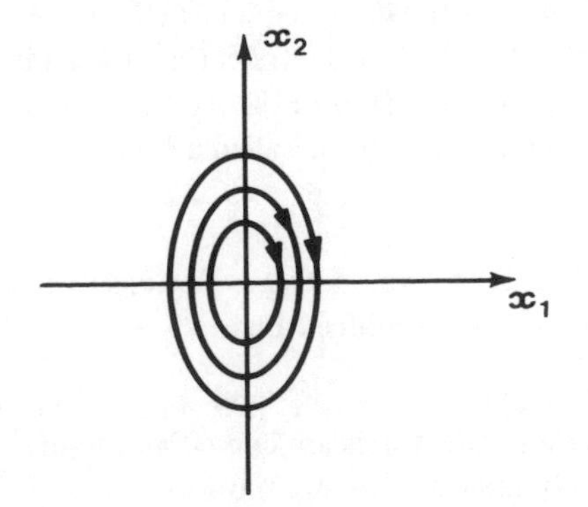

Larger values of C correspond to trajectories further from the origin.

Figure 3.3

This last equation may be integrated to obtain $4x_1^2 + x_2^2 = C$, where C is constant. The value of C may be determined by the initial conditions. For example, the graph of

Figure 3.2(a) could correspond to initial conditions $x_1 = 0$, $x_2 = u$ at $t = 0$ so that $C = u^2$. The corresponding phase plane trajectory is an ellipse: Why? It is possible for an infinite number of initial conditions to produce the same value of C and hence the same phase plane trajectories. It should also be clear that all possible trajectories for this case form a family of confocal ellipses. Some of these are shown in Figure 3.3.

If the initial velocity is in the positive x direction then $x_2 > 0$. The arrows on the trajectories indicate the direction of the motion with increasing time; try to reason why they are in the sense shown. Such a portrait is said to have a **centre** at (0, 0). Note that the eigenvalues for this particular matrix **A** are $\lambda_1 = 2i$, $\lambda_2 = -2i$.

Case (b) Take $n = 2$, $k = 1$ so that $\dot{x}_1 = x_2$, $\dot{x}_2 = -4x_1 - 2x_2$ and $dx_2/dx_1 = (-4x_1 - 2x_2)/x_2$.

In this instance the eigenvalues of **A** are $\lambda_1 = -1 + \sqrt{3}i$ and $\lambda_2 = -1 - \sqrt{3}i$. With the help of isoclines Figure 3.4 can be produced.

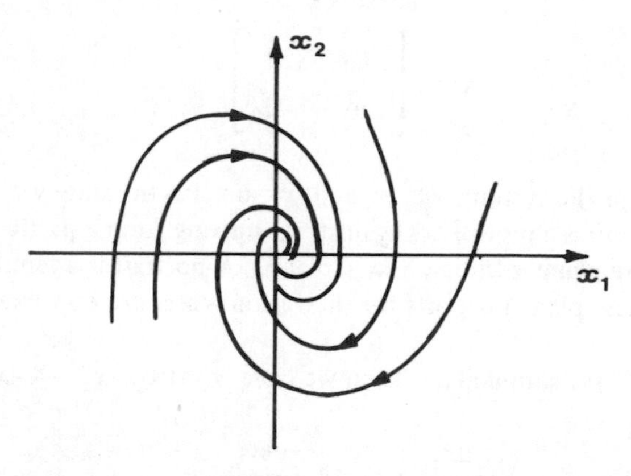

Figure 3.4

The link with the dynamical system is clear. As the damping takes effect the amplitude of the oscillations decays and the velocity falls; so the trajectory spirals in towards the origin but takes an infinite time to get there just as in theory the oscillations require an infinite time to be damped out. The point (0, 0) is called a **focus**. It represents a point of stable equilibrium.

Case (d) Take $n = 2$, $k = 2.5$ so that $\dot{x}_1 = x_2$, $\dot{x}_2 = -4x_1 - 5x_2$. The eigenvalues of **A** are $\lambda_1 = -1$, $\lambda_2 = -4$. With the aid of isoclines we produced Figure 3.5.

There are four rectilinear paths, lying on the lines $x_2 = -x_1$ and $x_2 = -4x_1$; the slopes of these lines are the eigenvalues. All other trajectories are curved and tend towards the line $x_2 = -x_1$. This corresponds dynamically to any motion $x = Ae^{-t} + Be^{-4t}$ being approximated by $x = Ae^{-t}$ for large t (unless of course $A = 0$).

Try to discover other links between the dynamics and the phase plane portraits so far mentioned.

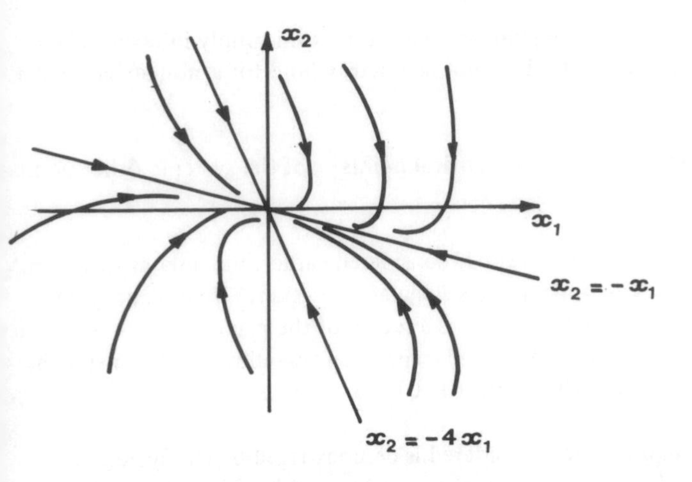

The point (0, 0) is a **node**. Notice that each trajectory **enters** the node and that it does so in a particular direction.

Figure 3.5

Also show that the phase portrait for case (c) also indicates a node at (0, 0). What is the difference from case (d)?

In general, we remark that the phase plane approach has two applications.

(i) it provides a general view of the types of transient response that may result from a variety of initial conditions;

(ii) it is possible to determine numerically a particular transient response corresponding to specific initial conditions (see Thaler and Brown (1960)).

Autonomous systems in the plane

Consider the system

$$\dot{\mathbf{x}} = \mathbf{f}(\mathbf{x}) \tag{3.7}$$

Since time does not appear **explicitly** on the right-hand side the system is called **autonomous.** In a two-dimensional system (3.7) is sometimes written

$$\begin{aligned} \dot{x} &= P(x, y) \\ \dot{y} &= Q(x, y) \end{aligned} \tag{3.8}$$

It is assumed that P and Q possess continuous first partial derivatives.† It follows that, given any t_0 and (x_0, y_0) there exists a unique solution satisfying the initial conditions. It is straightforward to show that if $x = x(t)$, $y = y(t)$ is a solution, then so

† Note that $P(0,0) = 0 = Q(0,0)$

is $x_1 = x(t+c)$, $y_1 = y(t+c)$ for a constant c. (What does this imply in terms of the phase plane trajectory?) This property does not necessarily hold for a non-autonomous system.

Points $\mathbf{x}$ at which $\mathbf{f}(\mathbf{x}) = 0$ are called **critical points**† of the system; other points are referred to as **regular**.

For linear systems $\dot{\mathbf{x}} = \mathbf{A}\mathbf{x}$, such as those considered earlier, the only critical point was the origin; this will be the case unless $\mathbf{A}$ is singular. In general we look at systems with **isolated** critical points $\mathbf{x}_c$ i.e. where there is a region of the plane $0 < |\mathbf{x} - \mathbf{x}_c| < \rho$ in which $\mathbf{x}_c$ is the *only* critical point. A simple translation will allow us to transfer the critical point under consideration to the point (0, 0).

One type of critical point not yet considered is demonstrated by the system $\dot{x}_1 = x_1 - x_2$, $\dot{x}_2 = -4x_1$. The eigenvalues are $\lambda_1 = -1$, $\lambda_2 = 2$. The solution is $x_1 = Ae^{-t} + Be^{2t}$, $x_2 = 2Ae^{-t} - Be^{2t}$; A and B constants. The phase plane portrait appears in Figure 3.6.

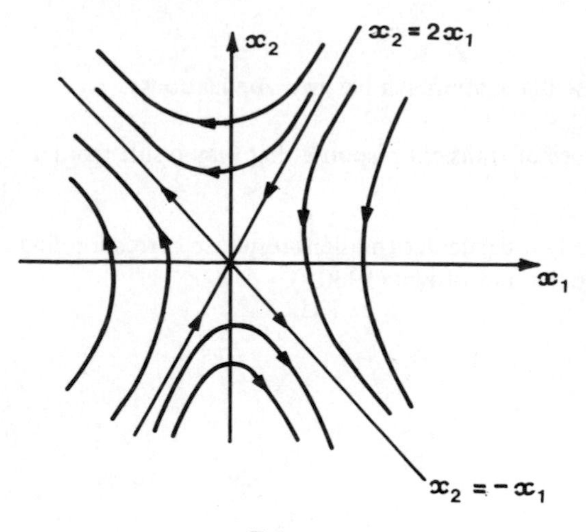

Figure 3.6

It is left to you to show that the phase trajectories have the equation

$$(x_1 + x_2)^2 (x_2 - 2x_1) = C$$

where C is constant.

Note that for large $t > 0$, $x \to Be^{2t}$, $y \to -Be^{2t}$ so that $y \sim -x$. Similarly, for large $t < 0$, $y \sim 2x$. Note that on $y = 2x$, the trajectory moves towards the origin, since $x = Ae^{-t}$, $y = 2Ae^{-t}$ whereas on $y = -x$ the trajectory moves away from the origin Note that other trajectories are horizontal on the y-axis and vertical on the x-axis. The critical point is called a **saddle point**. It is the *only* critical point crossed by a trajectory.

Intuitively, we expect the origin to be a stable critical point if nearby trajectories reach it and remain there. If they approach it, e.g. in the progress towards a focus, then

† Sometimes called **singular points** or **equilibrium points**.

we might refer to the critical point as stable. Could we refer to a centre as a stable point?

Classification of critical points

By solving the equations and plotting the phase plane portraits for linear systems where $\mathbf{A} = \begin{bmatrix} a_{11} & a_{12} \\ a_{21} & a_{22} \end{bmatrix}$, a_{ij} constants, it is found that the classification of critical points revolves around the nature of the eigenvalues of **A**. The main results are shown in Table 3.1. You are left to complete the table and to recast the results for the second-order dynamical systems studied earlier.

Table 3.1

Nature of eigenvalues	Stability of critical point	Nature of critical point
1) Imaginary conjugates	Stable?	Centre
2) Complex conjugates		
(i) Real part < 0	Stable	Focus
(ii) Real part > 0	Unstable	
3) Real, same sign		
(i) roots < 0	Stable	Node
(ii) roots > 0	Unstable	
4) Real, opposite sign	Unstable	Saddle Point

Three variables

Some results for a third-order system are now quoted. The system is $\dddot{x} + a\ddot{x} + b\dot{x} + c = 0$, where a, b and c are constants. State variables are $x_1 = x$, $x_2 = \dot{x}$ and $x_3 = \ddot{x}$.

- If all eigenvalues are real and of the same sign, the critical point is a node; if the common sign is negative the node is stable; if the common sign is positive the node is unstable.

- If the eigenvalues are real but of different sign, the critical point is a saddle point.

- If one eigenvalue is real and negative and the other two are purely imaginary and conjugate, the critical point is a centre.

- If one eigenvalue is real and the other two are complex conjugates then

(a) if the sum of the conjugates has the same sign as the real eigenvalue the critical point is a **focus** (Figure 3.7(a)).

(b) otherwise the critical point is a **saddle focus** (Figure 3.7(b)).

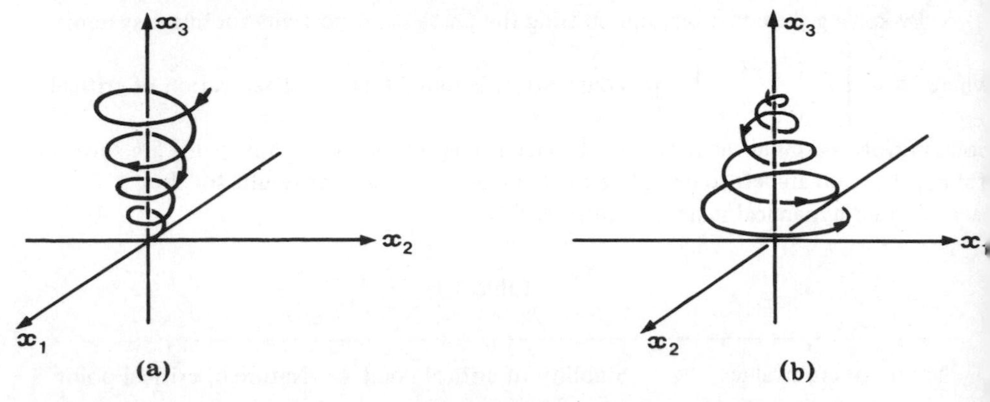

Figure 3.7

Note that if the real parts of the eigenvalues are negative the critical point is **stable.** If at least *one* real part is positive the critical point is **not stable**.

Linearisation

A non-linear system can be linearised near its (isolated) critical points, when their locations are known, to help determine the nature of these critical points. If we return to (3.8) we can expand $P(x, y)$ and $Q(x, y)$ in Taylor series centred on a critical point (x_c, y_c). Then the linearised approximation may be written

$$\frac{\mathrm{d}}{\mathrm{d}t}\begin{bmatrix} x - x_c \\ y - y_c \end{bmatrix} = \begin{bmatrix} \dot{x} \\ \dot{y} \end{bmatrix} = \mathbf{J}(x_c, y_c)\begin{bmatrix} x - x_c \\ y - y_c \end{bmatrix}$$

where

$$\mathbf{J}(x_c, y_c) = \begin{bmatrix} \dfrac{\partial P}{\partial x} & \dfrac{\partial P}{\partial y} \\[2ex] \dfrac{\partial Q}{\partial x} & \dfrac{\partial Q}{\partial y} \end{bmatrix} \tag{3.9}$$

and where all partial derivatives are evaluated at (x_c, y_c).

It can be shown that if any of the eigenvalues of (3.9) has positive real part, the critical point is unstable for the non-linear system. If the eigenvalues all have negative real parts the critical point is asymptotically stable.† Where some eigenvalues have zero real parts, it is not possible to distinguish between stability and instability of the critical point using the linearised approximation.

Some conclusions about the type of critical point can be drawn from consideration of the eigenvalues. Let the eigenvalues be λ_1, λ_2. Then

(a) If $\lambda_1 = \lambda_2$ are real and of the same sign — NODE

(b) If $\lambda_1 \neq \lambda_2$ are real and of opposite sign — SADDLE POINT

(c) If $\lambda_1 = \lambda_2$ are real and it is NOT true that $a_{11} = a_{22} \neq 0$, $a_{12} = a_{21} = 0$ — NODE

(d) If λ_1, λ_2 complex conjugate, real part $\neq 0$ — FOCUS

(e) If $\lambda_1 = \lambda_2$, real, $a_{11} = a_{22} \neq 0$, $a_{12} = a_{21} = 0$ — NODE OR FOCUS

(f) If λ_1, λ_2 purely imaginary — CENTRE OR FOCUS

Note that whereas the type of critical point may be the same as for the linearised system the actual phase plane portrait may be slightly different, for example, a non-linear saddle point may have two curved lines in place of the rectilinear trajectories corresponding to the eigenvectors.

Example 1

A relaxation oscillator is an electronic device which produces alternating current at a frequency determined by the time taken to charge or discharge a capacitor through a resistance. The governing equation for such a circuit with suitably chosen values of the parameters is

$$\frac{d^2 i}{dt^2} + 2\frac{di}{dt} - 4i + 9i^3 = 0$$

With the obvious state variables $x_1 = i$ and $x_2 = di/dt$ we find critical points where $\dot{x}_1 = \dot{x}_2 = 0$ i.e. $x_2 = 0$ and $4x_1 - 9x_1^3 - 2x_2 = 0$ i.e. $x_1 = 0$ or $x_1 = \pm 2/3$. The critical points are $(0, 0)$, $(2/3, 0)$, $(-2/3, 0)$.

The matrix of (3.9) is given by

$$\mathbf{J} = \begin{bmatrix} 0 & 1 \\ 4 - 27x_1^2 & -2 \end{bmatrix}$$

† See page 88 for a definition.

At (0, 0) $\mathbf{J} = \begin{bmatrix} 0 & 1 \\ 4 & -2 \end{bmatrix}$; its eigenvalues are $\lambda = -1 \pm \sqrt{5}$ and hence (0, 0) is a saddle point.

At (2/3, 0) and at (–2/3, 0) $\mathbf{J} = \begin{bmatrix} 0 & 1 \\ -8 & -2 \end{bmatrix}$. The eigenvalues are $\lambda = -1 \pm \sqrt{7}i$ and this shows that there is a stable focus at each point.

In Figure 3.8 is sketched the approximation to the phase plane portrait near the critical points. (Note that the non-linearity would distort the portrait there.) For other regions of the phase plane, we would need to use isoclines. Try to complete the portrait. (The oscillator switches from one stable state to the other.)

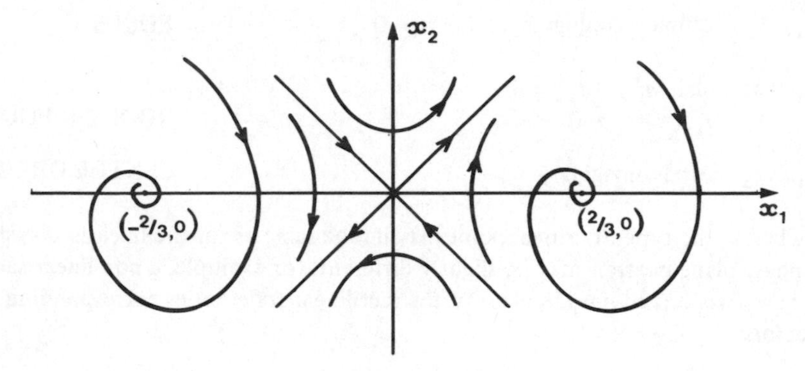

Figure 3.8

Example 2

Simple pendulum with no friction. The system in state space is $\dot{x}_1 = x_2$, $\dot{x}_2 = -\omega^2 \sin x_1$. First note that the critical points are at (0, 0), ($\pm\pi$, 0), ($\pm 2\pi$, 0) etc. Can we physically ignore most of these?

The matrix $\mathbf{J} = \begin{bmatrix} 0 & 1 \\ -\omega^2 \cos x_1 & 0 \end{bmatrix}$. At (0, 0), $\mathbf{J} = \begin{bmatrix} 0 & 1 \\ -\omega^2 & 0 \end{bmatrix}$ with eigenvalues $\lambda = \pm i\omega$. This indicates a centre or a focus. At ($\pm\pi$, 0), $\mathbf{J} = \begin{bmatrix} 0 & 1 \\ \omega^2 & 0 \end{bmatrix}$ with eigenvalues $\lambda = \pm\omega$.

This implies a saddle point. In fact, the governing differential equation can be integrated to obtain $x_2^2 = 2\omega^2 (C + \cos x_1)$ where C is a constant. If $C < -1$, x_2 ha no real value; if $C > 1$, the pendulum never comes to rest, whereas if $|C| < 1$ there are elliptic orbits representing oscillations about the downward vertical. When $C = 1$ there is a limiting case. What is it physically?

The relevant part of the phase plane portrait is shown in Figure 3.9(a).

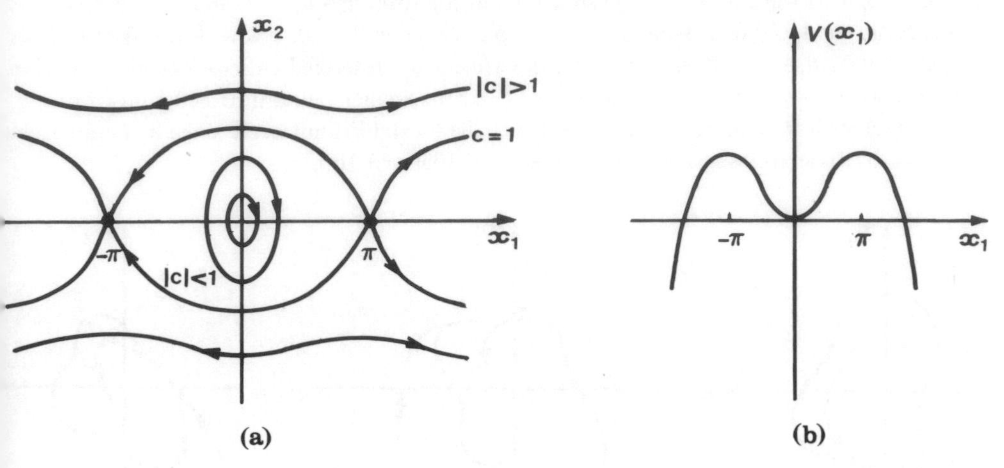

Figure 3.9

A trajectory which separates one type of motion from another type is known as a **separatrix**. Clearly, the cases $C = 1$ are separatrices. It is of interest to plot the potential energy $V(x_1)$ against x_1. Now $\frac{1}{2}m\,x_2^2$ is the kinetic energy, and since the total energy of the system is conserved, $\frac{1}{2}\,m\,x_2^2 + V(x_1) = A$, where A is constant. Therefore $V(x_1) = A - m\omega^2(C + \cos x_1)$ or $V(x_1) = B - m\omega^2 \cos x_1$ where B is constant.

If we define the constants such that $V(0) = 0$, then we obtain the graph of Figure 3.9(b). What suitable conclusions can you draw from this?

Limit cycles

Our study of phase portraits is concluded with a few remarks on closed paths in the portraits of non-linear systems. We have already studied the case of a centre, where there is an infinite number of closed paths in any region around the critical point. We now look at isolated closed paths.

Consider the following example

$$\begin{aligned} \dot{x}_1 &= x_2 + x_1(1 - x_1^2 - x_2^2) \\ \dot{x}_2 &= -x_1 + x_2(1 - x_1^2 - x_2^2) \end{aligned} \qquad (3.10)$$

By converting to polar coordinates we can eventually obtain the equations

$$\dot{r} = r(1 - r^2)$$
$$\dot{\theta} = -1 \tag{3.11}$$

It is straightforward to obtain the solutions in the form $\theta = \theta_0 - t$ and $r^2 = e^{2t}/(C + e^2$
where θ_0 and C are constants. For simplicity, take $\theta_0 = 0$. Now if $C = 0$ it follows that $r^2 = 1$, $\theta = -t$ which gives a circular trajectory described anticlockwise. For $C \neq$
note that as $t \to \infty$, r tends to the value 1, from smaller values if $C > 0$ and from larger values if $C < 0$. The curve $r = 1$ is called a **stable limit cycle** since it is approache
by nearby trajectories inside and outside. See Figure 3.10(a).

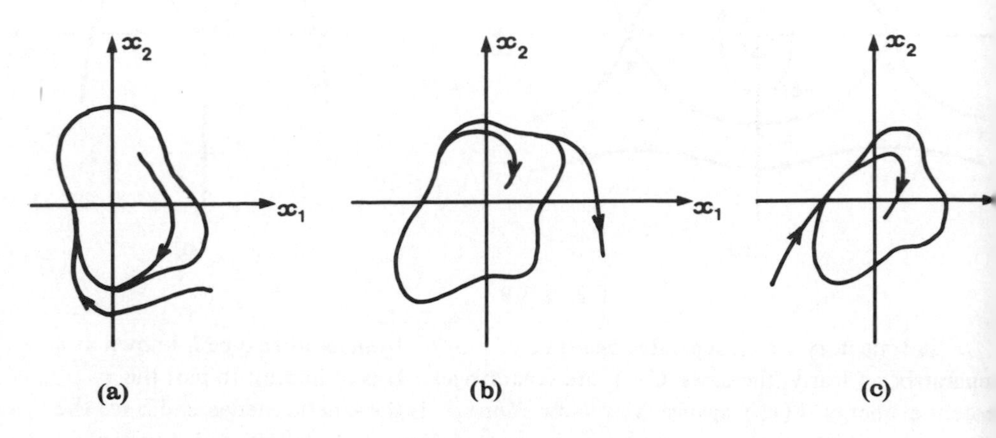

Figure 3.10

A physical example is provided by the Van der Pol equation (see AEM p. 159). In Figure 3.10(b) is shown an **unstable limit cycle**; nearby trajectories tend to move away from it. Figure 3.10(c) shows a **semi-stable limit cycle**; nearby trajectories on one side leave it, those on the other side approach it.

In some engineering applications limit cycles are undesirable, in others small oscillations are acceptable or even required. The prediction of the existence or non-existence of such cycles is important. We prove one result, known as Bendixson's **non-existence criterion**.

If $\left(\dfrac{\partial P}{\partial x} + \dfrac{\partial Q}{\partial y}\right)$ has the same sign throughout a region D in the xy plane then t
autonomous system (3.8) has no closed paths in D.

For the proof, let C be a closed curve in D and let R be the region, in D, bounded by C. Green's Theorem gives

$$\oint_C (P\mathrm{d}y - Q\mathrm{d}x) = \iint_R \left(\frac{\partial P}{\partial x} + \frac{\partial Q}{\partial y}\right) \mathrm{d}S$$

f C is a closed path of the system (3.8) then

$$\oint_C (P\mathrm{d}y - Q\mathrm{d}x) = \int_0^T (P\dot{y} - Q\dot{x})\mathrm{d}t = \int_0^T (PQ - QP)\mathrm{d}t = 0$$

Hence $\iint_R \left(\frac{\partial P}{\partial x} + \frac{\partial Q}{\partial y}\right) \mathrm{d}S = 0$. This contradicts the uniformity of sign of $\left(\frac{\partial P}{\partial x} + \frac{\partial Q}{\partial y}\right)$

nd the theorem follows.

Example

The relaxation oscillator circuit equation may be written in the form (3.8) with $P(x, y) \equiv y$, $Q(x, y) \equiv -4x + 4x^3 - 2y$; $\frac{\partial P}{\partial x} + \frac{\partial Q}{\partial y} = -2$ and therefore the system ossesses no closed paths.

Comment

It is worth restating the following.

(i) In a linear system $\dot{\mathbf{x}} = \mathbf{A}\mathbf{x}$, the eigenvalues of $\mathbf{A}$ must all have negative real parts for the system to be stable.

(ii) With a non-linear system $\dot{\mathbf{x}} = \mathbf{f}(\mathbf{x})$, we can linearise it and apply the result (i).

Problems

For each of the following systems, sketch the phase plane portrait and deduce the nature of the equilibrium point at the origin.

(a) $\dot{x}_1 = x_1 + x_2$, $\dot{x}_2 = 3x_1 - x_2$

(b) $\dot{x}_1 = x_1 - x_2$, $\dot{x}_2 = 2x_1 + 4x_2$

(c) $\dot{x}_1 = 3x_1 + 5x_2$, $\dot{x}_2 = -5x_1 - 3x_2$

(d) $\dot{x}_1 = -x_1 - x_2$, $\dot{x}_2 = x_1 - x_2$

(e) $\dot{x}_1 = 2x_1 + 4x_2$, $\dot{x}_2 = -2x_1 + 6x_2$

(f) $\dot{x}_1 = 2x_1 - 7x_2$, $\dot{x}_2 = 3x_1 - 8x_2$

Find the eigenvalues of the matrix $\mathbf{A}$ and verify your deductions for each case. Solve the equations explicitly for x_1 and x_2 as functions of time and verify your deductions a second time.

2. Given the linear autonomous system $\dot{\mathbf{x}} = \mathbf{A}\mathbf{x}$ where $\mathbf{A} = \begin{bmatrix} a & b \\ c & d \end{bmatrix}$, find the conditions on a, b, c and d which categorise the equilibrium point as a focus, a node, a saddle point and a centre. Further, obtain the conditions which separate stable foci and nodes from their unstable counter-parts. Recast these conditions in terms of the trace of $\mathbf{A}$ and its determinant. Sketch the results on a graph with ($-$trace $\mathbf{A}$) and $|\mathbf{A}|$ as the variables.

3. Linearise the following systems and deduce the nature of the equilibrium points at the origin. Sketch the phase plane portraits for both linear and non-linear cases.

(a) $\dot{x}_1 = x_1 + 4x_2 - x_1^2$

$\dot{x}_2 = 6x_1 - x_2 + 2x_1x_2$

(b) $\dot{x}_1 = \sin x_1 - 4x_2$

$\dot{x}_2 = \sin 2x_1 - 5x_2$

(c) $\dot{x}_1 = x_1 + x_1^2 - 3x_1x_2$

$\dot{x}_2 = -2x_1 + x_2 + 3x_2^2$

4. Find the critical points of the system governed by $\ddot{x} + 4x - 4x^3 = 0$.

Determine their nature and sketch the phase plane portrait.

5. Find the nature and location of the critical points of the system governed by

$$\ddot{x} + 6x - 5x^2 - 5x^3 + 5x^4 - x^5 = 0$$

6. Does the system

$$\dot{x}_1 = 2x_1 + x_2 + x_1^3$$
$$\dot{x}_2 = 3x_1 - x_2 + x_2^3$$

possess any limit cycles?

7. Does the system governed by

$$\ddot{x} + (x^2 + x^4)\dot{x} + (x + x^3) = 0$$

possess any limit cycles?

8. Draw the phase plane portrait for the system governed by $\ddot{x} + |\dot{x}|\dot{x} + x = 0$. What could the system be?

9. Sketch the phase plane portraits for Van der Pol's equation in the cases $\epsilon = 1$, $\epsilon = 10$. Comme

10. A feedback control system with a non-linear element is governed by

$$\ddot{x} + \dot{x} + x + \text{sign}(x) = 0$$

where $\text{sign}(x) = 1$ if $x > 0$, -1 if $x < 0$, 0 if $x = 0$. Sketch the phase plane portrait.

3.3 DEFINITIONS OF STABILITY

For the rest of this chapter we shall refer to critical points as **equilibrium points**.

Consider again the example of the simple pendulum without friction; there are two equilibrium positions, one with the pendulum hanging vertically downwards and one with it pointing vertically upwards. Clearly, the latter is an unstable position, by any reasonable definition; intuitively, we would say that the former is stable. However the local phase plane portrait indicates a centre and in theory the pendulum will never come to rest at the equilibrium position.

Further, there are problems in relating non-linear systems to linear autonomous systems. The latter have the equilibrium point at the origin (unless the matrix **A** is singular) and this will be stable or unstable whether or not a driving force exists. In a non-linear system there may be many equilibrium points and it is customary to speak of the stability of the states rather than the stability of the system.

For a linear stable system with no driving force, then whatever the initial state, the system will always return to equilibrium. With non-linear systems the initial state may affect the stability very markedly. Furthermore the driving force may stabilise a system (and may be the main purpose of a control (see Section 3.8)).

A precise definition of stability is needed. However, different kinds of engineers have different ideas of what stability means for them. In consequence, many different kinds of stability have been defined; this section examines a few of them. We shall assume the equilibrium point under consideration to be at the origin of state space.

The **Liapunov** concept of **stability** of an equilibrium point is that there is a region of state space around the point such that a trajectory starting from an initial state within the region will remain within the region for all time. This concept admits the downward vertical equilibrium position of the pendulum to be stable. It also allows 'stable' nodes, 'stable' foci, and embraces limit cycles. The damped dynamical systems studied in the ast section are both stable.

In Figure 3.11(a) we show in two-dimensional state space a possible trajectory in the egion of a stable equilibrium point and in diagram (b) we show the trajectory in 'motion pace'.

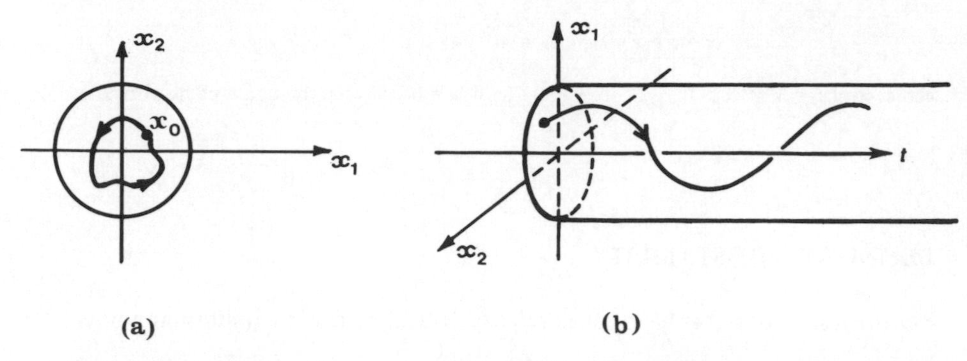

Figure 3.11

Note that the concept of Liapunov stability includes cases where the trajectory actually reaches the origin of state space. Such a special sub-class of cases merits a stronger definition. We say that an equilibrium point is **asymptotically stable** if it is stable and if also the trajectory either reaches the equilibrium point in an infinite time or reaches it in a finite time and stays there. The concept is represented graphically in Figure 3.12.

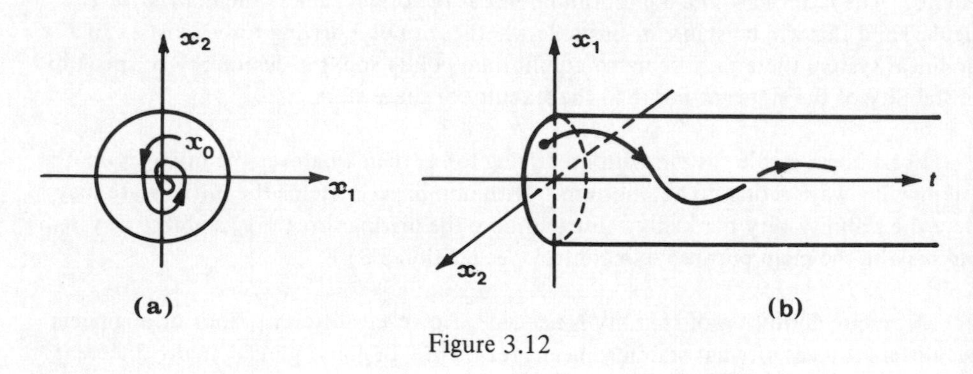

Figure 3.12

Thus stability alone implies a tendency towards orbital motion, asymptotic stability implies a tendency towards the origin and, obviously, instability implies a tendency to infinity, all as $t \to \infty$.

We now make precise mathematical definitions. Consider the system $\dot{\mathbf{x}} = \mathbf{f}(\mathbf{x}, t)$. The equilibrium point at the origin of state space is *stable* if for every $R > 0$ we can find an $r > 0$ (where the particular r chosen depends on the given R and on the initial time t_0) such that if we make $\|\mathbf{x}_0\| < r$ then $\|\mathbf{x}(t)\| < R$ for all $t > t_0$.† In other

† $\mathbf{x}_0$ is a notation for $\mathbf{x}(t_0)$. $\|\mathbf{x}_0\|$ may be interpreted for the moment as the magnitude of the vector $\mathbf{x}_0$; see Chapter 9. Note that some authors replace $\dot{\mathbf{x}}(t)$ by $\dot{\mathbf{x}}(t, \mathbf{x}_0, t_0)$ for emphasis.

words, by starting the trajectory sufficiently close to the origin it will remain within a required distance of the origin. This is illustrated for two dimensions in Figure 3.13(a).

The equilibrium point is further said to be **asymptotically stable** if it is stable and if $\lim_{t\to\infty} \|\mathbf{x}(t)\| = 0$ for $\|\mathbf{x}_0\| < r'$. See Figure 3.13(b).

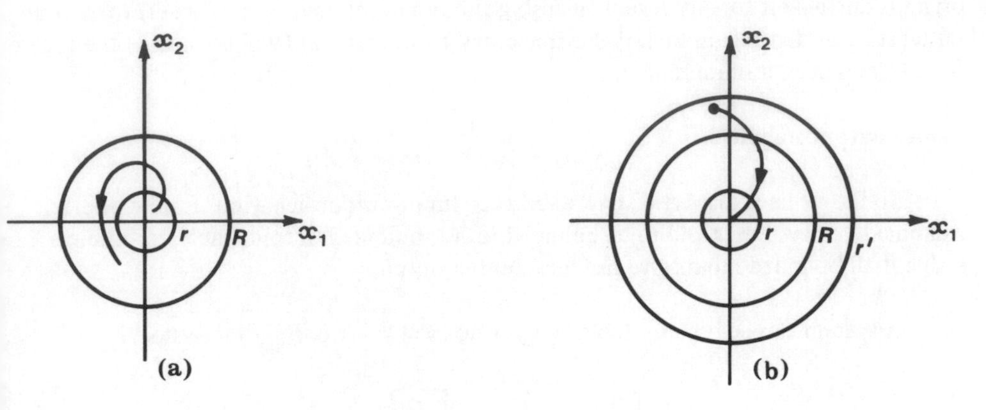

Figure 3.13

Note that r' does not have to be the same as r. Also note that the existence of limit cycles is precluded by asymptotic stability. If asymptotic stability occurs for all equilibrium states the system is said to exhibit **asymptotic stability in the large**, or **global asymptotic stability**. In this case r' can be made arbitrarily large. In general the hope is that the region in which asymptotic stability rules is large enough to contain any disturbance to the system. The largest region in which asymptotic stability holds (i.e. the largest value of r') is called the **domain of attraction**.

The initial state may have a strong influence on the output even inside a region of asymptotic stability. Zubov (1962) cites the following example. Let

$$\dot{x} = g(t)x \quad \text{where} \quad g(t) = \begin{cases} \log_e 10, & 0 \leqslant t \leqslant 10 \\ -1\ , & t > 10 \end{cases}$$

Also let $x = x_0$ at $t = 0$. Then the solution is given by

$$x(t) = \begin{cases} 10^t\, x_0 & , \quad 0 \leqslant t \leqslant 10 \\ 10^{10}\, e^{(10-t)}\, x_0, & \quad t > 10 \end{cases}$$

It is easy enough to show that the origin is asymptotically stable. However, a change in position of x_0 by only 10^{-5} will alter the value of x by 3 at $t = 20$.

Other forms of stability exist. We should remark that some systems declared unstable by the Liapunov criterion are in practice acceptable for design purposes. On the other hand systems declared stable may perform none too well from a practical point of view. Sometimes it is necessary not only to specify a permitted variation in the response of a system but also to specify a range of acceptable initial states.

A definition of instability is needed. We say that the equilibrium point at the origin is **unstable** if for any region enclosing the origin (no matter how small) there is an initial state in that region such that a trajectory from that state will lie outside the region for all times after a finite time t_0.

Input - output stability

So far we have discussed state stability. Input-output behaviour is now briefly examined. A system is said to be bounded input, bounded output stable, or b.i.b.o. stable if all bounded inputs give rise to bounded outputs.

A system can be b.i.b.o. stable and yet not stable. Consider the system

$$\dot{\mathbf{x}} = \begin{bmatrix} 0 & 6 \\ 1 & -1 \end{bmatrix} \mathbf{x} + \begin{bmatrix} -2 \\ 1 \end{bmatrix} u$$

with

$$\mathbf{y} = [0, \ 1]\mathbf{x}$$

The eigenvalues of the system matrix are $\lambda_1 = -3$, $\lambda_2 = 2$ and hence there is not a stable equilibrium point at the origin. On the other hand, due to cancellation effects, the unstable mode does not show up in the input-output analysis.

We make the remark that if a system is asymptotically stable, it is b.i.b.o. stable; however, if it is b.i.b.o. stable and also controllable and observable, it is asymptotically stable.

3.4 CRITERIA FOR CONSTANT LINEAR SYSTEMS

As might be anticipated, special stability criteria exist for linear systems. This section deals with one of the well-established methods.

Routh-Hurwitz Criteria

A continuous linear system is stable if all the roots λ_r of its characteristic equation have negative real parts. Let the characteristic equation be

$$a_0 \lambda^n + a_1 \lambda^{n-1} + \dots\dots + a_{n-1}\lambda + a_n = 0 \tag{3.12}$$

where a_r are real.

A **necessary** condition for stability is that all the coefficients should possess the same sign and be non-zero.†

For example, the systems with characteristic equations $\lambda^2 - a^2\lambda + \omega^2 = 0$ and $\lambda^7 + 1 = 0$ are both unstable.

In the case of higher degree equations, it is possible to have some roots λ_r with positive real parts and yet have the necessary conditions satisfied.

Where all coefficients are present and share the same sign a different criterion is needed. Assume without loss of generality that all the coefficients are positive. First we quote the Hurwitz criterion: The roots of (3.12) all have negative real parts if and only if all the leading principal minors of

$$H = \begin{vmatrix} a_1 & a_3 & a_5 & \cdot & \cdot & \cdot & a_{2n-1} \\ a_0 & a_2 & a_4 & \cdot & \cdot & \cdot & a_{2n-2} \\ 0 & a_1 & a_3 & \cdot & \cdot & \cdot & a_{2n-3} \\ 0 & 1 & a_2 & \cdot & \cdot & \cdot & a_{2n-4} \\ \cdot & \cdot & \cdot & \cdot & \cdot & \cdot & \cdot \\ \cdot & \cdot & \cdot & \cdot & \cdot & \cdot & \cdot \\ 0 & 0 & 0 & \cdot & \cdot & \cdot & a_n \end{vmatrix} \tag{3.13}$$

are positive, where $a_r = 0$ if $r > n$.

If $n = 3$, the requirement is that

$$H_3 = \begin{vmatrix} a_1 & a_3 & 0 \\ a_0 & a_2 & 0 \\ 0 & a_1 & a_3 \end{vmatrix} > 0, \quad H_2 = \begin{vmatrix} a_1 & a_3 \\ a_0 & a_2 \end{vmatrix} > 0, \quad H_1 = a_1 > 0$$

† If $a_0\lambda^2 + a_1\lambda + a_2 = 0 \equiv a_0(\lambda - \lambda_1)(\lambda - \lambda_2) = 0$ then $\lambda_1\lambda_2 = a_2/a_0$ and $(\lambda_1 + \lambda_2) = -a_1/a_0$. Hence if a_0, a_1 and a_2 are all > 0, both λ_1 and λ_2 are < 0.

i.e. $a_1 a_2 a_3 - a_3^2 a_0 > 0 \qquad a_1 a_2 - a_0 a_3 > 0, \quad a_1 > 0$

or $a_3(a_1 a_2 - a_3 a_0) > 0 \qquad a_1 a_2 - a_0 a_3 > 0, \quad a_1 > 0.$

Comparison of the first two inequalities requires a_3 to be positive. Further, in order for $a_1 a_2 - a_0 a_3 > 0$ we require $a_1 a_2 > 0$ and hence $a_2 > 0$. Therefore, the additional condition to $a_1 > 0$, $a_2 > 0$, $a_3 > 0$ is that $a_1 a_2 - a_0 a_3 > 0$.

Routh's extension

In order to avoid the evaluation of the determinants we employ an equivalent criterion due to Routh. Consider the matrix equation

$$\begin{bmatrix} 1 & 0 & 0 \\ -a_0/a_1 & 1 & 0 \\ \dfrac{-a_0 a_1}{(a_3 a_0 - a_1 a_2)} & \dfrac{a_1^2}{(a_3 a_0 - a_1 a_2)} & 1 \end{bmatrix} \begin{bmatrix} a_1 & a_3 & 0 \\ a_0 & a_2 & 0 \\ 0 & a_1 & a_3 \end{bmatrix} = \begin{bmatrix} r_{11} & r_{12} & r_{13} \\ 0 & r_{21} & r_{22} \\ 0 & 0 & r_{31} \end{bmatrix}$$

or

$$\mathbf{L}\,\mathbf{H} = \mathbf{R} \tag{3.14}$$

Note that $r_{22} = 0$, and that $r_{11} = H_1$, $r_{21} = H_2/H_1$, $r_{31} = H_3/H_2$. (You can verify this by direct multiplication.)

It follows that $r_{11}, r_{21}, r_{31} > 0$ if and only if $H_1, H_2, H_3 > 0$. Routh was able to develop his criterion from the following 'Routh' array. Each row and each column is terminated by zeros when we run out of non-zero coefficients.

$$\begin{array}{cccccc}
a_0 & a_2 & a_4 & a_6 & \cdot & \cdot \\
a_1 & a_3 & a_5 & a_7 & \cdot & \cdot \\
r_{21} & r_{22} & r_{23} & r_{24} & \cdot & \cdot \\
\cdot & \cdot & \cdot & \cdot & \cdot & \cdot \\
\cdot & \cdot & \cdot & \cdot & \cdot & \cdot \\
\cdot & \cdot & \cdot & \cdot & \cdot & \cdot \\
\cdot & \cdot & \cdot & \cdot & \cdot & \cdot \\
r_{i1} & r_{i2} & r_{i3} & r_{i4} & \cdot & \cdot \\
\cdot & \cdot & \cdot & \cdot & \cdot & \cdot \\
\cdot & \cdot & \cdot & \cdot & \cdot & \cdot \\
r_{n+1,1} & r_{n+1,2} & r_{n+1,3} & r_{n+1,4} & \cdot & \cdot
\end{array}$$

or

$$\begin{array}{cccccc}
r_{01} & r_{02} & r_{03} & r_{04} & \cdot & \cdot \\
r_{11} & r_{12} & r_{13} & r_{14} & \cdot & \cdot \\
r_{21} & r_{22} & r_{23} & r_{24} & \cdot & \cdot \\
\cdot & \cdot & \cdot & \cdot & \cdot & \cdot \\
\cdot & \cdot & \cdot & \cdot & \cdot & \cdot \\
\cdot & \cdot & \cdot & \cdot & \cdot & \cdot \\
r_{n+1,1} & r_{n+1,2} & r_{n+1,3} & r_{n+1,4} & \cdot & \cdot
\end{array} \qquad (3.15)$$

where $r_{01} = a_0$, $r_{02} = a_2$, $r_{03} = a_4$ etc., $r_{11} = a_{11}$ etc., and in general

$$r_{ij} = \frac{-1}{r_{i-1,1}} \begin{vmatrix} r_{i-2,1} & r_{i-2,\,j+1} \\ r_{i-1,1} & r_{i-1,\,j+1} \end{vmatrix} \qquad i = 2, 3, \ldots\ldots, n$$

The criterion is that all the eigenvalues have negative real parts if and only if the elements in the first column of the array (3.15) are all positive.

Example

$$\lambda^4 + 2\lambda^3 + 3\lambda^2 + 4\lambda + 5 = 0$$

Here $r_{01} = 1$, $r_{02} = 3$, $r_{03} = 5$, $r_{11} = 2$, $r_{12} = 4$ and all other r_{0j} and r_{1j} are zero. Now

$$r_{21} = \frac{-1}{r_{11}} \begin{vmatrix} r_{01} & r_{02} \\ r_{11} & r_{12} \end{vmatrix} = \frac{-1}{2} \begin{vmatrix} 1 & 3 \\ 2 & 4 \end{vmatrix} = 1$$

Similarly $r_{31} = -6$, $r_{41} = 5$, $r_{32} = 5$ and all other $r_{ij} = 0$. The Routh array is

$$\begin{array}{ccccc}
1 & 3 & 5 & \cdot & \cdot \\
2 & 4 & 0 & \cdot & \cdot \\
1 & 5 & 0 & \cdot & \cdot \\
-6 & 0 & 0 & \cdot & \cdot \\
5 & 0 & 0 & \cdot & \cdot \\
\cdot & \cdot & \cdot & \cdot & \cdot
\end{array}$$

Since all entries in the first column are not positive the eigenvalues do not all possess negative real parts and the system is unstable.

Further information can be gleaned from the array. The number of changes of sign of coefficient in the first column is equal to the number of roots with positive real part. In this example there are two sign changes.

(Note that if the second row were to be divided by 2 the qualitative conclusions would be unaltered.)

Further sophistications can be performed in the case of zero coefficients in the first column. See for example Ogata (1967).

Discrete systems

The solution of a continuous system essentially involves exponential functions of time, $e^{-\lambda_r t}$ where λ_r are the eigenvalues of the system matrix; this is why we require the real parts of the eigenvalues to be negative. Since the solution of a discrete system involves functions of the form λ_r^k for $k = 0, 1, 2, \ldots\ldots$ in this case the eigenvalues are required to lie within the unit circle in the complex z-plane. Via the bilinear mapping $w = (z + 1)/(z - 1)$ we can transform to the condition that the transformed eigenvalues μ lie in the negative real half of the complex w-plane. For example, the characteristic equation $a_0 \lambda^3 + a_1 \lambda^2 + a_2 \lambda + a_3 = 0$ is transformed into

$$a_0 \left(\frac{\mu + 1}{\mu - 1}\right)^3 + a_1 \left(\frac{\mu + 1}{\mu - 1}\right)^2 + a_2 \left(\frac{\mu + 1}{\mu - 1}\right)^1 + a_3 = 0$$

or

$$b_0 \mu^3 + b_1 \mu^2 + b_2 \mu + b_3 = 0$$

The Routh-Hurwitz technique may then be employed on this new polynomial. Alternatively the **Schur-Cohn** criterion may be used. A set of n determinants of which a typical member is Δ_r (of order $2r$) is given by (3.16). (See next page.)
Note that a_0 is assumed positive.

The condition due to Schur and Cohn is: The eigenvalues all have magnitude < 1 if and only if $\Delta_r < 0$ for r odd and $\Delta_r > 0$ for r even.

Example

The second order system $a_0 \lambda^2 + a_1 \lambda + a_2 = 0$. (continued after (3.16))

$$\Delta_r = \left|\begin{array}{cccc|cccc}
a_n & 0 & & & a_0 & a_1 & \cdots & a_{r-1} \\
a_{n-1} & a_n & & & 0 & a_0 & \cdots & a_{r-2} \\
\vdots & \vdots & \ddots & & & \ddots & & \vdots \\
a_{n-r+1} & a_{n-r+2} & \cdots & a_n & & & & a_0 \\
\hline
a_0 & 0 & & & a_n & a_{n-1} & \cdots & a_{n-r+1} \\
a_1 & a_0 & & & 0 & a_n & \cdots & a_{n-r+2} \\
\vdots & \vdots & \ddots & & & \ddots & & \vdots \\
a_{r-1} & a_{r-2} & \cdots & a_0 & & & & a_n
\end{array}\right| \qquad (3.16)$$

Example (continued) The relevant determinants are

$$\Delta_1 = \begin{vmatrix} a_2 & a_0 \\ a_0 & a_2 \end{vmatrix} = a_2^2 - a_0^2$$

$$\Delta_2 = \begin{vmatrix} a_2 & 0 & a_0 & a_1 \\ a_1 & a_2 & 0 & a_0 \\ a_0 & 0 & a_2 & a_1 \\ a_1 & a_0 & 0 & a_2 \end{vmatrix} = (a_2 - a_0)^2\left([a_2 + a_0]^2 - a_1\right)^2$$

It follows that $a_0^2 < a_2^2$ and $a_1^2 < (a_2 + a_0)^2$ are the requirements for the eigenvalues to have magnitude < 1.

Disadvantages of the criteria in this section

(i) The evaluation of some high-order determinants can be awkward.

(ii) Any changes in the system require completely new calculations to study their effects.

(iii) More information about a stable system may be needed.

A number of techniques have been developed to overcome these disadvantages. Among these we mention the **Nyquist diagram** which provides a graphical criterion for continuous negative feedback systems and the **root-locus method** for investigating roots of the characteristic equation $1 + G(s) = 0$ of feedback systems. These methods are discussed in Jacobs (1974) to which the reader is referred for further details.

Summary of eigenvalue criteria for linear constant systems

Assume that the eigenvalues are λ_i with real parts β_i.

(a) **Continuous systems** $\dot{\mathbf{x}} = \mathbf{A}\mathbf{x}$

(i) The system is **unstable** if $\beta_i > 0$ for *any* simple eigenvalue *or* $\beta_i \geqslant 0$ for *any* repeated eigenvalue.

(ii) The system is **stable** if $\beta_i \leqslant 0$ for *all* simple eigenvalues *and* $\beta_i < 0$ for *all* repeated eigenvalues.

(iii) The system is **asymptotically stable** if $\beta_i < 0$ for *all* eigenvalues.

(b) **Discrete systems** $\mathbf{x}(k + 1) = \mathbf{A}\mathbf{x}(k)$

(i) The system is **unstable** if $|\lambda_i| > 1$ for *any* simple eigenvalue *or* $|\lambda_i| \geqslant 1$ for *any* repeated eigenvalue.

(ii) The system is **stable** if $|\lambda_i| \leqslant 1$ for *all* simple eigenvalues *and* $|\lambda_i| < 1$ for *all* repeated eigenvalues.

(iii) The system is **asymptotically stable** if $|\lambda_i| < 1$ for *all* eigenvalues.

Problems

1. Discuss the stability of the continuous systems with characteristic equations as follows.

   (a) $\lambda^3 + \lambda^2 + 2\lambda + 24 = 0$

   (b) $\lambda^5 + \lambda^4 + 4\lambda^3 + 24\lambda^2 + 3\lambda + 63 = 0$

   (c) $\lambda^4 + 2\lambda^3 + 9\lambda^2 + \lambda + 4 = 0$

   (d) $\lambda^6 + 3\lambda^5 + 10\lambda^4 + 40\lambda^3 + 84\lambda^2 + 92\lambda + 40 = 0$

2. Discuss the stability of the discrete systems whose characteristic equations follow.

   (a) $2\lambda^3 + 4\lambda^2 - 5\lambda + 3 = 0$

   (b) $\lambda^4 + 2\lambda^3 + \lambda^2 + 3\lambda + 2 = 0$

3. Show that a necessary condition for the filtering system with characteristic equation

$$\lambda^3 + \left(\frac{1}{RC} + \frac{b}{J}\right)\lambda^2 + \frac{1}{RC}\,\frac{b}{J}\,\lambda + \frac{K}{JRC} = 0$$

to be stable is

$$0 < \frac{K}{JRC} < \left(\frac{1}{RC} + \frac{b}{J}\right)\frac{1}{RC}\cdot\frac{b}{J}$$

where R, C, b, J, K are constants.

3.5 LIAPUNOV STABILITY THEOREMS

The concept of asymptotic stability implies that the distance between a trajectory and the equilibrium point decreases to zero as $t \to \infty$. Liapunov was concerned with the problem of predicting the direction and destination of a trajectory without solving the differential equations; these properties were to be determined directly from the 'dynamics' of the system and the time behaviour of a positive definite function† of the state variables, called a **Liapunov function**. The requirement of positive definiteness means that there is some sense in which the function measures distance from the origin, as we shall see. To illustrate the background thinking a simple system is chosen.

Mass-spring-damping system

The system is governed by $\dot{x}_1 = x_2$, $\dot{x}_2 = -x_1 - x_2$; all the parameters are taken to be 1. The same equations will describe an LCR series electrical circuit and many other systems. The initial conditions $\mathbf{x}(0) = \begin{bmatrix} 2 \\ 0 \end{bmatrix}$ imply that the system is released from $x_1 = 2$ with zero velocity. The solution is found to be

$$x_1 = e^{-t/2}\left(\frac{2}{\sqrt{3}}\sin\frac{\sqrt{3}}{2}t + 2\cos\frac{\sqrt{3}}{2}t\right)$$

$$x_2 = \frac{-4}{\sqrt{3}}\,e^{-t/2}\sin\frac{\sqrt{3}}{2}t$$

† A function of two variables $f(x, y)$ is positive definite if $f(x, y) > 0$ for $(x, y) \neq (0, 0)$ and $f(0, 0) = 0$. Here, it is assumed that $f(x, y)$ does not contain an additive constant.

Figure 3.14(a) shows the phase plane portrait and Figure 3.14(b) shows the time variation of x_1 and x_2.

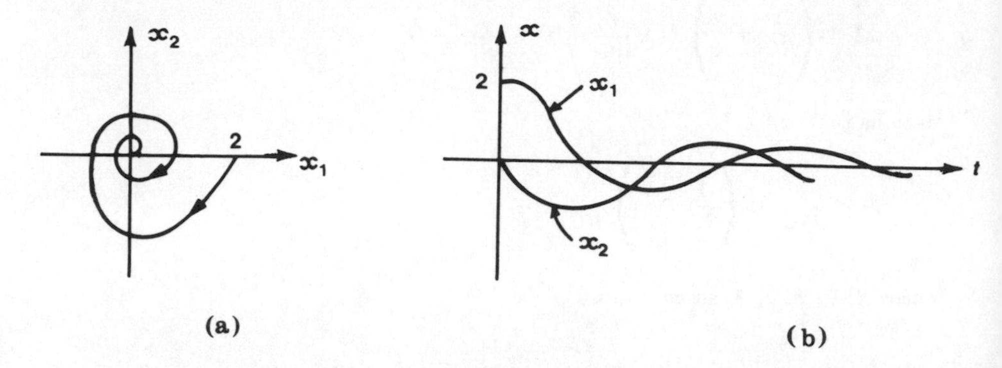

Figure 3.14

It is intuitively obvious that the origin is an asymptotically stable equilibrium point. Now consider the energy in the system. The kinetic energy of the mass is $\frac{1}{2} m\dot{x}_1^2 = \frac{1}{2} x_2^2$ and the potential energy stored in the spring is $\frac{1}{2} kx_1^2 = \frac{1}{2} x_1^2$, therefore the total energy, $E = \frac{1}{2} x_1^2 + \frac{1}{2} x_2^2$. Energy is dissipated as heat in the dashpot at a rate $\dot{E} = x_1\dot{x}_1 + x_2\dot{x}_2$. But along a trajectory in phase space, $\dot{x}_1 = x_2$ and $\dot{x}_2 = -x_1 - x_2$. Therefore, $\dot{E} = x_1x_2 - x_2x_1 - x_2^2$ i.e. $\dot{E} = -x_2^2$ along a trajectory. (Note that $\dot{E}$ has the dimensions of power.)

In Figure 3.15(a) are plotted the curves of constant E, with the trajectory of Figure 3.14(a) superimposed; in Figure 3.15(b) the graphs of E and $\dot{E}$ against time are depicted.

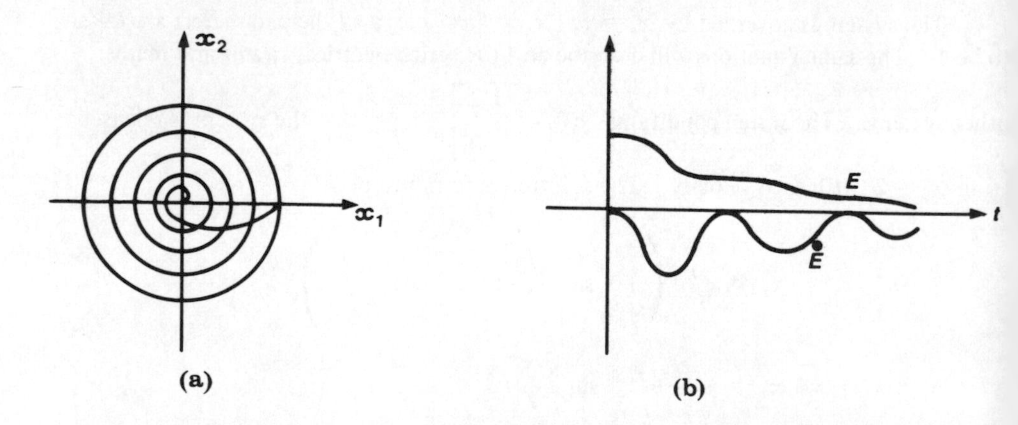

Figure 3.15

Write out the explicit epxressions for E and $\dot{E}$ and satisfy yourself that their time variation is as depicted. Think about what is physically happening to the system. Note that as the trajectory progresses it passes through decreasing values of E. In fact, the function $\dot{E}$, **along a trajectory**, is negative definite.*

No damping

Consider the damping-free system governed by $\dot{x}_1 = x_2$, $\dot{x}_2 = -k_1 x_1$.

The total energy $E = \frac{1}{2} k_1 x_1^2 + \frac{1}{2} x_2^2$. Along a trajectory, $\dot{E} = k_1 x_1 (x_2) + x_2(-k_1 x_1) = 0$. Clearly, $\dot{E}$ is not negative definite. In fact it can be regarded as a special case of a **negative semi-definite function.**†

In this example, we know that the origin is a stable equilibrium point (a centre). Systems for which $\dot{E} \equiv 0$ are called **conservative** systems.

Some authors require that functions should be continuous and possess continuous first partial derivatives in order to be classed for definiteness. We shall certainly require these properties in the rest of this chapter, so we shall assume that they hold.

Choice of Liapunov function

The energy of the system is not always the function to choose. Consider the system $\dot{x}_1 = x_2$, $\dot{x}_2 = -2x_1 - x_2$. Then $E = x_1^2 + \frac{1}{2} x_2^2$, but along a trajectory $\dot{E} = 2x_1 (x_2) + x_2(-2x_1 - x_2) = -x_2^2$. Now $\dot{E}$ is negative semi-definite since $\dot{E} = 0$ for all points $(x_1, 0)$; by analogy with the no-damping system we might conclude that the origin is a stable equilibrium point. However, direct solution of the equations (which we leave to you) shows that the origin is in fact asymptotically stable. For the moment consider the function $V(x_1, x_2) = ax_1^2 + bx_1 x_2 + cx_2^2$; this is a general quadratic form in two variables. Now $\dot{V} = 2ax_1 \dot{x}_1 + b(x_1 \dot{x}_2 + \dot{x}_1 x_2) + 2cx_2 \dot{x}_2$. Along a trajectory, $\dot{V} = 2ax_1 x_2 + bx_1(-2x_1 - x_2) + bx_2^2 + 2cx_2(-2x_1 - x_2)$ i.e. $\dot{V} = -2bx_1^2 + (2a - b - 4c) x_1 x_2 + (b - 2c)x_2^2$.

If we choose $a = 9$, $b = 2$, $c = 4$ then $\dot{V} = -4x_1^2 - 6x_2^2$ which is clearly negative definite and $V = 9x_1^2 + 2x_1 x_2 + 4x_2^2$ which, for example, $= 8x_1^2 + (x_1 + x_2)^2 + 3x_2^2$ and is therefore positive definite.†† We have found a function $V(x_1, x_2)$ which plays the same role for this system as the energy did for the first system studied in this section.

A suitable choice for the Liapunov function has given us more information. The key to success is to choose an appropriate function.

* A function $f(x, y)$ is negative definite if $f(x, y) < 0$ for $(x, y) \neq (0, 0)$ and $f(0, 0) = 0$.

† A function $f(x, y)$ is negative semi-definite if $f(0, 0) = 0$ and $f(x, y) \leqslant 0$ otherwise.

†† Other choices of a, b and c would have yielded the same result.

In Figure 3.16(a) and (b) are plotted contours of the functions E and V for this system and the trajectory from (2, 0) is superimposed on each. What conclusions can you draw?

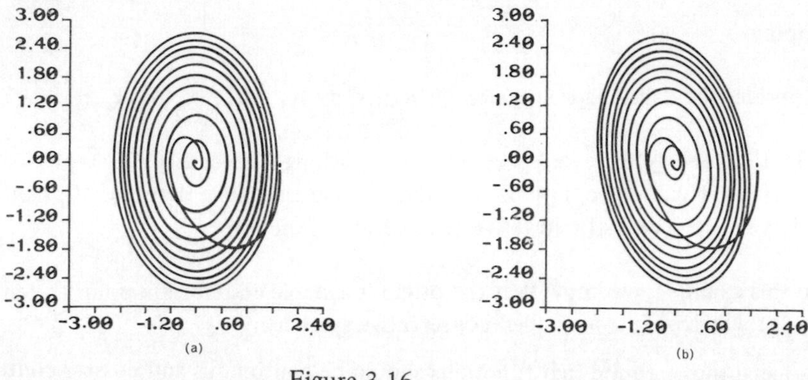

Figure 3.16

Liapunov stability theorems for autonomous systems

The systems so far studied in this section could have been treated via the Routh-Hurwitz criteria. The Liapunov theorems apply to linear or non-linear, autonomous or non-autonomous, deterministic or stochastic systems. We quote the relevant theorems for autonomous two-dimensional systems:

$$\begin{aligned} \dot{x}_1 &= P(x_1, x_2) \\ \dot{x}_2 &= Q(x_1, x_2) \end{aligned} \qquad \text{or} \qquad \dot{\mathbf{x}} = \mathbf{f}(\mathbf{x})$$

The generalisation to higher dimensions is clear.

Theorem 1

If it is posible to find a function $V(x_1, x_2)$ which has the properties near $\mathbf{x} = \mathbf{0}$ that

(i) V is continuous with continuous first partial derivatives

(ii) V is positive definite

(iii) $\dot{V}$ is negative semi-definite along any trajectory

then $\mathbf{x} = \mathbf{0}$ is stable. (This condition is sometimes called 'stable in the sense of Liapunov'.)

We assume that along a trajectory

$$\dot{V} \equiv \frac{\partial V}{\partial x_1}\dot{x}_1 + \frac{\partial V}{\partial x_2}\dot{x}_2$$

or

$$\dot{V} \equiv (\nabla V \,.\, \mathbf{f}(\mathbf{x})) = \frac{\partial V}{\partial x_1} \,.\, P(x_1, x_2) + \frac{\partial V}{\partial x_2} \,.\, Q(x_1, x_2)$$

V is called *a* **Liapunov function** for the system.

It is not unique, fortunately, and this allows a fair amount of freedom in the search for such a function. Failure to find such a function does not mean that one does not exist. (It is perhaps worth trying first the energy function, if this can be easily determined.)

Remember that $\dot{V} \leqslant 0$ means that the trajectories cannot get too far away from the origin.

Example

$$\dot{x}_1 = -2x_1x_2^2, \quad \dot{x}_2 = 2x_1^2x_2 - x_2$$

We try $V = x_1^2 + x_2^2$ which is positive definite. Now along a trajectory

$$\dot{V} = 2x_1(-2x_1x_2^2) + 2x_2(2x_1^2x_2 - x_2)$$

$$= -2x_2^2 = \mathbf{x}^{\mathrm{T}} \begin{bmatrix} 0 & 0 \\ 0 & -2 \end{bmatrix} \mathbf{x}$$

which is negative semi-definite. Hence the origin is stable.

We have seen that it may be possible, having proved stability, to find a V which will prove asymptotic stability.

Theorem 2

If in Theorem 1 condition (iii) is replaced by the condition

(iii) $\dot{V}$ is negative definite along any trajectory

then the origin is asymptotically stable.

Note that $\dot{V} < 0$ implies that trajectories cross the contours of constant V from outside to inside and reach the origin.

Example 1

$$\dot{x}_1 = -2x_1^3 - x_2, \quad \dot{x}_2 = x_1 - x_2$$

We try $V = x_1^2 + x_2^2$. Then

$$\dot{V} = 2x_1(-2x_1^3 - x_2) + 2x_2(x_1 - x_2)$$
$$= -4x_1^4 - 2x_2^2$$

which is negative definite. Hence the origin is asymptotically stable.

Example 2

A series LCR circuit can be modelled in state space by

$$\dot{x}_1 = x_2, \quad \dot{x}_2 = -\frac{1}{LC}x_1 - \frac{R}{L}x_2$$

We try $V = \dfrac{1}{2C}x_1^2 + \dfrac{1}{2}Lx_2^2$. Then

$$\dot{V} = \frac{1}{C}x_1x_2 + Lx_2\left(-\frac{1}{LC}x_1 - \frac{R}{L}x_2\right) = -Rx_2^2$$

which is negative semi-definite and yet by direct solution of the state equations it is known that the system is asymptotically stable. However, $\dot{V} = 0$ only when $x_2 = 0$; by the following theorem we can show that the origin is asymptotically stable. (Why is $x_2 = 0$ not a trajectory? Hint: what is the value of $\dfrac{dx_2}{dx_1}$ on $x_2 = 0$ and what does this imply?)

Theorem 3

If the conditions of Theorem 1 hold and if $\dot{V}$ is not identically zero on any trajectory other than $\mathbf{x} = \mathbf{0}$ then the origin is asymptotically stable.

Analogy

Consider a vessel shaped like a pudding basin with its lowest point at the origin and its axis of symmetry vertical. A small bead released anywhere on the inside surface of the vessel will eventually settle at the origin. 'Pudding basins' of many shapes can be constructed with the same result. We are effectively trying to construct one suitable pudding basin for the system being considered.

Asymptotic stability in the large

Consider the Van der Pol equation $\ddot{x} + \epsilon(x^2 - 1)\dot{x} + x = 0$ where $\epsilon < 0$.

Appropriate state space variables are $x_1 = x$, $x_2 = \dot{x}$, so that the state equations are $\dot{x}_1 = x_2$, $\dot{x}_2 = -x_1 - \epsilon(x_1^2 - 1)x_2$. We try $V = x_1^2 + x_2^2$ so that $\dot{V} = 2\epsilon(1 - x_1^2)x_2^2$ along a trajectory. We see that $\dot{V} \leqslant 0$ if $x_1^2 < 1$. It is reasonable to suppose that $-1 < x_1 < 1$ is a region of asymptotic stability. However it is possible for a trajectory which starts inside this region but outside the unit circle to move outside the strip and hence later diverge even whilst moving in the direction of decreasing V. Figure 3.17 may make this clear.

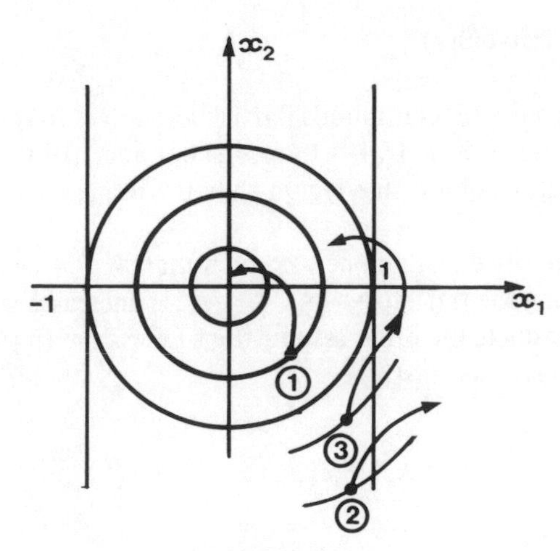

Figure 3.17

A trajectory such as (1) which starts inside the unit circle will be asymptotically stable. Trajectories (2) and (3) start outside the unit circle, but within $|x_1| < 1$; (2) moves in the direction of decreasing V, but leaves the region $|x_1| < 1$, then $\dot{V}$ becomes positive and quite large and it is taken further from that region, clearly demonstrating instability. However (3) leaves the region $|x_1| < 1$ near the x_1 axis and here $\dot{V}$ is fairly small; it actually re-enters the region $x_1^2 + x_2^2 < 1$ and exhibits stability.

Barnett (1975) chooses as state variables $x_1 = x$, $x_2 = \int_0^t x\,dt$ and with the same choice of V shows that the region of asymptotic stability can be widened to $x_1^2 + x_2^2 < 3$.

Theorem 4 (Asymptotic stability in the large)

If the conditions for asymptotic stability (Theorem 2) hold and if $V \to \infty$ as $\|\mathbf{x}\| \to \infty$ then the origin is asymptotically stable in the large, i.e. asymptotically stable regardless of the point $\mathbf{x}(0)$ from which the trajectory starts. Note that $\|\mathbf{x}\| \to \infty$ in any direction. The Example 1 for Theorem 2 possesses this additional property.

Theorem 5 (Instability)

If there is a function $V(x_1, x_2)$ which is continuous with continuous first partial derivatives and if V is positive definite in some region enclosing the origin and $\dot{V}$ is positive definite there also, then the origin is unstable.

Theorem 6 (Cetayev instability)

If V is continuous with continuous partial derivatives in some region enclosing the origin where $V > 0$ and $\dot{V} > 0$, if $V = 0$ on the boundary of the region and if the origin is itself a boundary point of the region then the origin is unstable.

Now there is a positive real number A such that $\dot{V} > A$ and therefore $V(t) - V(0) > At$ along the trajectory. As $t \to \infty$, V increases without limit along the trajectory which contradicts the other assumptions unless the trajectory cannot get near the origin no matter how close it starts.

Example

$$\dot{x}_1 = x_1^3 + x_2, \quad \dot{x}_2 = x_1 + x_2^3$$

Consider $V = x_1^2 - x_2^2$

Note that inside the region $|x_1| > |x_2|$, $V > 0$ and that

$$\dot{V} = 2x_1(x_1^3 + x_2) - 2x_2(x_1 + x_2^3) = 2(x_1^4 - x_2^4)$$

along a trajectory so that $\dot{V} > 0$ in this region. Further, at the boundary $|x_1| = |x_2|$, $V = 0$ and, finally, the origin is a boundary point. Hence the origin is unstable.

Liapunov functions can be used as a measure of the **rate** at which the trajectory approaches the origin. Such an application is, however, beyond our scope.

Non-autonomous systems

We briefly consider non-autonomous systems $\dot{\mathbf{x}} = \mathbf{f}(\mathbf{x}, t)$. First we assume a function $V(\mathbf{x}, t)$ is defined for $\|\mathbf{x}\| < h$ and $t \geq t_0$. It is positive definite if $V(\mathbf{0}, t) = 0$ and if for $\|\mathbf{x}\| < h$ there is a positive definite function $W(\mathbf{x})$ such that $V(\mathbf{x}, t) \geq W(\mathbf{x})$. If, however, $V(\mathbf{0}, t) = 0$ and $V(\mathbf{x}, t) \geq 0$ for $\|\mathbf{x}\| < h$ and $t \geq t_0$, V is said to be positive semi-definite.

It is left to you to deduce the conditions for negative definiteness and negative semi-definiteness.

The following theorem is typical.

The origin is stable if there is a (Liapunov) function $V(\mathbf{x}, t)$ which is positive definite in the 'region' $\|\mathbf{x}\| < h$, $t \geq t_0$ with derivative

$$\dot{V}(\mathbf{x}, t) \equiv \frac{\partial V}{\partial t} + \sum_{i=1}^{n} \frac{\partial V}{\partial x_i}\dot{x}_i = \frac{\partial V}{\partial t} + \sum_{i=1}^{n} \frac{\partial V}{\partial x_i} f_i$$

negative semi-definite in the 'region'. (Continuity of V is implied.)

For asymptotic stability $\dot{V}$ is required to be negative definite in the region and V to be, additionally, such that $\|V(x, t)\| \leq U(x)$ where $U(x)$ is positive definite.

Example

An LCR series circuit which is time varying may be described by the state space equations

$$\dot{x}_1 = \frac{x_2}{L(t)}, \quad \dot{x}_2 = -\frac{x_1}{C(t)} - x_2\frac{R(t)}{L(t)}$$

where $L(t)$, $C(t)$, $R(t) > 0$ for $t \geq 0$.

The Liapunov function $V = \left(R + \frac{2L}{RC}\right)x_1^2 + 2x_1x_2 + \frac{2}{R}x_2^2$ has the derivative along a trajectory

$$\dot{V} = \left(\dot{R} + \frac{2\dot{L}}{RC} - \frac{2L\dot{R}}{R^2C} - \frac{2L\dot{C}}{RC^2}\right)x_1^2 - \frac{2\dot{R}}{R^2}x_2^2 + 2x_1\left(R + \frac{2L}{RC}\right)\frac{x_2}{L}$$

$$+ \frac{2x_2^2}{L} + 2x_1\left(-\frac{x_1}{C} - \frac{x_2R}{L}\right) + \frac{4}{R}x_2\left(-\frac{x_1}{C} - \frac{x_2R}{L}\right)$$

i.e.

$$\dot{V} = \left(\dot{R} + \frac{2\dot{L}}{RC} - \frac{2L\dot{R}}{R^2C} - \frac{2L\dot{C}}{RC^2} - \frac{2}{C}\right)x_1^2 - \left(\frac{2}{L} + \frac{2\dot{R}}{R^2}\right)x_2^2$$

Deduce for yourselves sufficient conditions for $\dot{V}$ to be negative definite.

Problems

1. Assume the given forms for $\dot{V}$ and investigate the stability of the given systems.

   (a) $\dot{x} = \alpha x; \quad \dot{V} = -x^2$

   (b) $\dot{x}_1 = x_2, \ \dot{x}_2 = -\alpha x_1 - \beta x_2 \ ; \ \dot{V} = -x_2^2$

   (c) $\dot{x}_1 = x_2, \ \dot{x}_2 = x_3, \ \dot{x}_3 = -\alpha x_1 - \beta x_2 - \gamma x_3; \ \dot{V} = -x_3^2$

2. By choosing a suitable Liapunov function, show that the system

$$\dot{x}_1 = x_2 - \alpha x_1(x_1^2 + x_2^2), \quad \dot{x}_2 = -x_1 - \alpha x_2(x_1^2 + x_2^2)$$

   is asymptotically stable in the large if $\alpha > 0$.

3. Discuss the stability of the systems below, using the Liapunov functions suggested.

   (a) $\dot{x}_1 = x_2, \ \dot{x}_2 = -2x_1 - x_2; \ V = x_2^2 + 2x_1^2$

   (b) $\dot{x}_1 = -x_2, \ \dot{x}_2 = x_1; \ V = x_1^2 + x_2^2$

   (c) $\dot{x}_1 = -1.75x_1 + 0.25x_2, \ \dot{x}_2 = 0.75x_1 - 1.25x_2; \ V = 12x_1^2 + 12x_1x_2 + 20x_2^2$

   (d) $\dot{x}_1 = x_1 - x_2, \ \dot{x}_2 = -2x_1; \ V = x_1^2 - 2x_1x_2 - x_2^2$

   (e) $\dot{x}_1 = x_2 + x_1^3 + x_1x_2^2; \ \dot{x}_2 = -x_1 + x_1^2x_2 + x_2^3; \ V = x_1^2 + x_2^2$

3.6 CONSTRUCTION OF LIAPUNOV FUNCTIONS

It should be emphasised that the Liapunov stability theorems are theoretically important, but it is often very difficult to find a Liapunov function for non-linear systems.

It is possible to choose a tentative positive definite function $V(\mathbf{x})$, calculate $\dot{V}$ and examine it for definiteness. A more popular procedure is to choose a suitable $\dot{V}(\mathbf{x})$, calculate $V(\mathbf{x})$ and test its definiteness. In this section attention is confined to two of the many techniques developed for the construction of Liapunov functions.

Schultz and Gibson's Variable Gradient Method

Consider the autonomous system $\dot{\mathbf{x}} = \mathbf{f}(\mathbf{x})$ and note that if a Liapunov function $V(\mathbf{x})$ exists then its gradient $\dot{V} = \frac{\partial V_1}{\partial x_1}\dot{x}_1 + \ldots\ldots + \frac{\partial V_n}{\partial x_n}\dot{x}_n$ exists and is uniquely defined. Now

$$\dot{V} = (\nabla V) \, . \, \dot{\mathbf{x}} \tag{3.17}$$

and on integration we obtain

$$\int_0^t \dot{V}\, \mathrm{d}t = \int_0^{\mathbf{x}} (\nabla V) \, . \, \frac{\mathrm{d}\mathbf{x}}{\mathrm{d}t}\, \mathrm{d}t \tag{3.18}$$

where the limits t and $\mathbf{x}$ correspond and $\mathbf{x} = \mathbf{0}$ at $t = 0$. Therefore (3.18) becomes

$$V(\mathbf{x}) = \int_0^{\mathbf{x}} (\nabla V) \, . \, \mathrm{d}\mathbf{x} \tag{3.19}$$

The interpretation of the right-hand side of (3.19) is that the integral is a line integral to a point $\mathbf{x}$ in state space. For $V(\mathbf{x})$ to be uniquely determined from (3.19) it is necessary that the integral should be independent of the path of integration. Using vector analysis this condition implies that the n-dimensional curl of ∇V shall be zero, i.e.

$$\frac{\partial(\nabla V)_i}{\partial x_j} - \frac{\partial(\nabla V)_j}{\partial x_i} = 0, \; i \neq j; \; i, j = 1, 2, \ldots\ldots, n \tag{3.20}$$

where $(\nabla V)_i$ represents the ith component of ∇V.

A simple path is that which goes sequentially along the axes of state space i.e.

$$\int_0^{x_1} [\nabla V(u, 0, 0, \ldots\ldots, 0, 0]_1\, \mathrm{d}u + \int_0^{x_2} [\nabla V(x_1, u, 0, \ldots\ldots, 0, 0)]_2\, \mathrm{d}u$$

$$+ \ldots\ldots + \int_0^{x_n} [\nabla V(x_1, x_2, \ldots\ldots, x_{n-1}, u)]_n\, \mathrm{d}u \tag{3.21}$$

It can be shown that the conditions (3.20) are equivalent to requiring that the matrix

$$\begin{bmatrix} \frac{\partial(\nabla V)_1}{\partial x_1} & \cdots\cdots & \frac{\partial(\nabla V)_1}{\partial x_n} \\ \vdots & & \vdots \\ \frac{\partial(\nabla V)_n}{\partial x_1} & \cdots\cdots & \frac{\partial(\nabla V)_n}{\partial x_n} \end{bmatrix} \tag{3.22}$$

be symmetric.

The procedure is as follows

(1) Assume $\nabla V =$

$$\begin{bmatrix} a_{11}x_1 + a_{12}x_2 + \cdots + a_{1n}x_n \\ a_{21}x_1 + a_{22}x_2 + \cdots + a_{2n}x_n \\ \vdots \qquad \vdots \qquad \vdots \\ a_{n1}x_1 + a_{n2}x_2 + \cdots + a_{nn}x_n \end{bmatrix} \tag{3.23}$$

where the a_{ij} may be zero, constants, functions of time and/or the state variables. (Hence the name 'variable gradient'. There is a flexibility of choice which is tempered by intuition, conditions (3.20) and constraints on $\dot{V}$.)

(2) Determine $\dot{V}$.

(3) Constrain $\dot{V}$ to be at least negative semi-definite.

(4) Use (3.20) to determine the remaining a_{ij}.

(5) Check that $\dot{V}$ is still negative definite. If it is not, repeat the procedure.

(6) Obtain V from (3.19) via (3.21).

(Note that failure to find a suitable V by this method proves nothing about the stability of the equilibrium point.)

Example 1

$$\dot{x}_1 = x_2, \ \dot{x}_2 = -x_1 - Cx_2 - Dx_2^3; \ C, D > 0$$

Step (1) Assume $\nabla V = \begin{bmatrix} a_{11}x_1 + a_{12}x_2 \\ a_{21}x_1 + a_{22}x_2 \end{bmatrix}$

(2) $\dot{V} = (a_{11}x_1 + a_{12}x_2)\dot{x}_1 + (a_{21}x_1 + a_{22}x_2)\dot{x}_2$ and along a trajectory

$$\begin{aligned} \dot{V} &= (a_{11}x_1 + a_{12}x_2)\, x_2 + (a_{21}x_1 + a_{22}x_2)(-x_1 - Cx_2 - Dx_2^3) \\ &= -a_{21}x_1^2 + (a_{11} - a_{22} - Ca_{21})\, x_1 x_2 + (a_{12} - Ca_{22})x_2^2 \\ &\quad - Da_{21}x_1 x_2^3 - Da_{22}x_2^4 \end{aligned}$$

(3) Choose $a_{11} = 1$, $a_{12} = 0$, $a_{21} = 0$, $a_{22} = 1$ so that $\dot{V} = -Cx_2^2 - Dx_2^4$

which is clearly negative definite. Note that $\nabla V = \begin{bmatrix} x_1 \\ x_2 \end{bmatrix}$.

(4) We require that $\dfrac{\partial(\nabla V)_1}{\partial x_2} = \dfrac{\partial(\nabla V)_2}{\partial x_1}$ i.e. $\dfrac{\partial}{\partial x_2}(x_1) = \dfrac{\partial}{\partial x_1}(x_2)$ which is automatically satisfied.

(5) It is unnecessary to check $\dot{V}$ again.

(6) $V = \displaystyle\int_0^{x_1} x_1 \, dx_1 + \int_0^{x_2} x_2 \, dx_2 = \tfrac{1}{2}(x_1^2 + x_2^2)$.

Now V is positive definite and $V(0) = 0$. Hence the origin is asymptotically stable.

You may like to show that no other choice of numerical values for the a_{ij} will ensure $\dot{V}$ is negative definite. However, consider $a_{11} = a_{12}$, $a_{12} = x_1/x_2 = 1/a_{21}$ and follow through the working.

Example 2

$$\dot{x}_1 = x_2, \ \dot{x}_2 = x_3, \ \dot{x}_3 = -(x_1 + 2x_2)^3 - x_3$$

With ∇V as in (3.23) with $n = 3$ we evaluate $\dot{V}$ and choose $a_{13} = a_{31} = 0$, $a_{23} = a_{32} = 1$, $a_{22} = 1$, $a_{33} = 1$ so that

$$\nabla V = \begin{bmatrix} a_{11}x_1 + a_{12}x_2 \\ a_{21}x_1 + x_2 + x_3 \\ x_2 + x_3 \end{bmatrix}$$

and, along a trajectory

$$\dot{V} = (a_{11}x_1 + a_{12}x_2)x_2 + (a_{21}x_1 + x_2 + x_3)x_3$$
$$- (x_2 + x_3)(x_1 + 2x_2)^3 - (x_2 + x_3)x_3$$

This allows a partial cancellation of some terms and leaves three coefficients to be chosen. Conditions (3.20) are in this case

$$\begin{aligned} \frac{\partial}{\partial x_2}(a_{11}x_1 + a_{12}x_2) &= \frac{\partial}{\partial x_1}(a_{21}x_1 + x_2 + x_3) \\ \frac{\partial}{\partial x_3}(a_{11}x_1 + a_{12}x_2) &= \frac{\partial}{\partial x_1}(x_2 + x_3) = 0 \\ \frac{\partial}{\partial x_3}(a_{21}x_1 + x_2 + x_3) &= \frac{\partial}{\partial x_2}(x_2 + x_3) = 1 \end{aligned} \tag{3.24}$$

Remember that at this stage a_{11}, a_{12}, a_{21} could be functions of x_1, x_2, x_3 though we might hope to find suitable numerical values. From the last equation of (3.24), $x_1 \dfrac{\partial a_{21}}{\partial x_3} + 1 = 1$ and hence $\dfrac{\partial}{\partial x_3} a_{21} = 0$. From the form for $\dot{x}_3$ we are inspired to try $a_{21} = \dfrac{3}{x_1}(x_1 + 2x_2)^3$. This gives $\dfrac{\partial}{\partial x_1}(a_{21}x_1) = 3(x_1 + 2x_2)^2$ and by choosing $a_{12} = \dfrac{1}{2x_2}(x_1 + 2x_2)^3$ we find that $\dfrac{\partial}{\partial x_2}(a_{12}x_2) = 3(x_1 + 2x_2)^2$.

Suppose a_{11} is a function of x_1 only then all equations (3.24) are satisfied. We now have

$$\nabla V = \begin{bmatrix} a_{11}x_1 + \tfrac{1}{2}(x_1 + 2x_2)^3 \\ (x_1 + 2x_2)^3 + x_2 + x_3 \\ x_2 + x_3 \end{bmatrix}$$

where $a_{11} = a_{11}(x_1)$.

It can be deduced that along a trajectory

$$\dot{V} = -4x_2^4 - 3x_1^2x_2^2 - 6x_1x_2^3 - \tfrac{1}{2}x_2x_1^3 + a_{11}x_1x_2$$

The choice of a_{11} as $\tfrac{1}{2}x_1^2$ ensures $\dot{V}$ is somewhat simplified and capable of being rewritten as

$$\dot{V} = -x_2^2(4x_2^2 + 3x_1^2 + 6x_1x_2)$$

which is easily shown to be negative definite. Then

$$\nabla V = \begin{bmatrix} \tfrac{1}{2}x_1^3 + \tfrac{1}{2}(x_1 + 2x_2)^3 \\ (x_1 + 2x_2)^3 + x_2 + x_3 \\ x_2 + x_3 \end{bmatrix}$$

By (3.19) and (3.21) you can show that

$$V = \frac{1}{4}x_1^4 + \frac{1}{8}(x_1 + 2x_2)^4 + \frac{1}{2}(x_2 + x_3)^2$$

which is clearly positive definite.

This example demonstrates the ingenuity required sometimes You may care to repeat the example for the system with $\dot{x}_3 = -(x_1 + Cx_2)^3 - Dx_3$ and determine the conditions on C and D for asymptotic stability at the origin.

In some cases, more than one Liapunov function can be found, each leading to the conclusion of asymptotic stability. Which makes the more useful Liapunov function? The one which provides the larger region of stability. Clearly the engineer prefers a larger region to give him flexibility in design of the system. The search for a Liapunov function which widens the region is of value.

Liapunov Matrix Equation

The starting point is the linear autonomous system $\dot{\mathbf{x}} = \mathbf{Ax}$. Consider the function $V(\mathbf{x}) = \mathbf{x}^T\mathbf{Px}$ where $\mathbf{P}$ is a positive definite symmetric matrix with constant elements. Then, along a trajectory,

$$\dot{V} = \dot{\mathbf{x}}^T\mathbf{Px} + \mathbf{x}^T\mathbf{P}\dot{\mathbf{x}} = \mathbf{x}^T\mathbf{A}^T\mathbf{Px} + \mathbf{x}^T\mathbf{PAx} \qquad (\text{since } \dot{\mathbf{x}}^T = \mathbf{x}^T\mathbf{A}^T)$$

$$= \mathbf{x}^T(\mathbf{A}^T\mathbf{P} + \mathbf{PA})\mathbf{x}$$

If V is to be a Liapunov function, $\dot{V}$ must be at least negative semi-definite. If the concern is with asymptotic stability, $\dot{V}$ must be negative definite i.e. $\dot{V} = -\mathbf{x}^T\mathbf{Qx}$ where $\mathbf{Q}$ is a symmetric positive definite matrix. It is clear that we obtain

$$\mathbf{A}^T\mathbf{P} + \mathbf{P}\mathbf{A} = -\mathbf{Q} \tag{3.25}$$

which is the **Liapunov Matrix Equation.**

We may state the following theorem:

The origin is asymptotically stable if and only if there is a positive definite symmetric matrix $\mathbf{P}$ such that $\mathbf{P}$ is the unique solution of (3.25). (The appropriate Liapunov function is given by $V = \mathbf{x}^T\mathbf{P}\mathbf{x}$.)

A suitable matrix $\mathbf{Q}$ (often $\mathbf{I}$) is chosen in the hope that $\mathbf{P}$ turns out to be positive definite. For a unique $\mathbf{P}$ to exist it is necessary that $\mathbf{A}^T$ and $-\mathbf{A}$ have no common eigenvalues (alternatively, no two eigenvalues of $\mathbf{A}$, viz. λ_i and λ_j, are such that $\lambda_i + \lambda_j = 0$).

Example

$\mathbf{A} = \begin{bmatrix} -1 & -2 \\ 1 & -1 \end{bmatrix}$. In (3.25) take $\mathbf{Q} = \mathbf{I}$ and put $\mathbf{P} = \begin{bmatrix} p_{11} & p_{12} \\ p_{12} & p_{22} \end{bmatrix}$ so that

$$\begin{bmatrix} -1 & 1 \\ -2 & -1 \end{bmatrix}\begin{bmatrix} p_{11} & p_{12} \\ p_{12} & p_{22} \end{bmatrix} + \begin{bmatrix} p_{11} & p_{12} \\ p_{12} & p_{22} \end{bmatrix}\begin{bmatrix} -1 & -2 \\ 1 & -1 \end{bmatrix} = \begin{bmatrix} -1 & 0 \\ 0 & -1 \end{bmatrix}$$

i.e.

$$\begin{bmatrix} -p_{11} + p_{12} - p_{11} + p_{12} & -p_{12} + p_{22} - 2p_{11} - p_{12} \\ -2p_{11} - p_{12} - p_{12} + p_{22} & -2p_{12} - p_{22} - 2p_{12} - p_{22} \end{bmatrix} = \begin{bmatrix} -1 & 0 \\ 0 & -1 \end{bmatrix}$$

i.e.

$$\left.\begin{aligned} -2p_{11} + 2p_{12} \qquad &= -1 \\ -2p_{11} - 2p_{12} + p_{22} &= 0 \\ -4p_{12} - 2p_{22} \qquad &= -1 \end{aligned}\right\} \tag{3.26}$$

Hence $p_{11} = \dfrac{5}{12}$, $p_{12} = \dfrac{-1}{12}$, $p_{22} = \dfrac{8}{12}$

The matrix $\mathbf{P} = \begin{bmatrix} \frac{5}{12} & -\frac{1}{12} \\ -\frac{1}{12} & \frac{8}{12} \end{bmatrix}$ is positive definite; verify this. Therefore

$V = \frac{5}{12} x_1^2 - \frac{2}{12} x_1 x_2 + \frac{8}{12} x_2^2$ and, of course, $\dot{V} = -x_1^2 - x_2^2$. Hence the origin is asymptotically stable.

Because of the symmetry involved, the general case of $\mathbf{A}$ being $n \times n$ means the solution of $\frac{1}{2} n(n+1)$ equations. Write a computer program to effect the solution of (3.25) in the general case.

Liapunov further showed that non-linear systems of the form $\dot{\mathbf{x}} = \mathbf{Ax} + \mathbf{f}(\mathbf{x}, t)$ where $\mathbf{f}(\mathbf{x}, t) \to 0$ as $\|\mathbf{x}\| \to 0$ can be treated in the same way if $\mathbf{A}$ has no zero or purely imaginary eigenvalues.

That is, the stability properties of the non-linear system can be studied by omitting the term $\mathbf{f}(\mathbf{x}, t)$.

Infinite Series solution

The direct solution of the Liapunov matrix equation can be very slow if n is large. A method which is preferred for many problems relies on the result that if $\mathbf{A}$ is a stability matrix (i.e. all its eigenvalues have negative real parts) then the matrix $\mathbf{B} = (\mathbf{I} + \mathbf{A}^T)(\mathbf{I} - \mathbf{A}^T)^{-1}$ converges as an infinite series.

First we rearrange the equation for $\mathbf{B}$ in several ways. Now $\mathbf{B}(\mathbf{I} - \mathbf{A}^T) = \mathbf{I} + \mathbf{A}^T$ so that

$$\mathbf{B} - \mathbf{BA}^T = \mathbf{I} + \mathbf{A}^T \tag{3.27}$$

Hence $\mathbf{B} - \mathbf{I} = \mathbf{B}\mathbf{A}^T + \mathbf{A}^T = (\mathbf{B} + \mathbf{I})\mathbf{A}^T$

therefore

$$\mathbf{A}^T = (\mathbf{B} + \mathbf{I})^{-1}(\mathbf{B} - \mathbf{I}) \tag{3.28}$$

Taking the transpose of (3.27) we obtain

$$\mathbf{B}^T - \mathbf{AB}^T = \mathbf{I} + \mathbf{A} \tag{3.29}$$

so that $\mathbf{B}^T(\mathbf{I} - \mathbf{A}) = \mathbf{I} + \mathbf{A}$ and

$$\mathbf{B}^T = (\mathbf{I} + \mathbf{A})(\mathbf{I} - \mathbf{A})^{-1} \tag{3.30}$$

From (3.29) we find $\mathbf{B}^T - \mathbf{I} = \mathbf{A}\mathbf{B}^T + \mathbf{A} = \mathbf{A}(\mathbf{B}^T + \mathbf{I})$ and

$$\mathbf{A} = (\mathbf{B}^T - \mathbf{I})(\mathbf{B}^T + \mathbf{I})^{-1} \tag{3.31}$$

Again from (3.27) $\mathbf{B} - \mathbf{B}\mathbf{A}^T - \mathbf{A}^T = \mathbf{I}$, hence $\mathbf{B} + \mathbf{I} - \mathbf{B}\mathbf{A}^T - \mathbf{A}^T = 2\mathbf{I}$ or $(\mathbf{B} + \mathbf{I})(\mathbf{I} - \mathbf{A}^T) = 2\mathbf{I}$

so that
$$\mathbf{B} + \mathbf{I} = 2(\mathbf{I} - \mathbf{A}^T)^{-1} \tag{3.32}$$

See whether, starting from (3.29) you can deduce

$$(\mathbf{B}^T + \mathbf{I}) = 2(\mathbf{I} - \mathbf{A})^{-1} \tag{3.33}$$

Now we are ready to use this impressive array of results.

The equation $\mathbf{A}^T\mathbf{P} + \mathbf{P}\mathbf{A} = -\mathbf{Q}$ (3.25)

becomes
$$(\mathbf{B} + \mathbf{I})^{-1}(\mathbf{B} - \mathbf{I})\mathbf{P} + \mathbf{P}(\mathbf{B}^T - \mathbf{I})(\mathbf{B}^T + \mathbf{I})^{-1} = -\mathbf{Q}$$

Post-multiplying by $(\mathbf{B}^T + \mathbf{I})$ and premultiplying by $(\mathbf{B} + \mathbf{I})$ yields

$$(\mathbf{B} - \mathbf{I})\mathbf{P}(\mathbf{B}^T + \mathbf{I}) + (\mathbf{B} + \mathbf{I})\mathbf{P}(\mathbf{B}^T - \mathbf{I}) = -(\mathbf{B} + \mathbf{I})\mathbf{Q}(\mathbf{B}^T + \mathbf{I})$$

or

$$\mathbf{B}\mathbf{P}\mathbf{B}^T - \mathbf{P}\mathbf{B}^T + \mathbf{B}\mathbf{P} - \mathbf{P} + \mathbf{B}\mathbf{P}\mathbf{B}^T + \mathbf{P}\mathbf{B}^T - \mathbf{B}\mathbf{P} - \mathbf{P} = -(\mathbf{B} + \mathbf{I})\mathbf{Q}(\mathbf{B}^T + \mathbf{I})$$

Hence $\mathbf{P} - \mathbf{B}\mathbf{P}\mathbf{B}^T = \frac{1}{2}(\mathbf{B} + \mathbf{I})\mathbf{Q}(\mathbf{B}^T + \mathbf{I}) = \mathbf{M}$, say. It follows from (3.32) and (3.33) that

$$\mathbf{M} = 2(\mathbf{I} - \mathbf{A}^T)^{-1}\mathbf{Q}(\mathbf{I} - \mathbf{A})^{-1}$$

Suppose we write $\mathbf{B}\mathbf{P}\mathbf{B}^T$ as $\mathbf{\Phi}\mathbf{P}$ where $\mathbf{\Phi}$ is a matrix operator. Then

$$\begin{aligned}\mathbf{P} &= (\mathbf{I} - \mathbf{\Phi})^{-1}\mathbf{M} \\ &= (\mathbf{I} + \mathbf{\Phi} + \mathbf{\Phi}^2 + \ldots..)\mathbf{M}\end{aligned}$$

i.e.

$$\mathbf{P} = \mathbf{M} + \mathbf{B}\mathbf{M}\mathbf{B}^T + \mathbf{B}^2\mathbf{M}(\mathbf{B}^T)^2 + \ldots.. \tag{3.34}$$

Under the assumption that $\mathbf{A}$ is a stability matrix it can be shown that the series in (3.34) converges. The proof is beyond our scope; see Barnett and Storey (1970).

Then put $\mathbf{Q} = 2(\mathbf{B} + \mathbf{I})^{-1}(\mathbf{B}^T + \mathbf{I})^{-1}$ so that $\mathbf{M} = \mathbf{I}$. The solution $\mathbf{P}$ in this case is written as $\mathbf{T}$. Hence

$$\mathbf{T} = \mathbf{I} + \mathbf{B}\mathbf{B}^T + \mathbf{B}^2(\mathbf{B}^T)^2 + \ldots.. \tag{3.35}$$

Rate of convergence

If λ_i is an eigenvalue of $\mathbf{A}$, $\mu_i = (1 + \lambda_i)/(1 - \lambda_i)$ is an eigenvalue of $\mathbf{B}$. Convergence is most rapid for $\lambda_i = -1$ and slowest for $|\lambda_i|$ very large or very small.

We may improve the convergence by a nesting procedure.

Let the partial sums of the series in (3.34) be $\mathbf{T}_i$. Then $\mathbf{T}_1 = \mathbf{I}$, $\mathbf{T}_2 = \mathbf{T}_1 + \mathbf{B}\mathbf{T}_1\mathbf{B}^T$, $\mathbf{T}_3 = \mathbf{T}_2 + \mathbf{B}^2\mathbf{T}_2(\mathbf{B}^T)^2$, $\mathbf{T}_4 = \mathbf{T}_3 + \mathbf{B}^4\mathbf{T}_3(\mathbf{B}^T)^4$ etc. and, in general,

$$\mathbf{T}_{k+1} = \mathbf{T}_k + \mathbf{B}^{2^{(k-1)}}\mathbf{T}_k(\mathbf{B}^T)^{2^{(k-1)}}, \quad k = 1, 2, ...$$

Example

$$\mathbf{A} = \begin{bmatrix} -1 & -2 \\ 1 & -1 \end{bmatrix}; \quad \mathbf{B} = \begin{bmatrix} 0 & 1 \\ -2 & 0 \end{bmatrix}\begin{bmatrix} 2 & -1 \\ 2 & 2 \end{bmatrix}^{-1} = \begin{bmatrix} -\frac{1}{3} & \frac{1}{3} \\ -\frac{2}{3} & -\frac{1}{3} \end{bmatrix}. \text{ Then}$$

$$\mathbf{B}\mathbf{B}^T = \begin{bmatrix} \frac{2}{9} & \frac{1}{9} \\ \frac{1}{9} & \frac{5}{9} \end{bmatrix} \text{ and } \mathbf{T}_2 = \begin{bmatrix} \frac{11}{9} & \frac{1}{9} \\ \frac{1}{9} & \frac{14}{9} \end{bmatrix}. \text{ Similarly, } \mathbf{B}^2 = \begin{bmatrix} -\frac{1}{9} & -\frac{2}{9} \\ +\frac{4}{9} & -\frac{1}{9} \end{bmatrix},$$

$$(\mathbf{B}^T)^2 = \begin{bmatrix} -\frac{1}{9} & \frac{4}{9} \\ -\frac{2}{9} & -\frac{1}{9} \end{bmatrix}, \quad \mathbf{T}_3 = \frac{1}{729}\begin{bmatrix} 800 & -23 \\ -23 & 911 \end{bmatrix} \text{ and so on.}$$

Problems

1. Using the variable gradient method, investigate the stability of the system $\dot{x}_1 = x_2$, $\dot{x}_2 = x_2 f(x_1) - g(x_1)$.

2. Construct a Liapunov function for the system $\dot{x}_1 = -x_1 + 2x_1^2 x_2$, $\dot{x}_2 = -x_2$ using the variable gradient method and $\nabla V = \begin{bmatrix} a_{11}x_1 + a_{12}x_2 \\ a_{21}x_1 + 2x_2 \end{bmatrix}$

   (a) put $a_{11} = 1$, $a_{12} = 0$, $a_{21} = 0$

   (b) put $a_{11} = \dfrac{2}{(1 - x_1x_2)^2}$, $a_{12} = \dfrac{-x_1^2}{(1 - x_1x_2)} = -a_{21}$

In each case, sketch the region for which V is positive definite and V is negative definite.

3. Discuss the stability of the system $\dot{x}_1 = x_2$, $\dot{x}_2 = -4x_1 - 2x_2$ using the Liapunov matrix equation with $\mathbf{Q} = \mathbf{I}$.

4. Repeat Problem 3 for the linearised Van der Pol equation $\dot{x}_1 = x_2$, $\dot{x}_2 = -x_1 + kx_2$, $k > 0$.

5. Use the variable gradient method on the system

$$\dot{x}_1 = x_2, \ \dot{x}_2 = -x_1 - \alpha x_2 - \beta x_2^3; \qquad \alpha, \beta < 0$$

Take $a_{21} = a_{12} = 0$, $a_{11} = a_{22} = 1$. Show that the system is asymptotically stable.

6. Consider the stability of the system

$$\dot{x}_1 = -3x_2 - x_1^5, \ \dot{x}_2 = -2x_2 + x_1^5$$

Use the variable gradient method, taking $a_{12} = 0$, $a_{21} = 0$, $a_{11} = \dfrac{1}{3} a_{22} x_1^4$ to show that $a_{22} > 0$ guarantees asymptotic stability in the large. Why does linearisation not show this result?

3.7 DISCRETE SYSTEMS

Much of what has been said for continuous systems regarding Liapunov functions holds for discrete systems, except that $\dot{V}$ is replaced by $\Delta V = V[\mathbf{x}(k+1)] - V[\mathbf{x}(k)]$.

Consider the linear discrete system $\mathbf{x}(k+1) = \mathbf{A}\mathbf{x}(k)$. If $V = \mathbf{x}^{\mathrm{T}}\mathbf{P}\mathbf{x}$ then

$$\Delta V = \mathbf{x}^{\mathrm{T}}(k)\mathbf{A}^{\mathrm{T}}\mathbf{P}\mathbf{A}\mathbf{x}(k) - \mathbf{x}^{\mathrm{T}}(k)\mathbf{P}\mathbf{x}(k)$$

i.e.

$$\Delta V = \mathbf{x}^{\mathrm{T}}(\mathbf{A}^{\mathrm{T}}\mathbf{P}\mathbf{A} - \mathbf{P})\mathbf{x} = -\mathbf{x}^{\mathrm{T}}\mathbf{Q}\mathbf{x}$$

Then we may say by analogy that the origin is asymptotically stable if and only if there is a symmetric positive definite matrix $\mathbf{P}$ that uniquely satisfies the equation

$$\mathbf{A}^{\mathrm{T}}\mathbf{P}\mathbf{A} - \mathbf{P} = -\mathbf{Q} \tag{3.36}$$

where $\mathbf{Q}$ is symmetric and positive definite.

In general, the construction of Liapunov functions for discrete systems is very difficult.

8 STABILITY AND CONTROL

The inverted pendulum example of Section 1.2 may be subjected to a proportional
ntrol $u = K_1\theta$. In Section 1.3 it was found that the initial conditions
0) $= \dot{x}(0) = \dot{\theta}(0) = 0$ and $\theta(0) = \theta_0$ led to the solution

$$\theta = \theta_0 \cos \omega t \qquad (1.29)$$

r the angular deviation of the pendulum, where $\omega^2 = 3[K_1 - (m+M)g]/[(m+4M)l]$.

In Figure 3.18 are sketched the relationships between θ and t in the cases

$K_1 < K_{cr}$, (ii) $K_1 = K_{cr}$ (iii) $K_1 > K_{cr}$

ıere K_{cr} is that **critical** value of K_1 for which $\omega^2 = 0$. Strictly (1.29) applies only
case (iii). We leave you to show that the solution for case (ii) is $\theta = \theta_0$ and that for
se (i) is $\theta = \theta_0 \cosh |\omega| t$.

Try to interpret the graphs in terms of the physical set-up.

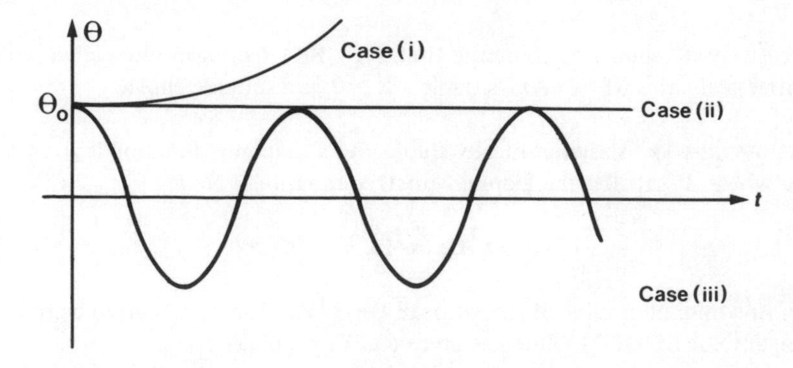

Figure 3.18

early, this form of control is unsatisfactory. We now consider the more likely
ndidate of proportional - plus - derivative (PPD) control $u = K_1\theta + K_2\dot{\theta}$. We leave
ɔu to derive the result

$$\theta = \frac{\theta_0 \omega}{\sqrt{\omega^2 - \alpha^2}} e^{-\alpha t} \sin\left(\sqrt{\omega^2 - \alpha^2}\, t + \tan^{-1} \frac{\sqrt{\omega^2 - \alpha^2}}{\alpha}\right)$$

ıere $\alpha = \dfrac{3K_2}{2(m+4M)l}$ and $\omega^2 \geqslant \alpha^2$.

The graph of θ versus t is shown in Figure 3.19.

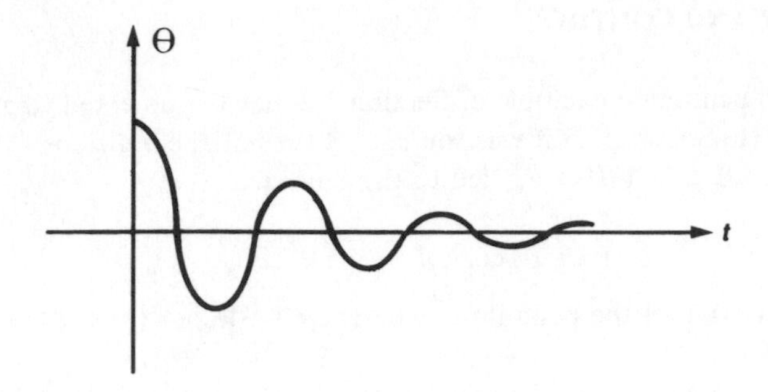

Figure 3.19

Again, we leave you to consider the case $\omega^2 < \alpha^2$. It should be clear that PPD control superior to the proportional control. Argue this on physical grounds.

In general, consider the linear system $\dot{\mathbf{x}} = \mathbf{Ax} + \mathbf{Bu}$, with input $\mathbf{u} = \mathbf{Kx}$. Then $\dot{\mathbf{x}} = (\mathbf{A} + \mathbf{BK})\mathbf{x}$. This means that the results derived earlier can be applied with the matrix $(\mathbf{A} + \mathbf{BK})$ replacing the matrix $\mathbf{A}$.

Effectively $\mathbf{K}$ should be chosen so that $(\mathbf{A} + \mathbf{BK})$ has eigenvalues all of which have negative real part. If $\dot{\mathbf{x}} = \mathbf{Ax}$ is stable, $\mathbf{K} = \mathbf{0}$ is a suitable choice.

Suppose $\dot{\mathbf{x}} = \mathbf{Ax}$ is asymptotically stable and a Liapunov function is given by $V = \mathbf{x}^T\mathbf{Px}$ where $\mathbf{P}$ satisfies the Liapunov matrix equation. Now,

$$\dot{V}/V = -\mathbf{x}^T\mathbf{Qx}/\mathbf{x}^T\mathbf{Px} \leqslant -\sigma, \text{ say} \tag{3.37}$$

(where σ, the minimum value of the ratio $\mathbf{x}^T\mathbf{Qx}/\mathbf{x}^T\mathbf{Px}$, can be shown to be the smallest eigenvalue of $\mathbf{QP}^{-1}$). Integration of (3.37) produces

$$V(\mathbf{x}) \leqslant e^{-\sigma t}V(\mathbf{x}_0) \tag{3.38}$$

where $\mathbf{x}_0$ is, as usual, the initial value of $\mathbf{x}$.

Now the larger the value of σ the more rapidly $\mathbf{x}(t) \to 0$. Suppose that we apply the control $\mathbf{u} = (\mathbf{S} - \mathbf{R})\mathbf{B}^T\mathbf{Px}$ where $\mathbf{P}$ is the solution of the Liapunov matrix equation, $\mathbf{S}$ is a skew-symmetric matrix and $\mathbf{R}$ is symmetric positive definite. Then the system i governed by $\dot{\mathbf{x}} = [\mathbf{A} + \mathbf{B}(\mathbf{S} - \mathbf{R})\mathbf{B}^T\mathbf{P}]\,\mathbf{x}$.

It can be shown that $\dot{V} = -\mathbf{x}^T\mathbf{Qx} - 2\mathbf{x}^T\mathbf{PB\,RB}^T\mathbf{Px}$ and since $\mathbf{PB\,RB}^T\mathbf{P} = (\mathbf{PB})\mathbf{R}(\mathbf{PB})^T$ is positive definite it follows that, for this control, $\dot{V} < -\mathbf{x}^T\mathbf{Qx}$, and hence σ is larger and in a sense the system is made more stable since trajectories tend to the origin more rapidly. (The reason for the somewhat peculiar choice of control will become clear in the next chapter.)

Problem

A rocket stands on a launching pad. The mass, m, of the rocket acts at its centre of gravity which is a distance l above the ground. Show that for a small perturbation θ of the system about its base, the equation of motion is

$$l\ddot{\theta} + u(t) = g\theta$$

where $u(t)$ is a horizontal linear acceleration per unit mass of the rocket which is applied in an attempt to counter the perturbation.

Show that the system may be modelled by the differential equation

$$\dot{\mathbf{x}} = \mathbf{A}\mathbf{x} + \mathbf{B}\mathbf{u}$$

Discuss the stability of the system when the control $u(t)$ is a linear function of θ.

Chapter Four

Introduction to Optimal Control

4.1 INTRODUCTION

It is clearly desirable to control a system in the best possible way, according to some suitable criterion. For example we may wish to return a disturbed system to its steady-state equilibrium condition in the shortest time possible. The problem of optimal control is to choose that control function which minimises the time taken, subject to the state differential equations and any end conditions on the state. The control may be subject to constraints: for example, an engine may not exert a thrust greater than a specified value or a steering device may not turn through an angle outside a particular range of values.

It is customary to define a **performance index**, I. In the example above, relating to a disturbed system, we can define the index as

$$I = \int_{t_0}^{t_f} 1 \, \mathrm{d}t = t_f - t_0$$

Here t_0 and t_f are the initial and final times at which the control is operating. This is an example of a **minimum-time** problem. To be specific, consider the simple example of the oven of Section 1.2 where we wish to heat the interior to a specified temperature as quickly as possible. (It should be noted at this early stage in the discussion that a system must be known to be controllable before any question of optimal control strategy is raised.)

The problem becomes to minimise $I = t_f - t_0$ subject to the equations

$$\dot{\mathbf{x}} = \mathbf{A}\mathbf{x} + \mathbf{B}\mathbf{u}$$

with $\mathbf{x}(t_0)$ and $\mathbf{x}(t_f)$ specified as the initial and final temperatures. Once the control $\mathbf{u}(t)$ has been determined† the state variables $\mathbf{x}(t)$ may be found for the optimal trajectory in state space. To make the problem realistic suppose that the heat input, u, cannot exceed a specified value u_{max}.

Other performance indices which may be used are

(i) **Minimum fuel**, where for a single control,

† The control may be open-loop or, if it involves the state variables, closed-loop.

$$I = \int_{t_0}^{t_f} |u| \, dt$$

As an example, consider the problem of soft-landing a lunar module. Let $z(t)$ be the height of the module above the moon's surface and let $z(0)$ be its initial hovering height.

Then the governing equation of motion is $\ddot{z} = -g - \frac{1}{m} u(t)$ where m is the mass of the module (partly fuel), g is the acceleration due to lunar gravity and $u(t)$ is the vertical rocket thrust given by

$$u(t) = -k\dot{m} \tag{4.1}$$

where $\dot{m}$ is the rate of burning of fuel and k is the velocity of exhaust gases relative to the module. (Check the validity of this expression.)
Introducing state variables† $x_1 = z$, $x_2 = \dot{z}$ we have the equations

$$\dot{x}_1 = x_2, \qquad \dot{x}_2 = -g - \frac{1}{m} u(t) \tag{4.2}$$

A suitable index of performance is

$$I = -\int_{t_0}^{t_f} \dot{m} \, dt = m(t_0) - m(t_f)$$

Via (4.1) it follows that

$$I = \frac{1}{k} \int_0^t u \, dt \tag{4.3}$$

Verify that the system is controllable.

It is clearly sensible to impose the condition $0 < u < u_{max}$. Further note that

$$\mathbf{x}(t_0) = \begin{bmatrix} z(0) \\ 0 \end{bmatrix}, \qquad \mathbf{x}(t_f) = \begin{bmatrix} 0 \\ 0 \end{bmatrix} \tag{4.4}$$

ii) **Terminal control**, where $I = [\mathbf{x}(t_f) - \mathbf{r}(t_f)]^T \mathbf{P} [\mathbf{x}(t_f) - \mathbf{r}(t_f)]$, the desired final state is $\mathbf{r}(t_f)$ and $\mathbf{P}$ is a real positive definite symmetric matrix.

These are measurable.

(iii) **Tracking problems**, where $I = \int_{t_0}^{t_f} [\mathbf{x}(t) - \mathbf{r}(t)]^{\mathrm{T}} \mathbf{P}[\mathbf{x}(t) - \mathbf{r}(t)]\,\mathrm{d}t$. Here the object is to follow as closely as possible a reference signal $\mathbf{r}(t)$ throughout $t_0 \leqslant t \leqslant t_f$.

In practice a more general index is needed to take account of the total effect of control and a combination of the indices in (i) and (iii) is employed.

In many simple problems the performance index is quadratic in its variables and if the system is linear then an analytical solution may be possible. However, other cases may lead to considerable difficulty.

In this chapter we first consider the calculus of variations which provides a classical approach to the solution of simple problems of optimal control. It is studied first in application to problems where $\mathbf{x}(t_0)$ and $\mathbf{x}(t_f)$ are determined, then to problems where $\mathbf{x}(t_f)$ is merely constrained to lie on a specified curve in state space. We then examine Pontryagin's maximum principle which generalises the calculus of variations to situations in which the control is bounded.† Essentially Pontryagin reformulates the problem in a manner analogous to the Hamiltonian formulation of classical dynamics. Finally we make a brief excursion into the topic of Dynamic Programming which is a numerical approach to the problem of optimal control.

It is necessary to emphasise that this chapter is merely an introductory account to establish the concepts of optimal control.

Two final remarks should be made. First, in minimising an integral, we should in practice need to allow for any *future* disturbances to the state, since these will affect the value of the integral. Second, many real-life problems require numerical methods for their solution and the requirements on computer storage and time may be very heavy.

4.2 A SIMPLIFIED EXAMINATION OF THE CALCULUS OF VARIATIONS

In this section the method of the calculus of variations is established. The control problems considered will have the state variable specified at the end points and will not place any restrictions on the control variable.

There are basically three kinds of problem by which optimal control may be classified according to the performance index

(i) $I = G(x_i, t_f)$ Mayer problem

† This case can also be handled by means of penalty functions in the classical calculus of variations.

(ii) $I = \int_{t_0}^{t_f} F(x_i, u_i, t)\mathrm{d}t$ Lagrange problem

(iii) $I = G(x_i, t_f) + \int_{t_0}^{t_f} F(x_i, u_i, t)\mathrm{d}t$ Bolza problem

These three types of problem are essentially equivalent and one form can be converted into another. In the next section a Lagrange problem is transformed into a Mayer problem.

Concept of a functional

As a function associates one scalar variable with another, for example, $y = f(x) = x^2 + 2$, so a **functional** associates a function with a scalar constant. For example, consider $I = \int_0^1 (2x - \dot{x}t^2)\mathrm{d}t$ where $x = x(t)$.

(a) $x = A$, constant : $I = \int_0^1 (2A)\mathrm{d}t = 2A$

(b) $x = t$: $I = \int_0^1 (2t - t^2)\mathrm{d}t = 1 - \frac{1}{3} = \frac{2}{3}$

(c) $x = t^2$: $I = \int_0^1 (2t^2 - 2t^3)\mathrm{d}t = \frac{2}{3} - \frac{1}{2} = \frac{1}{6}$

I is sometimes referred to as a **cost functional.**

Calculus of variations: scalar case

We wish to minimise $I = \int_{t_0}^{t_f} F(x, \dot{x}, t)\mathrm{d}t$ by choosing a suitable function $x(t)$. Suppose that $\dot{x}(t_0)$ and $x(t_f)$ are given.

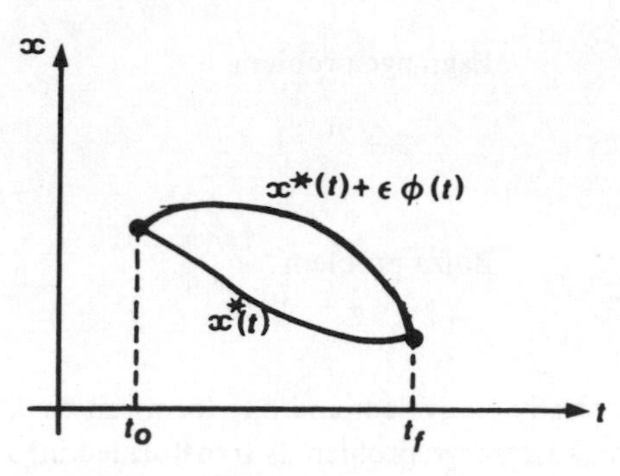

Figure 4.1

In Figure 4.1 is shown the optimal trajectory $x^*(t)$ (which minimises I) and a 'neighbouring' trajectory $x(t) = x^*(t) + \epsilon\phi(t)$ where ϵ is a small scalar parameter.

Assume that variations in the slope $\dot{x}$ are kept small. Then, since the values of $x(t)$ at $t = t_0$ and $t = t_f$ are fixed,

$$\phi(t_0) = 0 = \phi(t_f) \tag{4.5}$$

Also

$$\dot{x}(t) = \dot{x}^*(t) + \epsilon\dot{\phi}(t) \tag{4.6}$$

Consider now the integral

$$J(\epsilon) = \int_{t_0}^{t_f} F(x^* + \epsilon\phi, \ \dot{x}^* + \epsilon\dot{\phi}, \ t)\mathrm{d}t$$

By Taylor's expansion

$$F(x^* + \epsilon\phi, \dot{x} + \epsilon\dot{\phi}, t) = F(x^*, \dot{x}^*, t) + \epsilon\left(\phi\frac{\partial F}{\partial x} + \dot{\phi}\frac{\partial F}{\partial \dot{x}}\right) + 0(\epsilon^2)$$

$$\text{Let } \Delta J = \int_{t_0}^{t_f} \left\{F(x^* + \epsilon\phi, \dot{x}^* + \epsilon\dot{\phi}, t) - F(x^*, \dot{x}^*, t)\right\}\mathrm{d}t$$

$$\text{Then } \Delta J = \epsilon \int_{t_0}^{t_f} \left\{\phi\frac{\partial F}{\partial x} + \dot{\phi}\frac{\partial F}{\partial \dot{x}}\right\}\mathrm{d}t$$

The right-hand side is called the **first variation** in $J(\epsilon)$. If the trajectory $x^*(t)$ is optimal then $\dfrac{\partial J}{\partial \epsilon} = 0$ at $\epsilon = 0$. Therefore a **necessary** condition for the trajectory x^* to produce an extreme value of I (we cannot yet specify maximum or minimum) is that

$$\int_{t_0}^{t_f} \left(\phi\frac{\partial F}{\partial x} + \dot{\phi}\frac{\partial F}{\partial \dot{x}}\right)\mathrm{d}t = 0 \tag{4.7}$$

But

$$\int_{t_0}^{t_f} \frac{\partial F}{\partial \dot{x}} \dot{\phi} \, dt = \left[\frac{\partial F}{\partial \dot{x}} \phi \right]_{t_0}^{t_f} - \int_{t_0}^{t_f} \frac{d}{dt} \left(\frac{\partial F}{\partial \dot{x}} \right) \phi(t) dt$$

However, $\phi = 0$ at $t = t_0$ and $t = t_f$. Therefore

$$\int_{t_0}^{t_f} \frac{\partial F}{\partial \dot{x}} \dot{\phi} \, dt = - \int_{t_0}^{t_f} \frac{d}{dt} \left(\frac{\partial F}{\partial \dot{x}} \right) \phi(t) dt$$

Hence (4.7) becomes

$$\int_{t_0}^{t_f} \phi(t) \left\{ \frac{\partial F}{\partial x} - \frac{d}{dt} \left(\frac{\partial F}{\partial \dot{x}} \right) \right\} dt = 0$$

However, the variation $\phi(t)$ is arbitrary, It follows that a **necessary** condition for the minimisation of I is that

$$\frac{\partial F}{\partial x} - \frac{d}{dt} \left(\frac{\partial F}{\partial \dot{x}} \right) \equiv 0 \qquad (4.8)$$

The function $x(t)$ which satisfies (4.8) is the optimal trajectory. Equation (4.8) is referred to as the **Euler-Lagrange** equation . Note that it is a necessary and sufficient condition for the trajectory to be **extremal.**

Example 1

Minimise the performance index

$$I = \int_{t_0}^{t_f} (\alpha^2 x^2 + \dot{x}^2) dt$$

where α is a scalar and t_0, t_f, $x(t_0)$, $x(t_f)$ are given. In this instance $F = \alpha^2 x^2 + \dot{x}^2$. Therefore

$$\frac{\partial F}{\partial x} = 2\alpha^2 x \quad \text{and} \quad \frac{\partial F}{\partial \dot{x}} = 2\dot{x}$$

Equation (4.8) becomes $2\alpha^2 x = 2\ddot{x}$. The solution of this equation is $x = Ae^{\alpha t} + Be^{-\alpha t}$ where A and B are constants to be determined by the end conditions. If, to be specific, $t_0 = 0$, $t_f = 1$, $x(t_0) = 0$, $x(t_f) = 1$ then $x = \sinh \alpha t/\sinh \alpha$.

This problem can be formulated in terms of a control function u if the system equation is $\dot{x} = u$ and then we seek to minimise

$$I = \int_{t_0}^{t_f} (\alpha^2 x^2 + u^2)\mathrm{d}t$$

The quantity α^2 may be interpreted as a weighting of the relative importance of x and u in the performance index. We see that u is a function of t only and therefore this is a case of open loop control.

Example 2

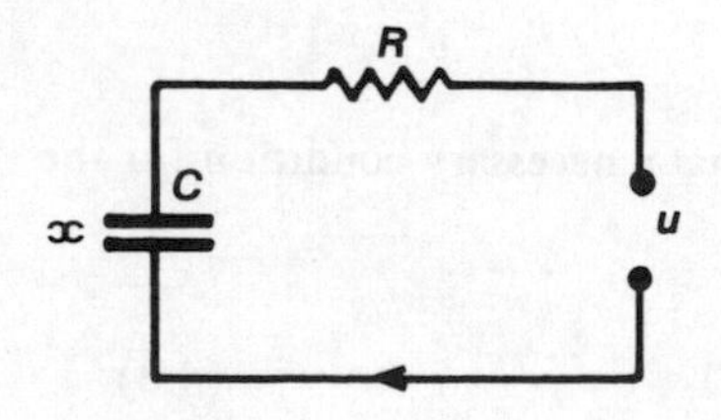

Figure 4.2

Figure 4.2 shows the circuit for the charging of a capacitor from a voltage x_0 at time t_0 to a final value x_f at t_f. The energy dissipated in the resistor is

$$\int_{t_0}^{t_f} RC^2\dot{x}^2\,\mathrm{d}t$$

The state equation is $\dot{x} = \dfrac{-1}{RC}x + \dfrac{1}{RC}u$ where u is the source of voltage. The system can easily be shown to be controllable. We require to choose that control strategy which minimises the dissipated energy. It should be clear that we need to minimise

$$I = \int_{t_0}^{t_f} (u - x)^2\,\mathrm{d}t \tag{4.9}$$

For the moment we minimise $I = \int_{t_0}^{t_f} \dot{x}^2\,\mathrm{d}t$. (Why? Refer to the state equation.) Then we have from equation (4.8) $\ddot{x} = 0$. Hence $x(t) = a + bt$. From the end conditions we have

$$x(t) = x_0 + (x_f - x_0)(t - t_0)/(t_f - t_0) \tag{4.10}$$

Hence the capacitor voltage increases at a constant rate.

Special case

In the case that F does not contain t explicitly the Euler -Lagrange equation has a first integral

$$F - \dot{x}\frac{\partial F}{\partial \dot{x}} = C, \text{ constant} \tag{4.11}$$

To see this, note that

$$\frac{\mathrm{d}}{\mathrm{d}t}F = \frac{\partial F}{\partial t} + \frac{\partial F}{\partial x} \cdot \frac{\mathrm{d}x}{\mathrm{d}t} + \frac{\partial F}{\partial \dot{x}}\frac{\mathrm{d}\dot{x}}{\mathrm{d}t} = 0 + \dot{x}\frac{\partial F}{\partial x} + \ddot{x}\frac{\partial F}{\partial \dot{x}}$$

so that $\frac{\mathrm{d}}{\mathrm{d}t}$ (4.11) becomes

$$\dot{x}\frac{\partial F}{\partial x} + \ddot{x}\frac{\partial F}{\partial \dot{x}} - \ddot{x}\frac{\partial F}{\partial \dot{x}} - \dot{x}\frac{\mathrm{d}}{\mathrm{d}t}\left(\frac{\partial F}{\partial \dot{x}}\right) = 0$$

which is automatically satisfied.

For Example 1, $F = \alpha^2 x^2 + \dot{x}^2$ therefore (4.11) becomes

$$\alpha^2 x^2 + \dot{x}^2 - \dot{x}\, 2\dot{x} = C$$

or

$$\alpha^2 x^2 - \dot{x}^2 = C$$

You can easily verify that this equation is satisfied by the function $x = \sinh \alpha t/\sinh \alpha$ obtained earlier.

Reformulation in terms of a control variable

Now suppose that we are given the state equation $\dot{x} = f(x, u, t)$ and the initial condition $x(t_0) = x_0$ and it is required to minimise

$$I = \int_{t_0}^{t_f} g(x, u, t)\mathrm{d}t$$

A technique analogous to Lagrange multipliers is employed. Consider

$$F(x, \dot{x}, u, \dot{u}, t) = g(x, \dot{x}, u, \dot{u}, t) + \lambda(t)h(x, \dot{x}, u, \dot{u}, t)$$

where $$h(x, \dot{x}, u, t) = \dot{x} - f(x, u, t)$$

The minimisation of $J = \int_{t_0}^{t_f} F \, dt$ will give the required result for the minimum performance index. Consider the optimum state and control as x^*, u^* and let $x(t) = x^*(t) + \epsilon\phi(t)$ and $u(t) = u^*(t) + \epsilon\psi(t)$ where ϕ and ψ are suitable arbitrary functions. A development similar to the preceding case gives a necessary condition for minimisation as

$$\int_{t_0}^{t_f} \left[\phi \left\{ \frac{\partial F}{\partial x} - \frac{d}{dt}\left(\frac{\partial F}{\partial \dot{x}}\right) \right\} + \psi \left\{ \frac{\partial F}{\partial u} - \frac{d}{dt}\left(\frac{\partial F}{\partial \dot{u}}\right) \right\} \right] dt = 0$$

Hence, allowing for the case where $\dot{u}$ does not appear in the state equations or in the original performance index we derive the necessary conditions (4.8) as before and, additionally,

$$\frac{\partial F}{\partial u} = 0 \tag{4.12}$$

Note that the use of Lagrange multipliers allows the constrained problem in $x(t)$, $u(t)$ to be treated as an unconstrained problem in $x(t)$, $u(t)$, $\lambda(t)$.

Example 1

An inertia wheel is to be brought to a new stationary angle θ_2 from the initial position $\theta = \theta_0$. The control is $u(t) = \ddot{\theta}$ and the performance index is

$I = \frac{1}{2} \int_0^2 [u(t)]^2 \, dt$. The end conditions are $\theta(0) = \theta_0$, $\theta(2) = \theta_2$, $\dot{\theta}(0) = 1$, $\dot{\theta}(2) = 0$. Let $x_1 = \theta$, $x_2 = \dot{x}_1$; hence $u = \dot{x}_2$. Then we form

$$J = \int_0^2 \left\{ \tfrac{1}{2}[u(t)]^2 + \lambda_1(t)[x_2(t) - \dot{x}_1] + \lambda_2(t)[u(t) - \dot{x}_2] \right\} dt$$

(We can, for consistency label u as x_3, noting that $\frac{\partial F}{\partial \dot{x}_3} \equiv 0$)

The Euler-Lagrange equations are then

$$\frac{\partial F}{\partial x_1} - \frac{d}{dt}\left(\frac{\partial F}{\partial \dot{x}_1}\right) = 0; \text{ i.e. } \dot{\lambda}_1 = 0 \tag{4.13}$$

$$\frac{\partial F}{\partial x_2} - \frac{d}{dt}\left(\frac{\partial F}{\partial \dot{x}_2}\right) = 0; \text{ i.e. } \lambda_1(t) + \dot{\lambda}_2 = 0 \tag{4.14}$$

$$\frac{\partial F}{\partial x_3} - \frac{d}{dt}\left(\frac{\partial F}{\partial \dot{x}_3}\right) = 0 \text{ or } \frac{\partial F}{\partial u} = 0;$$

$$\text{i.e. } u(t) + \lambda_2(t) = 0 \tag{4.15}$$

From (4.13) $\lambda_1 = a$, constant

From (4.14) $\lambda_2 = -at + b$, b constant

From (4.15) $u = at - b$

To determine a and b return to the equation $\dot{x}_2 = u$ and hence $x_2 = \frac{1}{2}at^2 - bt + c$, c constant. Since $\dot{x}_1 = x_2$, $x_1 = \frac{1}{6}at^3 - \frac{1}{2}bt^2 + ct + d$, d constant.

Furthermore, $x_2 = 1$ at $t = 0$ and $x_1 = \theta_0$ at $t = 0$; hence $c = 1$, $d = \theta_0$. Also $x_2 = 0$ at $t = 2$ therefore $2a - 2b + 1 = 0$ and $x_1 = \theta_2$ at $t = 2$ so that $\frac{4}{3}a - 2b + 2 + \theta_0 = \theta_2$. It follows that $a = \frac{3}{2}(\theta_0 - \theta_2 + 1)$, $b = \frac{3}{2}(\theta_0 - \theta_2) + 2$.

It is left to you to write down the formulae for x_1, x_2, u and sketch the graphs of these functions against time.

Bang-Bang control (Amplitude control)

Consider the problem of moving a vehicle on a horizontal road from rest to rest a distance a by accelerating or decelerating only. We shall assume that the governing equation of motion is

$$\ddot{x} = u(t)$$

where $u(t)$, the control, is related to vehicular thrust. The aim is to achieve the movement in minimum time. Further suppose that $|u| \leqslant u_{max}$, which is specified. Then we *could* augment the function F by the term $\left(\frac{u}{u_{max}}\right)^{2M}$ where M is a large positive integer. If $|u| \leqslant u_{max}$ this term is very small, whereas if $|u| > u_{max}$ the term is very large. By using this penalty function the constraint is effectively included. An alternative approach is developed.

Consider $h(x) = \frac{1}{2}\dot{x}^2$ then $h' = \frac{dh}{dx} = \dot{x}\,\frac{d\dot{x}}{dx} = u$ with $h(0) = 0 = h(a)$; the task is to minimise

$$I = \int_0^a \left(\frac{1}{2h}\right)^{1/2} dx = \int_0^{t_f} dt$$

For simplicity assume $|u| \leqslant 1$. This can be expressed by the equation $\alpha^2 = (u+1)(1-u)$ where α is a real scalar. Now consider

$$J = \int_0^a \left(\left[\frac{1}{2h}\right]^{1/2} + \lambda_1\left[\frac{\partial h}{\partial x} - u\right] + \lambda_2\left[\alpha^2 - (u+1)(1-u)\right]\right) dx$$

Here h is the state variable, u and α are control variables and x is the independent variable; λ_1 and λ_2 are now functions of x. The Euler-Lagrange equations are

$$\frac{\partial F}{\partial h} - \frac{d}{dx}\left(\frac{\partial F}{\partial h'}\right) = 0; \quad -\left[\frac{1}{2h}\right]^{3/2} - \frac{d\lambda_1}{dx} = 0 \tag{4.16}$$

$$\frac{\partial F}{\partial u} - \frac{d}{dx}\left(\frac{\partial F}{\partial u'}\right) = 0; \quad -\lambda_1 + \lambda_2(2u) = 0 \tag{4.17}$$

$$\frac{\partial F}{\partial \alpha} - \frac{d}{dx}\left(\frac{\partial F}{\partial \alpha'}\right) = 0; \quad 2\lambda_2\alpha = 0 \tag{4.18}$$

From (4.18) we could take $\lambda_2 = 0$; the implications of this are that $\lambda_1 = 0$ and hence $\frac{1}{2h} = 0$ which is impossible.

Therefore $\alpha = 0$ and so $(u+1)(1-u) = 0$ which means that u takes only its extreme values ± 1. (Bang-bang control.) In practical terms this implies maximum acceleration followed by maximum deceleration (we are entitled to assume one instantaneous switch-over). Now

$$\ddot{x} = u = \begin{cases} 1 & \text{for } 0 \leqslant t \leqslant t_1 \\ -1 & \text{for } t_1 < t \leqslant t_f \end{cases}$$

Then

$$\dot{x} = \begin{cases} t & \text{for } 0 \leqslant t \leqslant t_1 \\ -(t - t_f) & \text{for } t_1 < t \leqslant t_f \end{cases}$$

since $\dot{x} = 0$ at $t = 0$ and at $t = t_f$. Also

$$x = \begin{cases} \tfrac{1}{2}t^2 & \text{for } 0 \leqslant t \leqslant t_1 \\ -\tfrac{1}{2}(t - t_f)^2 + a & \text{for } t_1 < t \leqslant t_f \end{cases}$$

since $x = 0$ at $t = 0$ and $x = a$ at $t = t_f$.

We require x and $\dot{x}$ to be continuous at $t = t_1$, therefore $t_1 = -(t_1 - t_f)$ and $-\tfrac{1}{2}(t_1 - t_f)^2 + a = \tfrac{1}{2}t_1^2$. Then $-\tfrac{1}{2}t_1^2 + a = \tfrac{1}{2}t_1^2$ so that the switch-over time is $t_1 = a^{1/2}$ and, as a bonus, $t_f = 2a^{1/2}$.

The control is graphed with x and $\dot{x}$ in Figure 4.3.

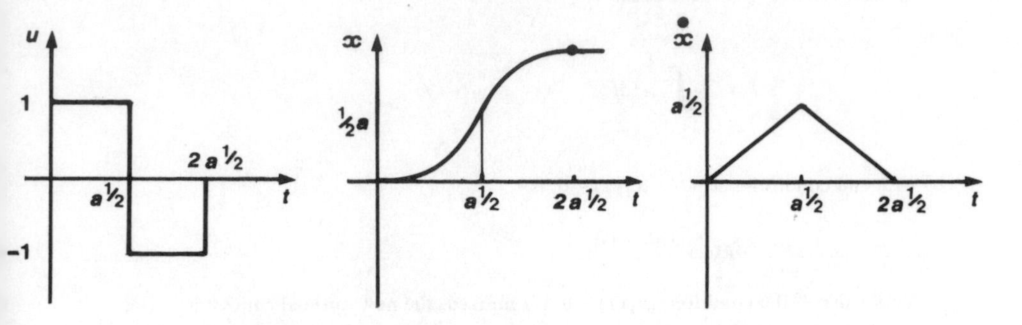

Figure 4.3

Problems

1. Find the shortest path connecting the points $(0, 0)$ and (a, b).

2. A curve $y = y(x)$ is rotated about the x-axis. Consider the curved surface generated by that part of the curve between $x = 0$ and $x = a$. Show that the curve which has minimum surface area is a catenary. Interpret this physically.

3. The ends of a given straight line are to be joined by a plane curve of a specified length so as to enclose the maximum possible area with the straight line. Show that the curve is an arc of a circle.

   (Hint: Assume that the line has end points $(0, 0)$ and $(a, 0)$)

4. Evaluate $\displaystyle\int_0^1 \left\{ y^2 + \left[\frac{dy}{dx}\right]^2 \right\} dx$ along the paths

   (i) $y = x$
   (ii) $y = x^2$

5. A system is governed by the equations $\dot{x}_1 = -x_1 + u$, $x_2 = x_1$. The control variable u is to be chosen to minimise

$$I = \int_0^\infty (x_2^2 + \frac{16}{3}u^2)\mathrm{d}t$$

The boundary conditions are $x_1(0) = \alpha$, $x_2(0) = \beta$; $x_1(\infty) = 0$, $x_2(\infty) = 0$. Show that the optimal control is given by

$$u_{\text{opt}} = -0.37x_1(t) - 0.43x_2(t)$$

6. A system is governed by $\dot{x}_1 = x_2$, $\dot{x}_2 = u$ where $x_1(0) = 1$, $x_2(0) = 1$. Show that the optimal control which minimises

$$I = \int_0^1 u^2\,\mathrm{d}t$$

with end conditions $x_1(1) = x_2(1) = 0$ is

$$u_{\text{opt}} = 18t - 10$$

Verify that if the conditon $x_2(1) = 0$ is removed, the new optimal control is

$$u_{\text{opt}} = 6t - 6$$

Compare the two minimised values of I and comment.

4.3 FURTHER FEATURES OF CALCULUS OF VARIATIONS

This section starts with an examination of the case where not both end points are fixed.

Transversality conditions

Consider the general problem of minimising

$$I = \int_{t_0}^{t_f} F(x, \dot{x}, t)\,\mathrm{d}t$$

where t_0 and $x(t_0)$ are fixed, but t_f and $x(t_f)$ are allowed to vary; See Figure 4.4.

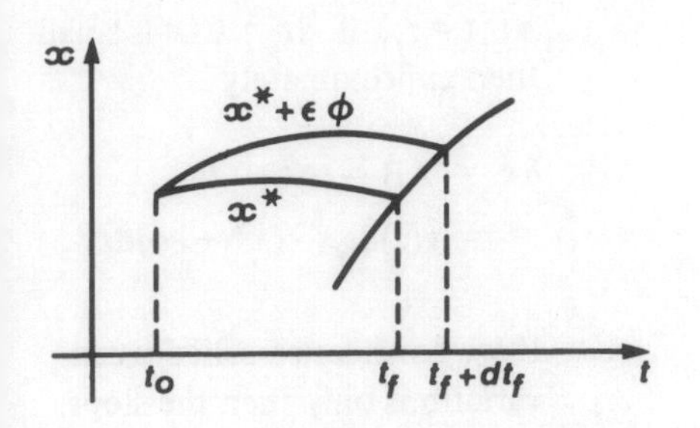

Figure 4.4

Consider

$$J(\epsilon) = \int_{t_0}^{t_f+dt_f} F(x^* + \epsilon\phi, \dot{x}^* + \epsilon\dot{\phi}, t)dt$$

Proceeding as in the previous section the first variation in $J(\epsilon)$ now becomes

$$\int_{t_0}^{t_f} \left[\phi \frac{\partial F}{\partial x} + \dot{\phi} \frac{\partial F}{\partial \dot{x}}\right] dt + F(x, \dot{x}, t_f)\, dt_f$$

By the argument of that section, a necessary condition for the trajectory x^* to be optimal is that

$$\int_{t_0}^{t_f} \phi \left\{\frac{\partial F}{\partial x} - \frac{d}{dt}\left[\frac{\partial F}{\partial \dot{x}}\right]\right\} dt + \left[\frac{\partial F}{\partial \dot{x}}\phi\right]_{t_0}^{t_f} + F(x, \dot{x}, t_f)dt_f = 0$$

We therefore recover the Euler-Lagrange equations (4.8) and the **transversality condition**

$$\left[F\, dt + \frac{\partial F}{\partial \dot{x}}\phi\right]_{t=t_f} = 0^\dagger$$

or

$$\left[F\, dt + \frac{\partial F}{\partial \dot{x}} dx\right]_{t=t_f} = 0 \qquad (4.19)$$

If t_f is specified then we have the condition that

$$\left[\frac{\partial F}{\partial \dot{x}} dx\right]_{t=t_f} = 0 \qquad (4.20)$$

We can satisfy this equation by specifying that

$$\frac{\partial F}{\partial \dot{x}} = 0 \quad \text{at} \quad t = t_f \qquad (4.21)$$

Suppose the end point $x(t_f)$ is constrained to lie along a curve $x = g(t)$; see Figure 4.5.

† Note that $\phi = 0$ at $t = t_0$ since $x(t_0)$ is specified.

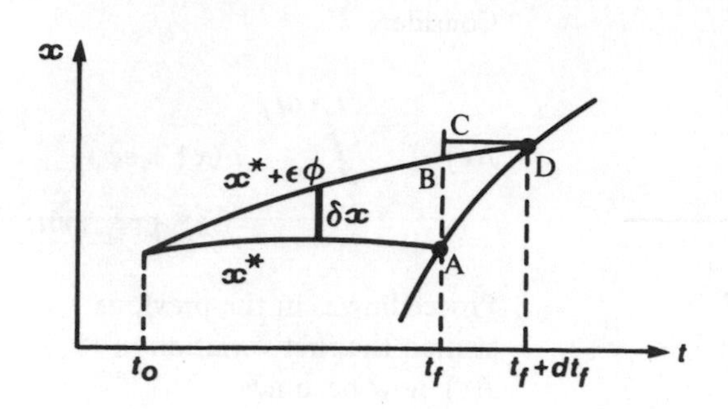

Figure 4.5

At $t = t_f$, if $dt_f = CD$ is small then approximately

$$\delta x = AB = AC - BC$$
$$= \dot{g}(t)dt_f - (\dot{x}^* + \epsilon\dot{\phi})dt_f$$

If we consider so-called **weak variations** only then the slope variation is negligible , i.e. $\dot{\phi} = 0$ so that

$$\delta x = [\dot{g}(t) - \dot{x}(t)]\,dt_f$$

Substituting into (4.19) we obtain

$$F\,dt_f + \frac{\partial F}{\partial \dot{x}}[\dot{g}(t)dt_f - \dot{x}(t)dt_f] = 0$$

or

$$F + [\dot{g}(t) - \dot{x}(t)]\frac{\partial F}{\partial \dot{x}} = 0 \quad \text{at} \quad t = t_f \tag{4.22}$$

This is the **transversality condition.** It expresses the condition which is obeyed when the end state $x(t_f)$ is not specified.

Example

Minimise $I = \int_0^{t_f} (1 + \dot{x}^2)^{1/2}\,dt$ where $x(0) = 1$ and $x(t_f)$ lies on the curve $x = 2 - t$. Here $F = (1 + \dot{x}^2)^{1/2}$ and the Euler-Lagrange equation is

$$0 = \frac{d}{dt}\left[\frac{\dot{x}}{(1 + \dot{x}^2)^{1/2}}\right]$$

i.e. $\dot{x} = c(1 + \dot{x}^2)^{1/2}$ where c is constant or $\dot{x} = c^2/(1 - c^2) = a$, where a is constant.

Then $x(t) = at + b$, where b is constant. Now $x(0) = 1$. $\therefore$ $x = at + 1$. Also $g(t) = 2 - t$. $\therefore$ $\dot{g}(t) = -1$. Hence condition (4.22) is

$$(1 + \dot{x}^2)^{1/2} + (-1 - \dot{x})\frac{\dot{x}}{(1 + \dot{x}^2)^{1/2}} = 0 \text{ at } t = t_f$$

This reduces to $\dot{x} = 1$ at $t = t_f$.

But $\dot{x} = a$ and therefore $a = 1$ to give the optimal trajectory $x = t + 1$. This trajectory intersects $g(t)$ at $t = t_f$ whence

$$t_f + 1 = 2 - t_f \quad \text{or} \quad t_f = \tfrac{1}{2}$$

Note that at t_f the slope of the optimal trajectory is perpendicular to that of the given curve, hence the term '*transversality*'.

Mayer Formulation

Now consider the problem of minimising $I = \int_{t_0}^{t_f} (x^2 + \dot{x}^2)\mathrm{d}t$ where the system constraint is $\dot{x} = u$ with end conditions $x(t_0) = x_0$, $x(t_f) = x_1$.

Method 1

$F = x^2 + \dot{x}^2$ and the Euler-Lagrange equation is $2x = \frac{\mathrm{d}}{\mathrm{d}t}(2\dot{x})$ i.e. $\ddot{x} = x$ for which the general solution is $x = A\mathrm{e}^t + B\mathrm{e}^{-t}$ and the values of A and B can be found via the boundary conditions. [If $x(t_f)$ were free but t_f were specified the condition to be used would be (4.21) viz. $\partial F/\partial\dot{x} = 0$ at $t = t_f$ so that $\dot{x}(t_f) = 0$. (Note that the slope of x is perpendicular to the curve $t = t_f$)]. We find u from the constraint $u = \dot{x}$.

Method 2

$$I = \int_{t_0}^{t_f} (x^2 + u^2)\,\mathrm{d}t \tag{4.23}$$

Now $\dot{x} = u$ so we form the augmented functional $J = \int_{t_0}^{t_f} [x^2 + u^2 + \lambda(\dot{x} - u)]\,\mathrm{d}t$.

The Euler-Lagrange equations are

$$2x = \dot{\lambda}, \qquad 2u - \lambda = 0, \qquad \dot{x} - u = 0$$

Again we find $x = A\mathrm{e}^t + B\mathrm{e}^{-t}$. The transversality condition would in general be

$$\left[F - \dot{x}\frac{\partial F}{\partial\dot{x}} - \dot{u}\frac{\partial F}{\partial\dot{u}} - \dot{\lambda}\frac{\partial F}{\partial\dot{\lambda}}\right]\mathrm{d}t + \frac{\partial F}{\partial\dot{x}}\,\mathrm{d}x + \frac{\partial F}{\partial\dot{u}}\,\mathrm{d}u + \frac{\partial F}{\partial\dot{\lambda}}\,\mathrm{d}\lambda = 0 \text{ at } t = t_f$$

However, F does not contain $\dot{u}$ or $\dot{\lambda}$ so that the condition becomes

$$\left(F - \dot{x}\,\frac{\partial F}{\partial \dot{x}}\right) \mathrm{d}t + \frac{\partial F}{\partial \dot{x}}\,\mathrm{d}x = 0 \tag{4.24}$$

Method 3

To formulate the example as a Mayer problem we use the state variable notation $x_1 = x$, $u = \dot{x}_1$. The problem is to drive the system from the state $x_1(t_0)$ to the state $x_1(t_f)$, so as to minimise

$$I = \int_{t_0}^{t_f} (x_1^2 + u^2)\,\mathrm{d}t \tag{4.25}$$

We introduce a new state variable $x_0 = \int_{t_0}^{t_f} (x_1^2 + u^2)\,\mathrm{d}t$ so that $\dot{x}_0 = x_1^2 + u^2$ with $x_0(t_0) = 0$.

Now we consider $\Lambda = \lambda_0(t)(\dot{x}_0 - x_1^2 - u^2) + \lambda_1(t)(\dot{x}_1 - u)$ where λ_0 and λ_1 are auxiliary variables.

The Euler-Lagrange equations for the five variables are

$$\frac{\partial \Lambda}{\partial x_0} - \frac{\mathrm{d}}{\mathrm{d}t}\left(\frac{\partial \Lambda}{\partial \dot{x}_0}\right) = 0; \quad \dot{\lambda}_0 = 0 \tag{4.26}$$

$$\frac{\partial \Lambda}{\partial x_1} - \frac{\mathrm{d}}{\mathrm{d}t}\left(\frac{\partial \Lambda}{\partial \dot{x}_1}\right) = 0; \quad -2\lambda_0 x_1 - \dot{\lambda}_1 = 0 \tag{4.27}$$

$$\frac{\partial \Lambda}{\partial u} - \frac{\mathrm{d}}{\mathrm{d}t}\left(\frac{\partial \Lambda}{\partial \dot{u}}\right) = 0; \quad -2\lambda_0 u - \lambda_1 = 0 \tag{4.28}$$

$$\frac{\partial \Lambda}{\partial \lambda_0} - \frac{\mathrm{d}}{\mathrm{d}t}\left(\frac{\partial \Lambda}{\partial \dot{\lambda}_0}\right) = 0; \quad \dot{x}_0 - x_1^2 - u^2 = 0 \tag{4.29}$$

$$\frac{\partial \Lambda}{\partial \lambda_1} - \frac{\mathrm{d}}{\mathrm{d}t}\left(\frac{\partial \Lambda}{\partial \dot{\lambda}_1}\right) = 0; \quad \dot{x}_1 - u = 0 \tag{4.30}$$

From (4.26) we see that λ_0 is constant.

The transversality condition *can be shown to be*

$$dx_0 + \left(\Lambda - \dot{x}_0 \frac{\partial \Lambda}{\partial \dot{x}_0} - \dot{x}_1 \frac{\partial \Lambda}{\partial \dot{x}_1} \right) dt + \frac{\partial \Lambda}{\partial \dot{x}_0} dx_0 + \frac{\partial \Lambda}{\partial \dot{x}_1} dx_1 = 0 \text{ at } t = t_f$$

Now $dt_f = 0$ and $dx_1 = 0$ at $t = t_f$ by the specification of the problem. Hence

$$dx_0 + \frac{\partial \Lambda}{\partial \dot{x}_0} dx_0 = 0 \quad \text{at} \quad t = t_f$$

i.e.

$$dx_0 + \lambda_0 dx_0 = 0 \quad \text{at} \quad t = t_f$$

or

$$\lambda_0(t) = -1 \quad \text{at} \quad t = t_f$$

Since λ_0 is constant, we conclude that $\lambda_0 = -1$ for all t. Then, from (4.27), (4.28) and (4.30) it follows that $2x_1 = \dot{\lambda}_1 = 2\dot{u} = 2\ddot{x}_1$ and we proceed as before.

The advantage of the Mayer formulation is that it leads via the Hamiltonian formulation of the next section into Pontryagin's principle.

Disadvantages of the approach

When the state equations are linear and the performance index involves quadratic functions then the Euler-Lagrange equations are linear and time-invariant and are easy to solve. However, a performance index of the form $\int_{t_0}^{t_f} (|x| + |u|)\, dt$ leads to a dead end via the Euler-Lagrange approach (Why?) and other methods need to be found. In addition the Euler-Lagrange equations generally give rise to open-loop control strategy and only under certain conditions can the strategy be framed as a closed-loop one. Also, for purposes of analytical solution it is desirable to have few variables. Note that the Euler-Lagrange equations can be solved by a numerical method (e.g. Runge-Kutta).

In general the solution of the problem via calculus of variations becomes that of a two-point boundary-value problem which is not easy to solve, even numerically.

If, as is often the case, the controls are piecewise continuous† only, a new approach is needed. This is provided in the next section;

† A function is piecewise continuous over an interval I if there are at most a finite number (includes the case zero: a continuous function) of points in the interval at which the function is discontinuous and at these is only allowed to have finite jumps.

Problems

1. A system is governed by $\dot{x}_1 = x_3 - x_1$, $\dot{x}_2 = x_1$, $\dot{x}_3 = u$. It is required to minimise

$$I = \int_0^T (x_2^2 + ax_3^2)\,dt,$$

$a > 0$. Consider the special case $T \to \infty$ and $a = 0.1$ and show that

$$u(0) \simeq -1.7x_1(0) - 3.2x_2(0)$$

2. Show that the control which drives the system $\dot{x} = -2x_1 + u$ from $(1, 1)$ so as to minimise $I = \int_0^1 u^2\,dt$ is given by $u_{opt} = \dfrac{-4e^{2t}}{e^4 - 1}$

3. Minimise $I = \int_0^1 (\dot{x}^2 - \beta^2 x^2)\,dt$ with $x(0) = 0$, $x(1) = 0$ and show that either $x = 0$ or $x = \alpha \sin n\pi t$ for $\beta = n\pi$ gives $I = 0$. Show that $x = t(1-t)$ gives $I = \dfrac{1}{3} - \dfrac{1}{30}a^2$ and noting that $a^2 > 10 \Rightarrow I < 0$, explain the paradox.

4.4 PONTRYAGIN'S MAXIMUM PRINCIPLE

In this section we first reformulate the Mayer problem of the last section via a Hamiltonian function, then we proceed to the statement of the Maximum Principle and finally we give some examples of its application.

Hamiltonian formulation

In the Mayer problem we define the *Hamiltonian*

$$H = \sum_{i=1}^{n} \lambda_i f_i \tag{4.31}$$

Note that

$$H = \Lambda + \sum_{i=0}^{n} \lambda_i \dot{x}_i \tag{4.32}$$

where

$$\Lambda = \sum_{i=0}^{n} \lambda_i (f_i - \dot{x}_i)$$

Using the general multi-dimensional case of the Mayer problem, it can be shown that in terms of the function H the equations become (on the assumption that H does not contain terms in $\dot{u}_i$)

$$\frac{d\lambda_i}{dt} = -\frac{\partial H}{\partial x_i} \; ; \qquad i = 0, \; ....., \; n \tag{4.33}$$

$$\frac{dx_i}{dt} = \frac{\partial H}{\partial \lambda_i} \; ; \qquad i = 0, \; ....., \; n \tag{4.34}$$

$$\frac{\partial H}{\partial u_j} = 0 \quad ; \qquad j = 0, \; ....., \; r \tag{4.35}$$

The transversality condition becomes

$$dx_0 - H dt + \sum_{i=0}^{n} \lambda_i \, dx_i = 0 \quad \text{at} \quad t = t_f \tag{4.36}$$

Note that if the final time, t_f, is not specified it is necessary that $H = 0$ at $t = t_f$.

It can be shown that the Hamiltonian is constant along the optimal trajectory. This follows since

$$\frac{dH}{dt} = \frac{\partial H}{\partial x_1}\dot{x}_1 + ... + \frac{\partial H}{\partial x_n}\dot{x}_n + \frac{\partial H}{\partial \lambda_1}\dot{\lambda}_1 + ... + \frac{\partial H}{\partial \lambda_n}\dot{\lambda}_n + \frac{\partial H}{\partial u_1}\dot{u}_1 + ... + \frac{\partial H}{\partial u_r}\dot{u}_r$$

Along the optimal trajectory we may substitute (4.33), (4.34) and (4.35). The terms of $\frac{dH}{dt}$ containing $\dot{u}_j$ disappear and the remainder cancel in pairs, e.g.

$$\frac{\partial H}{\partial x_1}\dot{x}_1 + \frac{\partial H}{\partial \lambda_1}\dot{\lambda}_1 = \frac{\partial H}{\partial x_1}\frac{\partial H}{\partial \lambda_1} + \frac{\partial H}{\partial \lambda_1}\left(-\frac{\partial H}{\partial x_1}\right) = 0$$

Hence $\frac{dH}{dt} = 0$ and H is constant along the optimal trajectory. It has been assumed that H does not contain t explicitly. Note that if t_f is free then $H = 0$ for $t_0 \leqslant t \leqslant t_f$.

Example

Given the system $\dot{x}_1 = u = f_1$, the aim is to minimise $\int_{t_0}^{t_f} (x_1^2 + u^2)\, dt$, given $x_1(t_0)$ and $x_1(t_f)$.

As before we introduce $x_0 = \int_{t_0}^{t_f} (x_1^2 + u^2)\, dt$ so that $\dot{x}_0 = x_1^2 + u^2 = f_0$, $x_0(t_0) = 0$.

The Hamiltonian becomes

$$H = \lambda_0 (x_1^2 + u^2) + \lambda_1 u$$

Then

$$\text{(4.33)} \quad \text{gives} \quad \dot{\lambda}_0 = -\frac{\partial H}{\partial x_0} = 0; \; \dot{\lambda}_1 = -2x_1 \lambda_0$$

$$\text{(4.35)} \quad \text{gives} \quad 2\lambda_0 u + \lambda_1 = 0 \text{ for optimal control}$$

Equations (4.34) merely state the formulae for $\dot{x}_0$ and $\dot{x}_1$. Therefore, these are the same equations as Method 3 of the last section and hence the solution procedure is the same.

The Maximum Principle

In the Hamiltonian we replace λ_i by p_i† and restate the problem as that of minimising $I = \int_{t_0}^{t_f} f_0(\mathbf{x}, \mathbf{u})\, dt$ where $\dot{x}_0 = f_0(\mathbf{x}, \mathbf{u})$ and $x_0(t_0) = 0$.

The state variables $x_0, x_1, \ldots, x_n$ are augmented by the adjoint or **co-state variables**†† $p_0, p_1, \ldots, p_n$ and a Hamiltonian is defined by

$$H = \sum_{i=0}^{n} p_i f_i(\mathbf{x}, \mathbf{u})$$

† In Hamiltonian dynamics, the Newtonian coordinates of position and velocity are replaced by position and momentum.

†† For a linear system $\dot{\mathbf{x}} = \mathbf{Ax} + \mathbf{Bu}$ the co-state system is $\dot{\mathbf{p}} = -\mathbf{A}^{\mathrm{T}}\mathbf{p}$.

where the system equations are

$$\dot{x}_i = f_i(\mathbf{x}, \mathbf{u}) \quad i = 1, \ldots, n$$

The equations to be solved are

$$\frac{\partial H}{\partial \mathbf{u}} = 0; \qquad \frac{\partial H}{\partial x_i} = -\frac{dp_i}{dt}; \quad \frac{\partial H}{\partial p_i} = \frac{dx_i}{dt}$$

Pontryagin's principle, which we state without proof, is that for the above system *a necessary and sufficient condition for minimising I is that at all times H should be maximised with respect to the control vector* $\mathbf{u}$.

At any instant of time H must achieve the maximum value possible that can be attained by any permitted control subject to the state conditions existing at that instant.

The following first reactions are worthy of note.

1. Instead of minimising the *functional* I we are now maximising a *function* H. Furthermore, the function depends on the instantaneous value of $\mathbf{u}$.

2. The nature of the optimum control strategy can be deduced by inspection of the Hamiltonian. Obviously we first ignore those terms which do not involve $\mathbf{u}$. Suppose this has been done and also that there is a scalar control u.

   (a) **Continuous control** (with or without saturation). H involves u^2. If $\frac{\partial H}{\partial u} = 0$ gives a result within $[u_{min}, u_{max}]$ then this is the optimal control, otherwise u takes its minimum and/or maximum values (saturation).

   (b) **Bang-Bang control.** $H = g(t) \cdot u$. The strategy is $u = u_{max}$ where $g(t) > 0$ and $u = u_{min}$ where $g(t) < 0$.

   (c) **Bang-Coast-Bang control.** $H = \alpha^2 |u| + g(t)u$. The strategy is $u = u_{max}$ where $\alpha^2 < g(t)$, $u = 0$ where $-\alpha^2 < g(t) < \alpha^2$, $u = u_{min}$ where $g(t) < -\alpha^2$

3. The exact strategy, for example when to switch, must be determined by further analysis.

4. We still have a two-point boundary value problem.

5. The information obtained for the control is of the open-loop variety.

Example 1

As a first example we return to the problem on page 129. We see that the optimal control strategy for $0 \leqslant t \leqslant t_1$ is $u = -1$ or $u = +1$ where t_1 is the as yet unspecified time at which saturation ends. After this time we may use the criterion $\frac{\partial H}{\partial u} = 0$.

Suppose the initial control is $u = -1$† then it can easily be found that for $0 \leqslant t \leqslant t_1$, $x_1(t) = -t + A$, where A is constant, $\dot{p}_1 = +2x_1$ and hence $p_1(t) = -t^2 + 2At + B$, B constant.

At $t = t_0 = 0$, $x_1 = 4$ so that $A = 4$. Also,

$$H \equiv 0 \Rightarrow -[x_1(0)]^2 - [u(0)]^2 + p_1(0)\,u(0) = 0$$

i.e. $-16 - 1 - p_1(0) = 0$ hence $p_1(0) = -17 = B$

Therefore, for $0 \leqslant t \leqslant t_1$, $x_1(t) = -t + 4$, $p_1(t) = -t^2 + 8t - 17$. Immediately after t_1, $\frac{\partial H}{\partial u} = 0$ so that $p_1(t_1) = 2u(t_1) = -2$. Equating the two expressions for $p(t)$ at $t = t_1$ we have $-t_1^2 + 8t_1 - 17 = -2$ so that $t_1^2 - 8t_1 + 15 = 0$ giving $t_1 = 3$ or 5.

Now if $t_1 = 5$, $x_1(t_1) = -1$ whereas if $t_1 = 3$, $x_1(t_1) = 1$. The less drastic chang from $x_1(0) = 4$ is accomplished by choosing $t_1 = 3$, and this we do.

For the period $t_1 < t$ we have $\frac{dx_1}{dt} = u = \frac{1}{2} p_1$ and $\frac{dp_1}{dt} = 2x_1$. Therefore $\ddot{x}_1 = x_1$ so that $x_1(t) = Ae^t + Be^{-t}$ and $u(t) = \dot{x} = Ae^t - Be^{-t}$.

Fit the boundary conditions on x, and u at $t = t_1$ and complete the solution.

Example 2

Suppose we wish to drive the system $\dot{x}_1 = -x_1 + u$ from the initial state $x_1(0) = 4$ to $x_1(t_f) = x_1(4) = 0$. The control force is constrained by $|u| \leqslant 2$ and we seek to minimise the effort $I = \int_0^4 |u|\,dt$.

Defining $\dot{x}_0 = |u|$, $x_0(0) = 0$ we find the Hamiltonian to be $H = p_0|u| - p_1 x_1 + p_1 u$. Ignoring the term $-p_1 x_1$, the aim is to maximise $H = H'$

† You can show that a choice of $u = 1$ here leads to negative t_1.

ith respect to u. Now, as usual, $p_0 = -1$ so that $H' = -|u| + p_1 u$. From the eneral case quoted earlier we expect to employ bang-coast-bang strategy. From the ature of the problem we reject the control $u = 2$ and employ $u = 0$ followed by $= -2$.

When $u = 0$ the system equation is $\dot{x}_1 = -x_1$ so that $x_1 = Ae^{-t}$; since $x_1(0) = 4$, then $x_1 = 4e^{-t}$. Let t_1 be the switching time. Since $u = -2$ the system quation is $\dot{x}_1 = -x_1 - 2$ so that $x_1 = Be^{-t} - 2$. But x_1 is continuous at $t = t_1$ o that $4e^{-t_1} = Be^{-t_1} - 2$. Substituting for B we find that for $t \geqslant t_1$, $x_1 = (4 + 2e^{t_1})e^{-t} - 2$. Now $x_1(4) = 0$ so that $(4 + 2e^{t_1}) \cdot e^{-4} - 2 = 0$. Hence we ind $t_1 = \log_e(e^4 - 2) = 3.96$ (3 s.f.).

Show that the value of H is constant but non-zero over the time interval $[0, 4]$.

Minimum-time problem

Consider the problem of driving an inertia wheel from the initial state $x_1(0) = x_0$, $x_2(0) = \dot{x}_1(0) = v_0$ to a final state $x_1(t_f) = 0$, $x_2(t_f) = 0$ in minimum time. We uppose that the dynamic equation is $\ddot{x}_1 = u$ i.e. $\dot{x}_2 = u$ and that $|u| \leqslant 1$. The

ndex of performance $I = \int_0^{t_f} 1\, dt$ so that the Hamiltonian is $H = p_0 + p_1 x_2 + p_2 u$

where $\dot{x}_0 = 1$, $x_0(0) = 0$. Now $p_0 = -1$ as usual so that $H = -1 + p_1 x_2 + p_2 u$ and herefore the relevant part of the Hamiltonian is $H' = p_2 u$. This is maximised when $u = -1$ for $p_2 < 0$ and $u = +1$ for $p_2 > 0$†. This, of course, is bang-bang control. Returning to the full Hamiltonian we have the equations

$$\dot{p}_1 = -\frac{\partial H}{\partial x_1} = 0 \qquad \dot{p}_2 = -\frac{\partial H}{\partial x_2} = -p_1$$

$$\dot{x}_1 = x_2 \qquad \dot{x}_2 = u = \text{sign}(p_2)$$

Hence $p_1 = \alpha$, a constant and $p_2 = -\alpha t + \beta$ where β is a constant. Hence the sign of p_2 changes at most once in $[0, t_f]$.

We expect to use the control $u = -1$ for $0 \leqslant t < t_1$ and then follow with the control $u = 1$ for $t_1 < t < t_f$. Then for $0 \leqslant t \leqslant t_1$, $u = -1$, $\dot{x}_2 = -1$ so that $x_2 = -t + A$, for A constant and $x_1 = -\dfrac{t^2}{2} + At + B$, B constant. Now $x_2(0) = v_0$, $x_1(0) = x_0$ so that $A = v_0$ and $B = x_0$. Hence $x_2 = -t + v_0$, $x_1 = \dfrac{-t^2}{2} + v_0 t + x_0$, for $0 \leqslant t \leqslant t_1$.

† We may say the same thing by writing $u = \text{sign}(p_2)$.

Now suppose that $u = 1$ so that $x_2 = t + C$, $x_1 = \frac{1}{2}t^2 + Ct + D$. We know that $x_2(t_f) = 0$ so that $C = -t_f$ and $x_1(t_f) = 0$ so that $D = -\frac{1}{2}t_f^2 - (-t_f)t_f = \frac{1}{2}t_f^2$. Then $x_2 = t - t_f$, $x_1 = \frac{1}{2}t^2 - t_f t + \frac{1}{2}t_f^2$ for $t_1 < t \leqslant t_f$.

We require x_1 to be continuous at $t = t_1$ so that

$$-\tfrac{1}{2}t_1^2 + v_0 t_1 + x_0 = \tfrac{1}{2}t_1^2 - t_f t_1 + \tfrac{1}{2}t_f^2$$

and we *could* obtain an expression for t_1 in terms of t_f. However, from a design point of view we would prefer to express the control in terms of the state variables x_1 and x_2, in order to derive a feedback control.

Let us first examine the trajectories in state space under the controls $u = -1$ and $u = 1$.

If $u = -1$, $x_2 = -t + A$, $x_1 = -\dfrac{t^2}{2} + At + B = -\dfrac{1}{2}x_2^2 + E$ where E is a constant. The family of trajectories for varying E is shown in Figure 4.6(a). Similarly, if $u = +1$, $x_1 = \frac{1}{2}x_2^2 + F$; the family of trajectories for varying F is shown in Figure 4.6(b). Justify the direction of the arrows.

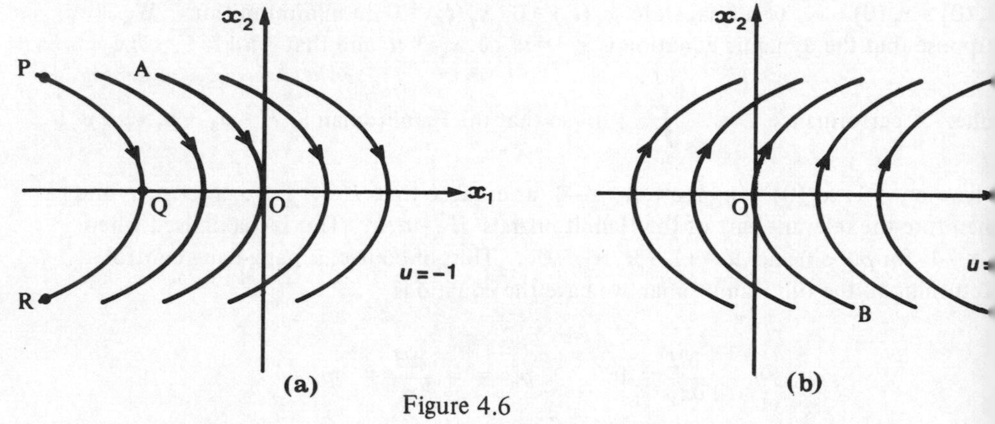

Figure 4.6

Consider Figure 4.6(a). A point in state space corresponds to initial conditions on $x_1 = x$ and $x_2 = \dot{x}$. An initial point P will move in state space along the trajectory PQ and will clearly miss the origin (where $x = \dot{x} = 0$). For initial point P, then, it is useless to use the strategy $u = -1$ for all time. However, the point A will make its way to the origin along the trajectory AO, as will any point on this trajectory. We conclude that to reach the origin, from points not on AO we must at some stage switch to the control $u = +1$.

Similarly, examining Figure 4.6(b) we see that points on the trajectory BO will reach the origin; others will require a switch to the control $u = -1$.

The equation of AO is $x_1 = -\frac{1}{2}x_2^2$ and that of BO is $x_1 = \frac{1}{2}x_2^2$. In Figure 4.7 we superimpose the diagrams of Figures 4.6(a) and (b).

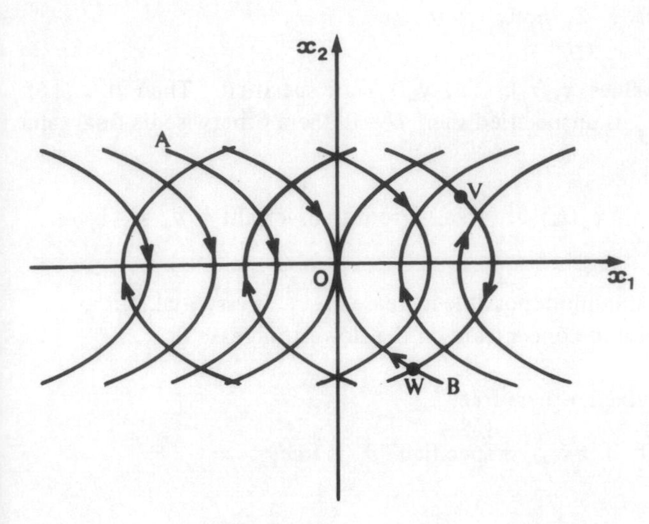

Figure 4.7

Consider, for example, the initial state V with initial control $u = -1$. The optimum policy would be to maintain this control until the curve BO is reached (at W) and then switch the control to $u = +1$ to reach the origin.

In Figure 4.8 we have removed the unnecessary curves.

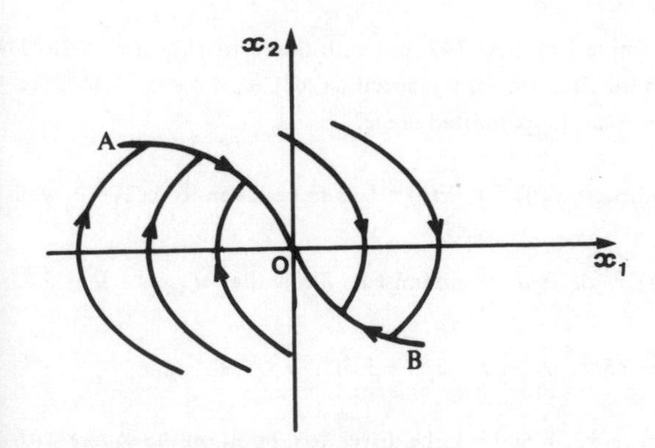

Figure 4.8

The curve AOB is called a **switching curve.** The control policy is $u = -1$ if the state (x_1, x_2) is above AOB or on AO, but $u = +1$ if the state (x_1, x_2) is below AOB or on BO.

'e can specify AOB by the single equation $2x_1 + x_2 |x_2| = 0$
Verify that this is so.)

Then we can frame the control policy as $u = -\text{sign}\,(2x_1 + x_2 |x_2|)$, and the roblem is fully solved.

Sage (1968) points out that some minimum-time problems have a singular olution.

emarks on the transversality condition

The general condition is

$$dx_0 - H\,dt + \sum_{i=0}^{n} p_i\,dx_i = 0 \quad \text{at} \quad t = t_f$$

(a) Assume that the final values $x_1(t_f)$, ..., $x_n(t_f)$ are specified. Then p_1, ..., p_n are free at $t = t_f$. If t_f is unspecified then $H = 0$ there otherwise its final value is unspecified.

(b) t_f fixed but $x_1(t_f)$, ..., $x_n(t_f)$ are free. From transversality, $p_0 = -1$, $p_1 = p_2 = \ldots\ldots = p_n = 0$.

At the last instant, the minimum possible increase in x_0 is essential, but at earlier times it is not wise to concentrate on the slowest increase in x_0.

(c) Some values of $x_i(t_f)$ fixed, others free.

If $x_i(t_f)$ is free, $p_i = 0$; if $x_j(t_f)$ is specified, p_j is free.

Problems

1. Consider the same problem as Example 1 on page 142, but with the restriction $|u| < 1$. The initial condition is $x_1(0) = 4$ and the final time is not specified, but $x_1(t_f) = 0$. Why does $H_{max} = 0$ give a value of $u(0)$ outside the permitted range?

2. The system $\ddot{x} = u$ with initial conditions $x(0) = 1$, $\dot{x}(0) = 1$ is to be taken to $x(2) = 0$ whilst

   the quantity $J = 0.5 \displaystyle\int_0^2 [u(t)]^2\, dt$ is to be minimised. Show that $u_{opt} = 3t - 3.5$.

   Verify that $x = 1 + t - 1.75t^2 + 0.5t^3$, $\dot{x} = 1 - 3.5t + 1.5t^2$.

   The system $\ddot{x} = u$ starts from (x_{10}, x_{20}) and is to be driven to a point on the square with vertices $(\pm 1, \pm 1)$ in the state-space plane in minimum time. Show that the switching curves are $x_1 = 0.5x_2^2 + 0.5$ to the vertex $(1, -1)$ and $x_1 = -0.5x_2^2 - 0.5$ to $(-1, 1)$.

3. A missile moves in a vertical plane under a constant thrust F: it is subject to a constant gravity force. Introduce state variables x_1, x_3 as horizontal components of displacement and velocity with x_2, x_4 their vertical counterparts. Let $u(t)$ be the angle of thrust, measured from the horizontal. Show that the optimum thrust angle $u_{opt}(t)$ is given by $\tan u_{opt} = at + b$ where a and b are constants. (Consider the performance index $J = -x_3(t_f)$.)

4. Consider the system $\dot{x} - x = u$ with $x(0) = 2$. Find the optimal control, subject to $-1 \leqslant u \leqslant 1$ which takes $x(t)$ to zero in the least time.

5. The system $\dot{x}_1 = x_2$, $\dot{x}_2 = 10u$, $\dot{x}_3 = 0.5$ with performance index

$$J = \int_0^\infty [(x_3 - x_1)^2 + k^2u^2]\, dt$$

is claimed to have optimal control $u_{opt} = 5p_2/k^2$; verify this.

. Consider the system $\ddot{x} = u$ with $|u| \leqslant 1$. It is required to reach the final state $x_2 = 0$ in minimum time. Sketch some optimal trajectories.

. (Minimum fuel problem). In the system of **Problem** 6 it is required to minimise

$$I = \int_{t_0}^{t_f} |u|\, dt$$

Sketch some optimal trajectories. Show that of two trajectories, that which uses least fuel takes the longer time.

.5 DYNAMIC PROGRAMMING: THE DISCRETE PROBLEM

The dynamic programming method, which is a numerical multi-stage decision rocess, was developed by Bellman in the nineteen-fifties. In this and the next section e shall look at a few simple examples to show its application.

In practice the formulation of problems for dynamic programming solution is ften very difficult. A selection of problems treated by the approach is given in Stark nd Nicholls (1972).

We shall first solve two typical problems via dynamic programming and then xtract the essential features of the problem to study the underlying basic principle of he approach.

The essence of a multi-stage decision process is that, at discrete times during the ransition in state space, decisions are taken which affect the outcome of the process. ince the process (and any control) varies with time, we may use the term *dynamic*. The *rogram* is the selection of the sequence of decisions. The set of rules which governs he decision-making is called the **policy**.

A large class of problems are those where a total cost or total time has to be inimised. Bellman and Dreyfus (1962) cite the example of the minimum-time-to-climb f an aircraft. Instead of a continuous variation of flight path we are in effect recognising hat decisions can in practice only be made at discrete intervals and we are discretizing he problem so that the theoretical continuous curve for the flight path is approximated y straight line segments which join mesh points. We now consider simpler examples hich are in their nature discrete.

Example 1

In Figure 4.9 we depict a network of major trunk routes from town A to town B via suitable intermediate towns. On each route is marked the distance in tens of miles. Let us assume (somewhat rashly) that the routes carry sufficiently similar traffic loads that time of travel is approximately proportional to distance and also that it is required to get from A to B in minimum time. Therefore, the distance travelled is to be minimised. For simplicity assume that at each stage the general direction of travel is from left to right.

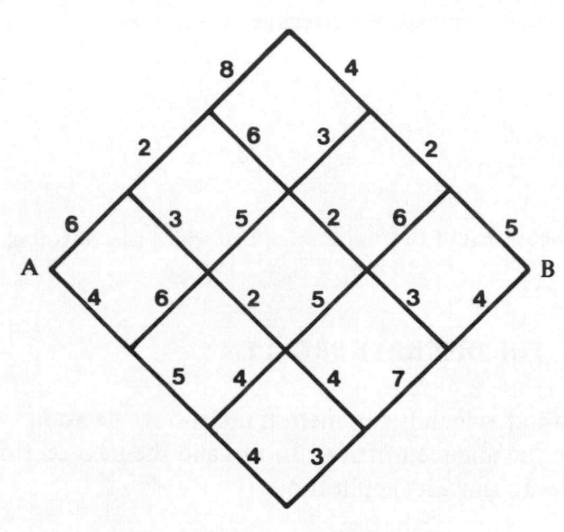

Figure 4.9

The idealised networ shown, though suitable for a computer, may be objected to for its symmetry. If certain routes did no exist in practice we could associate a cos of 10^6 with them sc that any automatic computer routine could exclude them. The procedure is as follows. At each node of the mesh mark in a circle the shortest distance of that node from B. The two points to the immediate south-west and north-west of B are easily seen to be at distances 4 and 5 from B respectively. The nearest point to the west of B has possible distances (6 + 5) or (3 + 4); clearly its minimum distance is 7.

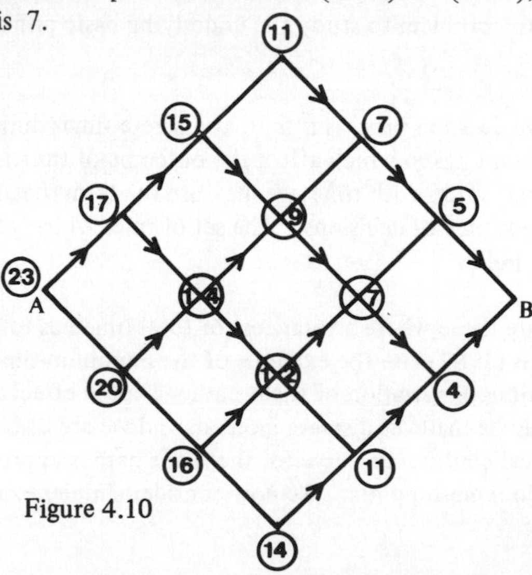

Figure 4.10

Other distances are calculated similarly. The results are show in Figure 4.10. Verify them for yourselves. The routes used in calculating these minimum distances are indicated by arrows. Now proceed from A to B following any route which is mar

y arrows. Any one of them will yield a minimum distance of 23 units (i.e. 230 miles).
ı this problem there was a fixed end-point. The second example
as a freedom of choice of end-point.

xample 2

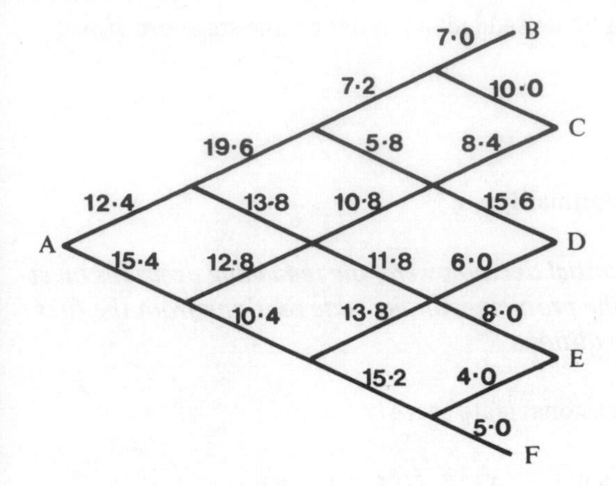

Figure 4.11

The network shown in Figure 4.11 is on the same principle as that of Figure 4.9 . On this occasion we want to choose the nearest of five potential destinations, B, C, D, E or F. Again the symmetry can be provided in practical examples by adding non-existent routes with costs of 10^6 .

Figure 4.12 shows the results of a similar computation.

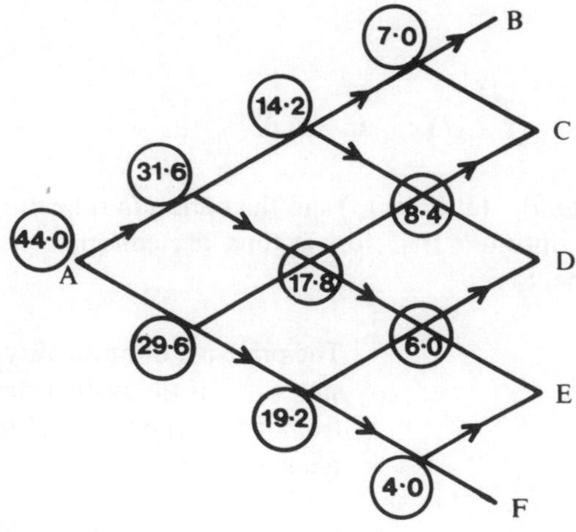

Figure 4.12

This example corresponds to a free-end point problem. Note that on this occasion there is only one optimum route. (Is this a general result for free-end problems?)

One point that does arise is that we have worked backwards in time to select those utes which are optimal from each node (or, equivalently, to reject some unsuitable utes). We then work forwards to pick up the overall optimal route (or routes). owever, the last example shows that we do not always choose the shortest route *at each ıge* from one node to the next.

What is clear is that at each stage of the optimal route, the path taken from a particular node is the optimal path *from that node*. For the example in Figure 4.12, once we have progressed two stages from A, so that we are now 17.8 units from our destination, we are in effect solving a new problem which is to find the optimum route from that point to one of the destinations. (A process in which the remaining decisions after the kth stage depend only on the state of the system at the kth stage is referred to as a **Markov process.**)

The original problem is said to be **imbedded** in a series of one-stage processes.

Principle of Optimality

Bellman stated a **principle of optimality**:

Whatever the initial state and initial decision were, the remaining decisions must constitute an optimal sequence for the problem with the state resulting from the first decision now considered as initial conditions.

Consider a system in two-dimensional state space

$$\dot{x}_1 = f_1(x_1, x_2, u); \qquad \dot{x}_2 = f_2(x_1, x_2, u)$$

where $u(t)$ is chosen to minimise

$$C = \int_{t_0}^{t_f} F(x_1, x_2, u)\, dt$$

Figure 4.13 shows the initial state P_1 (at time t_0) and the optimum trajectory to P_2 (at time t_f). P* is a point on the optimum trajectory at time t^*; consider a trajectory from P* to reach Q at time t_f.

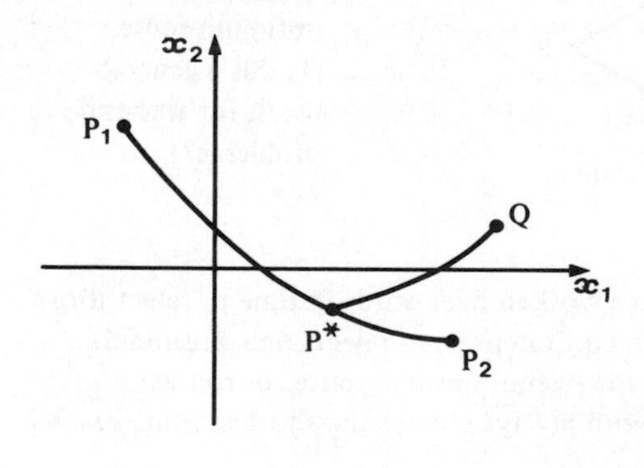

Figure 4.13

The principle of optimality states that if the system started from P* at time t^* and it was required to minimise

$$C^* = \int_{t^*}^{t_f} F(x_1, x_2, u)\, dt$$

then the optimum trajectory would be P^*P_2.

Let

$$C = \int_{t_0}^{t_f} F(x_1, x_2, u)\mathrm{d}t = \int_{t_0}^{t^*} F(x_1, x_2, u)\mathrm{d}t + \int_{t^*}^{t_f} F(x_1, x_2, u)\mathrm{d}t$$

$$= C_1 + C^*, \text{ say.}$$

Suppose the cost from P* to Q was C_2.

Then since the trajectory $P_1 P^* P_2$ is optimal, its cost is least and the cost of $P_1 P^* Q$ is greater. Hence $C_1 + C^* < C_1 + C_2$ so that $C^* < C_2$. Hence $P^* P_2$ is the optimal trajectory from P_2. This justifies our 'backwards approach'.

Application to a control problem

Let a system be governed by the equation

$$x_{k+1} = g(x_k, u_k) \tag{4.37}$$

where the accumulated cost after n stages is

$$C = \sum_{i=0}^{n-1} C_i(x_i, u_i) \tag{4.38}$$

Given an initial stage x_0 a sequence of control inputs $u_0, u_1, \ldots\ldots, u_{n-1}$ is to be chosen to minimise C.

(i) Suppose the system is at state x_{n-1} and one stage remains. Let $S_1(x_{n-1})$ denote the minimum cost possible for the last stage, and $C_{n-1}(x_{n-1}, u_{n-1})$ be the actual cost using control u_{n-1}. Then

$$S_1(x_{n-1}) = \min_{u_{n-1}} \left\{ C_{n-1}(x_{n-1}, u_{n-1}) \right\} \tag{4.39}$$

(Note that the right-hand side must be a function of x_{n-1} only.)

(ii) Suppose the system is at state x_{n-2} and two stages remain. Then

$$S_2(x_{n-2}) = \min_{u_{n-2}, u_{n-1}} \left\{ C_{n-2}(x_{n-2}, u_{n-2}) + C_{n-1}(x_{n-1}, u_{n-1}) \right\}$$

However, $x_{k+1} = g(x_k, u_k)$ so that

$$S_2(x_{n-2}) = \min_{u_{n-2}} \left\{ C_{n-2}(x_{n-2}, u_{n-2}) + \min_{u_{n-1}} \left\{ C_{n-1} \left[g(x_{n-2}, u_{n-2}), u_{n-1} \right] \right\} \right\}$$

by the principle of optimality; whichever u_{n-2} is selected the optimal policy thereafter is to minimise $C_{n-1}\left[g(x_{n-2}, u_{n-2}), u_{n-1}\right]$. Now $g(x_{n-2}, u_{n-2})$ is the new state which results from the first stage. Hence

$$S_2(x_{n-2}) = \min_{u_{n-2}} \left\{ C_{n-2}(x_{n-2}, u_{n-2}) + S_1 \left[g(x_{n-2}, u_{n-2}) \right] \right\} \tag{4.40}$$

Note that since S_1 is now a function of x_{n-2}, S_2 is a function of x_{n-2} only.

The computational approach is

(a) Using (4.39) calculate S_1 for each possible x_{n-1}

(b) For each possible u_{n-2} calculate x_{n-1} from (4.37) using the given x_{n-2}

(c) Then calculate the cost of the term $\left\{ C_{n-2} + \min_{u_{n-1}} \{C_{n-1}\} \right\}$ for each u_{n-2} using (a) and (b). The lowest cost is found by inspection.

(iii) Now suppose the system is at state x_{n-3} and three stages remain.

You should be able to argue that

$$S_3(x_{n-3}) = \min_{u_{n-3}} \left\{ C_{n-3}(x_{n-3}, u_{n-3}) + S_2 \left\{ g(x_{n-3}, u_{n-3}) \right\} \right\} \tag{4.41}$$

Having found all the S_2 by application of the computational approach above, we would then calculate $x_{n-2} = g(x_{n-3}, u_{n-3})$ using each possible u_{n-3} and the given x_{n-3}. The cost $\{\}$ in (4.41) is calculated and the least cost identified.

Example 1

Consider the system $x_{k+1} = x_k + u_k$ where the cost is

$$C = \sum_{i=1}^{n-1} (x_i^2 + u_i^2) = C_0 + C_1 + \dots + C_{n-1}$$

Note the continuous analogue $\dot{x} = u$, $C = \int_{t_0}^{t_f} (x^2 + u^2)\, du$

Suppose $x_n = 0$. The problem is to find the optimal sequence of controls $u_0, u_1, \dots, u_{l}$

) **One-stage process**

$$0 = x_n = x_{n-1} + u_{n-1}$$

Hence we choose $u_{n-1} = -x_{n-1}$. Then

$$S_1(x_{n-1}) = x^2_{n-1} + u^2_{n-1} = 2x^2_{n-1}$$

i) **Two-stage process**

$$S_2(x_{n-2}) = \min_{u_{n-2}} \left\{ (x^2_{n-2} + u^2_{n-2}) + S_1(x_{n-1}) \right\} = \min_{u_{n-2}} \left\{ (x^2_{n-2} + u^2_{n-2}) + 2x^2_{n-1} \right\}$$

Now $x_{n-1} = x_{n-2} + u_{n-2}$ so that

$$S_2(x_{n-2}) = \min_{u_{n-2}} \left\{ (x^2_{n-2} + u^2_{n-2}) + 2(x_{n-2} + u_{n-2})^2 \right\}$$

A computer program would employ a search strategy like that of Fibonacci search, but we shall use calculus to find the minimum. It is easy to see that the minimising value of $u_{n-2} = -\frac{2}{3} x_{n-2}$ with cost $S_2(x_{n-2}) = \frac{5}{3} x^2_{n-2}$

ii) **Three-stage process**

$$S_3(x_{n-3}) = \min_{u_{n-3}} \left\{ (x^2_{n-3} + u^2_{n-3}) + S_2(x_{n-2}) \right\}$$

Using the result for $S_2(x_{n-2})$ and the formula $x_{n-2} = x_{n-3} + u_{n-3}$ we find that

$$S_3(x_{n-3}) = \min_{u_{n-3}} \left\{ (x^2_{n-3} + u^2_{n-3}) + \frac{5}{3} (x_{n-3} + u_{n-3})^2 \right\}$$

Hence $u_{n-3} = -\frac{5}{8} x_{n-3}$, $S_3(x_{n-3}) = \frac{13}{8} x^2_{n-2}$

v) **General process**

Earlier mention of Fibonacci may have caused you to view the fractions, -1, $-\frac{2}{3}$, $-\frac{5}{8}$ for u_i and $\frac{2}{1}$, $\frac{5}{3}$, $\frac{13}{8}$ for S_i with recognition. It is worth conjecturing (and can be shown to be the case) that $u_{n-k} = -\frac{F(2k-1)}{F(2k)} x_{n-k}$ and $S_k(x_{n-k}) = \frac{F(2k+1)}{F(2k)} x_{n-k}^2$ where $F(n)$ is the nth number of the Fibonacci sequence.

Let the final time $t_f = 5$, reached after 5 stages, and let $x_0 = 1$. Compute and graph the control sequence and compare it with the solution for the continuous problem computed via calculus of variations.

Example 2

The system of Example 1 with x_n free. Since x_n is not fixed, u_{n-1} must be zero. Hence $x_{n-1} = x_n$, $u_{n-1} = 0$ and $S_1(x_{n-1}) = x_{n-1}^2$. Now

$$S_2(x_{n-2}) = \min_{u_{n-2}} \left\{ (x_{n-2}^2 + u_{n-2}^2) + x_{n-1}^2 \right\}$$

and since $x_{n-1} = x_{n-2} + u_{n-2}$

$$S_2(x_{n-2}) = \min_{u_{n-2}} \left\{ (x_{n-2}^2 + u_{n-2}^2) + (x_{n-2} + u_{n-2})^2 \right\}$$

Hence $u_{n-2} = -\frac{1}{2} x_{n-2}$ and $S_2(x_{n-2}) = \frac{3}{2} x_{n-2}^2$. Similarly, $u_{n-3} = -\frac{3}{5} x_{n-3}$, $S_3(x_{n-3}) = \frac{8}{5} x_{n-3}^2$ and, in general,

$$u_{n-k} = -\frac{F(2k-2)}{F(2k-1)} x_k$$

$$S_k(x_{n-k}) = \frac{F(2k)}{F(2k-1)} x_{n-k}^2$$

Again, compare with the continuous case.

Problems

1. Find the shortest route between A and J for the network shown in Figure 4.14.

2. Find the shortest path which starts at A and ends at one of the points D, G, I or J for the network shown in Figure 4.15.

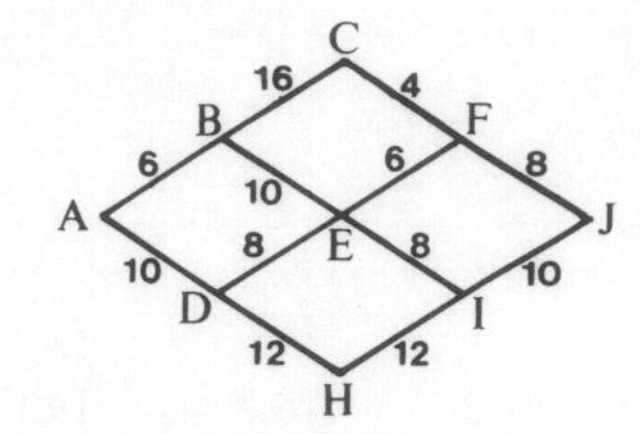

Figure 4.14

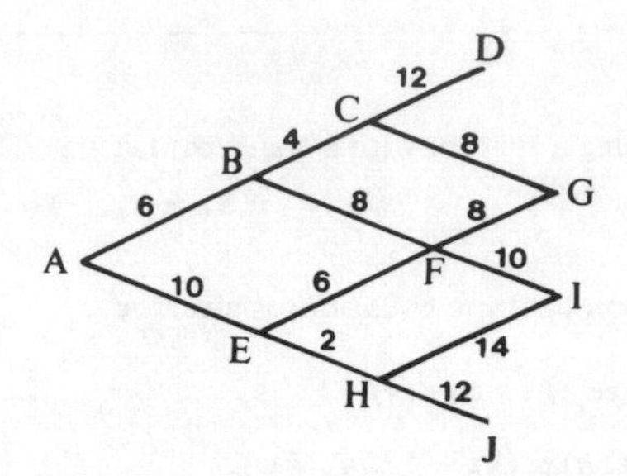

Figure 4.15

Find the control sequence $u(0)$, $u(1)$, $u(2)$ which minimises the cost

$$C = \sum_{k=0}^{2} \left\{ [u(k)]^2 + 1 \right\}$$

for the system $x(k + 1) = x(k) + u(k)$ with boundary conditions $x(0) = 0$, $x(3) = 3$. Comment.

The system $x(k + 1) = x(k) + u(k)$ is to be driven from $x(0) = 10$ under controls $u(0)$ and $u(1)$ so as to minimise $C = [x(2) - 20]^2 + \sum_{k=0}^{1} \left\{ [x(k)]^2 + [u(k)]^2 \right\}$. Show that the control sequence is $u(0) = -2$, $u(1) = 6$, that $x(1) = 8$, $x(2) = 14$ and that the minimum cost is 540.

Find a control sequence which minimises

$$C = [x(2) + 2]^2 + [u(0)]^2 + [u(1)]^2$$

for the system $x(k + 1) = 0.5x(k) + u(k)$ where $x(0) = 10$. Show that the value of $x(2)$ obtained is 0 and the minimised cost is 9. Verify that the uncontrolled system would give a cost of 20.25. Interpret.

6. An operator has eight hours to allocate amongst four projects. In the Table below is shown his earnings for various extra times spent on each project, based on previous work. Decide the optimal allocation of time to earn the most, assuming that he must work an integer number of hours (includes zero) on each project.

Hours Spent \ Project	*A*	*B*	*C*	*D*
0	12	14	14	18
1	14.5	14.5	15.2	19.1
2	16.5	15.7	16.2	19.5
3	17.5	16.1	17.1	19.7
4	18.3	16.9	17.8	19.8

7. The differential equations governing a system with a transport lag are $\dot{x}_1 = -x_1 + u_1$, $\dot{x}_2 = 0.5(u_2 - x_2)$, $\dot{x}_3 = x_1 + x_2 - x_3$, $\dot{x}_4 = 0$, $\dot{x}_5 = x_4 - x_5$, $y(t) = x_3(t - 0.3)$.

Show that a discrete approximation to these equations is given by

$$x_1(k+1) = (1-h)x_1(k) + hu_1(k)$$

$$x_2(k+1) = (1-0.5h)x_2(k) + 0.5hu_2(k)$$

$$x_3(k+1) = hx_1(k) + hx_2(k) + (1-h)x_3(k)$$

$$x_4(k+1) = x_4(k)$$

$$x_5(k+1) = hx_4(k) + (1-h)x_5(k)$$

To allow for the time lag, introduce $0.3/h$ extra state variables. Thus, if $h = 0.1$, define

$$x_6(k+1) = x_5(k),\ x_7(k+1) = x_6(k),\ x_8(k+1) = x_7(k),\ y = x_8$$

Suppose that the control has to minimise

$$\int_0^6 ([x_6 + x_8]^2 + \lambda_1 u_1^2 + \lambda_2 u_2^2)dt$$

Show that a suitable approximation is given by

$$\sum_{k=0}^{N} ([x_6(k) + x_8(k)]^2 + \lambda_1 [u_1(k)]^2 + \lambda_2 [u_2(k)]^2) . h$$

.6 DYNAMIC PROGRAMMING: CONTINUOUS CASE

The continuous problem is very difficult to handle via dynamic programming. As n example consider the system $\dot{x}_1 = x_2$, $\dot{x}_2 = u$ and the cost functional

$C = \int_{t_0}^{t_f} (\alpha^2 x_1^2 + u^2)dt$. We seek to minimise C given initial values $x_1(0)$, $_2(0)$. Define

$$S\left[x_1(t), x_2(t), t\right] = \min_u \left\{\int_{t_0}^{t_f} (\alpha^2 x_1^2 + u^2)dt\right\} \tag{4.42}$$

vhere $t_0 \leqslant t \leqslant t_f$.

Assume that the values of x_1, x_2 at time t, achieved during the process, act as nitial conditions to help determine the cost from that time onwards.

Since the optimisation from time t onwards depends only on the state at time t and t itself) then S may be regarded as being a function of the three independent ariables: $x_1(t)$, $x_2(t)$ and t.

By the principle of optimality,

$$S(x_1, x_2, t) = \min_u \left\{\int_t^{t+\delta t} (\alpha^2 x_1^2 + u^2)dt + \min_u \left\{\int_{t+\delta t}^{t_f} (\alpha^2 x_1^2 + u^2)dt\right\}\right\}$$

$$= \min_u \left\{\int_t^{t+\delta t} (\alpha^2 x_1^2 + u^2)dt + S(x_1 + \delta x_1, x_2 + \delta x_2, t + \delta t)\right\}$$

: is assumed that $x_1 + \delta x_1$ and $x_2 + \delta x_2$ describe the state which results from pursuing the ptimum trajectory from t to $t + \delta t$. If δt is small, then by the Mean Value Theorem,

$$\int_t^{t+\delta t} (\alpha^2 x_1^2 + u^2)dt = (\alpha^2 x_1^2 + u^2)\delta t + 0[(\delta t)^2]$$

follows from a Taylor series expansion that

$$S(x_1 + \delta x_1, x_2 + \delta x_2, t + \delta t) \simeq S(x_1, x_2, t) + \frac{\partial S}{\partial x_1}\delta x_1 + \frac{\partial S}{\partial x_2}\delta x_2 + \frac{\partial S}{\partial t}\delta t$$

$$= S(x_1, x_2, t) + \left(\frac{\partial S}{\partial x_1}\dot{x}_1 + \frac{\partial S}{\partial x_2}\dot{x}_2 + \frac{\partial S}{\partial t}\right)\delta t$$

Therefore,

$$S = \min_u \left\{(\alpha^2 x_1^2 + u^2)\delta t + S + \left(\frac{\partial S}{\partial x_1}\dot{x}_1 + \frac{\partial S}{\partial x_2}\dot{x}_2 + \frac{\partial S}{\partial t}\right)\delta t + 0[(\delta t)^2]\right\}$$

Subtract S from both sides, substitute for $\dot{x}_1$ and $\dot{x}_2$ from the state equations, divide throughout by δt and take the limit as $\delta t \to 0$. Then

$$0 = \min_u \left\{\alpha^2 x_1^2 + u^2 + \frac{\partial S}{\partial x_1}x_2 + \frac{\partial S}{\partial x_2}u + \frac{\partial S}{\partial t}\right\} \tag{4.43}$$

In the next section we consider a special class of control problems which lead to a simplification of (4.43).

4.7 LINEAR SYSTEMS WITH QUADRATIC PERFORMANCE INDICES (LQP CONTROL)

This section deals with systems which give rise to a particular form of differential equation, known as the Riccati equation, which is relatively straightforward to solve. We first consider an example leading to a one-variable equation and then generalise to a matrix equation.

Example

The system $\dot{x} = -ax + bu$ with performance index

$$I = \int_{t_0}^{t_f} (\alpha x^2 + \beta u^2)\,\mathrm{d}t$$

where α, $\beta > 0$ and t_0, t_f are specified in advance.

We follow the procedure described in the last section to obtain

$$\min_u \left\{\alpha x^2 + \beta u^2 + \dot{x}\frac{\partial S}{\partial x} + \frac{\partial S}{\partial t}\right\} = 0 \tag{4.44}$$

Assume that $S(x,\ t) = P(t)x^2$ [†]. Then (4.44) becomes

$$\min_u \left\{ \alpha x^2 + \beta u^2 + (-ax + bu)\,2Px + \dot{P}x^2 \right\} = 0 \tag{4.45}$$

Via partial differentiation with respect to u we find that

$$2\beta u + 2bPx = 0$$

Hence the optimum control is given by

$$u_{\text{opt}} = -\frac{bP}{\beta}x \tag{4.46}$$

Substitution of this expression into (4.45) produces

$$\left(\alpha + \frac{b^2P^2}{\beta} - 2aP - \frac{2b^2P^2}{\beta} + \dot{P}\right)x^2 = 0 \tag{4.47}$$

We can satisfy (4.47) if the coefficient of x^2 is zero.

The following Riccati differential equation results:

$$\dot{P} = -\alpha + 2aP + \frac{b^2P^2}{\beta} \tag{4.48}$$

[The general Riccati equation will be $\dot{P} = A(t) + B(t)P + C(t)P^2$.]

Since this is a first-order equation, one boundary condition is needed for its solution. In his example, the free-end-point at t_f implies that $S(x,\ t_f) = 0$ since it represents the minimum cost for a process of zero time. Hence $P(t_f) = 0$.

For simplicity, consider the case where $\alpha = \beta = 1$, $a = 0$, $b = 1$. Then the optimum control is $u_{\text{opt}} = -Px$ and the Riccati equation becomes $\dot{P} = P^2 - 1$ with oundary condition $P(t_f) = 0$. It is left to you to show the following results:

$$P = -\tanh(t - t_f) \tag{4.49}$$

$$\dot{x} = x\tanh(t - t_f)$$

$$\therefore \quad x = C\cosh(t - t_f) \tag{4.50}$$

$$u_{\text{opt}} = C\sinh(t - t_f) \tag{4.51}$$

ou may care to compare these results with those on page 142 where the solution was pressed in terms of exponential functions.

[†] This is a reasonable assumption by analogy with the discrete case.

If the end-point is fixed the problem is not so straightforward. Refer to McCausland (1969) p. 193.

If the final time, t_f, is infinite then it can be argued that P is a constant; if we examine the discrete case on page 154 the ratio $\dfrac{F(2k-1)}{F(2k)}$ tends to a definite limit as $k \to \infty$ and u is directly proportional to x. Set $\dot{P} = 0$ in the Riccati equation and solve an algebraic equation for P.

Matrix Riccati equation

Consider the system $\dot{\mathbf{x}} = \mathbf{A}(t)\mathbf{x} + \mathbf{B}(t)\mathbf{u}$ where $\mathbf{A}$ is an $n \times n$ matrix and $\mathbf{B}$ is an $n \times m$ matrix. We seek to minimise the cost

$$C = \int_{t_0}^{t_f} (\mathbf{x}^{\mathrm{T}} \mathbf{Q}\mathbf{x} + \mathbf{u}^{\mathrm{T}} \mathbf{R}\mathbf{u})\, \mathrm{d}t$$

where $\mathbf{Q}$ and $\mathbf{R}$ are symmetric and positive definite matrices.
The control $\mathbf{u}$ is chosen to minimise C given $\mathbf{x}(t_0) = \mathbf{x}_0$. Define

$$f(\mathbf{x}, t) = \min_{\mathbf{u}} \left\{ \int_{t}^{t_f} (\mathbf{x}^{\mathrm{T}} \mathbf{Q}\mathbf{x} + \mathbf{u}^{\mathrm{T}} \mathbf{R}\mathbf{u})\, \mathrm{d}t \right\}$$

for $t_0 \leqslant t \leqslant t_f$.

By the principle of optimality

$$f(\mathbf{x}, t) = \min_{\mathbf{u}} \left\{ \int_{t}^{t+\delta t} (\mathbf{x}^{\mathrm{T}} \mathbf{Q}\mathbf{x} + \mathbf{u}^{\mathrm{T}} \mathbf{R}\mathbf{u})\, \mathrm{d}t + f(\mathbf{x}+\delta\mathbf{x},\ t+\delta t) \right\}$$

$$= \min_{\mathbf{u}} \left\{ (\mathbf{x}^{\mathrm{T}} \mathbf{Q}\mathbf{x} + \mathbf{u}^{\mathrm{T}} \mathbf{R}\mathbf{u})\, \delta t + f(\mathbf{x}, t) + \left(\frac{\partial f}{\partial \mathbf{x}}\right)^{\mathrm{T}} \times \left(\mathbf{A}(t)\mathbf{x} + \mathbf{B}(t)\,\mathbf{u}\right) \delta t + \frac{\partial f}{\partial t}\, \delta t + 0\,[(\delta t)^2] \right\}$$

by Taylor's theorem.

Cancel $f(\mathbf{x}, t)$, divide throughout by δt and take the limit as $\delta t \to 0$ to obtain

$$-\frac{\partial f}{\partial t} = \min_{\mathbf{u}} \left\{ \mathbf{x}^T \mathbf{Q}\mathbf{x} + \mathbf{u}^T \mathbf{R}\mathbf{u} + \left(\frac{\partial f}{\partial \mathbf{x}}\right)^T (\mathbf{A}\mathbf{x} + \mathbf{B}\mathbf{u}) \right\} \qquad (4.52)$$

ssume that $f(\mathbf{x}, t) = \mathbf{x}^T \mathbf{P}(t)\mathbf{x}$, i.e. that f can be expressed as a quadratic form here $\mathbf{P}$ is an $n \times n$ matrix which we may assume, without loss of generality, is ymmetric. It is left to you to show that

$$\frac{\partial f}{\partial \mathbf{x}} = 2\mathbf{P}(t)\mathbf{x} \qquad (4.53)$$

nd

$$\frac{\partial f}{\partial t} = \mathbf{x}^T \frac{d\mathbf{P}}{dt} \mathbf{x} \qquad (4.54)$$

Ience,

$$-\mathbf{x}^T \frac{d\mathbf{P}}{dt} \mathbf{x} = \min_{\mathbf{u}} \left\{ \mathbf{x}^T \mathbf{Q}\mathbf{x} + \mathbf{u}^T \mathbf{R}\mathbf{u} + 2\mathbf{x}^T \mathbf{P}(\mathbf{A}\mathbf{x} + \mathbf{B}\mathbf{u}) \right\} \qquad (4.55)$$

)ifferentiating $\{\ \}$ partially with respect to $\mathbf{u}$ and equating to zero gives $\mathbf{R}\mathbf{u} + 2\mathbf{B}^T \mathbf{P}\mathbf{x} = \mathbf{0}$ so that

$$\mathbf{u}_{opt} = -\mathbf{R}^{-1} \mathbf{B}^T \mathbf{P}\mathbf{x} \qquad (4.56)$$

'he fact that $\mathbf{R}$ is positive definite guarantees the existence of $\mathbf{R}^{-1}$. Hence, the omponents of the optimal control vector are linear functions of the state vector omponents (linear feedback). Substituting (4.56) into (4.55) gives

$$-\mathbf{x}^T \frac{d\mathbf{P}}{dt} \mathbf{x} = \{ \mathbf{x}^T \mathbf{Q}\mathbf{x} + \mathbf{x}^T \mathbf{P}\mathbf{B}\mathbf{R}^{-1} \mathbf{B}^T \mathbf{P}\mathbf{x} + \mathbf{x}^T \mathbf{P}\mathbf{A}\mathbf{x} + \mathbf{x}^T \mathbf{A}^T \mathbf{P}\mathbf{x} - 2\mathbf{x}^T \mathbf{P}\mathbf{B}\mathbf{R}^{-1} \mathbf{B}^T \mathbf{P}\mathbf{x} \}$$

ote that the third and fourth terms on the right-hand side are scalars, so that it is ermissible to express $2\mathbf{x}^T \mathbf{P}\mathbf{A}\mathbf{x}$ as their sum. (Verify this.) Therefore,

$$-\mathbf{x}^T \frac{d\mathbf{P}}{dt} \mathbf{x} = \mathbf{x}^T (\mathbf{Q} + \mathbf{P}\mathbf{A} + \mathbf{A}^T \mathbf{P} - \mathbf{P}\mathbf{B}\mathbf{R}^{-1} \mathbf{B}^T \mathbf{P}) \mathbf{x}$$

nce $\mathbf{x}$ is arbitrary,

$$-\frac{d\mathbf{P}}{dt} = \mathbf{Q} + \mathbf{P}\mathbf{A} + \mathbf{A}^T\mathbf{P} - \mathbf{P}\mathbf{B}\mathbf{R}^{-1} \mathbf{B}^T\mathbf{P} \qquad (4.57)$$

ıis is the **matrix Riccati equation**. Since

$$\mathbf{x}^T(t_f)\mathbf{P}(t_f)\mathbf{x}(t_f) = f(\mathbf{x}, t_f) = \int_{t_f}^{t_f} \ldots\ldots\ldots \, dt = 0$$

it follows that the appropriate boundary condition is

$$\mathbf{P}(t_f) = 0 \tag{4.58}$$

The minimised cost is

$$C_{\text{opt}} = \mathbf{x}_0^T \mathbf{P}(t_0) \mathbf{x}_0 \tag{4.59}$$

Note that when **A**, **B**, **Q** and **R** are constant matrices and $t_f = \infty$ then it can be shown that the control is independent of t and $\frac{d\mathbf{P}}{dt} = \mathbf{0}$ so that the Riccati equation becomes

$$\mathbf{Q} + \mathbf{PA} + \mathbf{A}^T\mathbf{P} - \mathbf{PBR}^{-1}\mathbf{B}^T\mathbf{P} = \mathbf{0} \tag{4.60}$$

(The control now depends on the state of the system and not on time left – always ∞ – hence it is time-independent.) Confirm that these results reduce to the scalar case.

It is of interest to note further that if **R** is the zero matrix then the derivation of (4.60) would lead to the equation

$$\mathbf{A}^T\mathbf{P} + \mathbf{PA} = -\mathbf{Q} \tag{4.61}$$

which is the Liapunov matrix equation encountered in Chapter 3. Of course, $\mathbf{R} \equiv \mathbf{0}$ implies that control considerations have not been taken into account.

Example 1

$$\mathbf{A} = \begin{bmatrix} 0 & 1 \\ 0 & -4 \end{bmatrix}, \; \mathbf{B} = \begin{bmatrix} 0 \\ 1 \end{bmatrix}, \; \mathbf{C} = \int_0^\infty \left\{ \mathbf{x}^T \begin{bmatrix} 4 & 0 \\ 0 & 1 \end{bmatrix} \mathbf{x} + u^2 \right\} dt.$$ Now

$\mathbf{Q} = \begin{bmatrix} 4 & 0 \\ 0 & 1 \end{bmatrix}$, $\mathbf{R} = 1$ so that from (4.56)

$$u_{\text{opt}} = -\begin{bmatrix} 0 \\ 1 \end{bmatrix}^T \mathbf{Px} = -\begin{bmatrix} 0 \\ 1 \end{bmatrix}^T \begin{bmatrix} p_{11} & p_{12} \\ p_{12} & p_{22} \end{bmatrix} \begin{bmatrix} x_1 \\ x_2 \end{bmatrix}$$

.e.

$$u_{opt} = -[\, p_{12} \quad p_{22} \,] \begin{bmatrix} x_1 \\ x_2 \end{bmatrix} = -(p_{12}x_1 + p_{22}x_2) \qquad (4.62)$$

From (4.60)

$$\begin{bmatrix} 4 & 0 \\ 0 & 1 \end{bmatrix} + \begin{bmatrix} p_{11} & p_{12} \\ p_{12} & p_{22} \end{bmatrix} \begin{bmatrix} 0 & 1 \\ 0 & -4 \end{bmatrix} + \begin{bmatrix} 0 & 0 \\ 1 & -4 \end{bmatrix} \begin{bmatrix} p_{11} & p_{12} \\ p_{12} & p_{22} \end{bmatrix}$$

$$- \begin{bmatrix} p_{11} & p_{12} \\ p_{12} & p_{22} \end{bmatrix} \begin{bmatrix} 0 \\ 1 \end{bmatrix} [0 \quad 1] \begin{bmatrix} p_{11} & p_{12} \\ p_{12} & p_{22} \end{bmatrix} = \mathbf{0}$$

You should show that this leads to the equations

$$p_{12}^2 = 4$$

$$p_{11} - 4p_{12} - p_{12}p_{22} = 0$$

$$-2p_{12} + 8p_{22} + p_{22}^2 = 1$$

and that these have possible solutions

$$\left.\begin{aligned} p_{12} &= 2 \\ p_{22} &= -4 \pm \sqrt{21} \\ p_{11} &= \pm 2\sqrt{21} \end{aligned}\right\} \quad \text{or} \quad \left.\begin{aligned} p_{12} &= -2 \\ p_{22} &= -4 \pm \sqrt{13} \\ p_{11} &= \mp 2\sqrt{13} \end{aligned}\right\}$$

The optimal control can be found from (4.62) and the least cost via (4.59).

Example 2

$$\dot{x}_1 = x_2; \quad \dot{x}_2 = 5u; \quad \dot{x}_3 = 0; \quad x_3(0) = 1.0$$

We seek to minimise $\int_0^{t_f} [(x_3 - x_1)^2 + \alpha^2 u^2]\, dt$. Here

$$\mathbf{A} = \begin{bmatrix} 0 & 1 & 0 \\ 0 & 0 & 0 \\ 0 & 0 & 0 \end{bmatrix}, \quad \mathbf{B} = \begin{bmatrix} 0 \\ 5 \\ 0 \end{bmatrix}, \quad \mathbf{Q} = \begin{bmatrix} 1 & 0 & -1 \\ 0 & 0 & 0 \\ -1 & 0 & 1 \end{bmatrix} \text{ with } \mathbf{R} = \alpha^2$$

Then (4.57) becomes

$$-\frac{d\mathbf{P}}{dt} = \begin{bmatrix} 1 & 0 & -1 \\ 0 & 0 & 0 \\ -1 & 0 & 1 \end{bmatrix} + \begin{bmatrix} 0 & 0 & 0 \\ p_{11} & p_{12} & p_{13} \\ 0 & 0 & 0 \end{bmatrix} + \begin{bmatrix} 0 & p_{11} & 0 \\ 0 & p_{12} & 0 \\ 0 & p_{13} & 0 \end{bmatrix}$$

$$- \frac{25}{\alpha^2} \begin{bmatrix} p_{12}^2 & p_{12}p_{22} & p_{12}p_{23} \\ p_{12}p_{22} & p_{22}^2 & p_{22}p_{23} \\ p_{12}p_{23} & p_{22}p_{23} & p_{23}^2 \end{bmatrix}$$

It is left to you to write out the six component differential equations and to show that

$$u_{\text{opt}} = -\frac{5}{\alpha^2}\left\{p_{12}x_1 + p_{22}x_2 + p_{23}x_3\right\}$$

We can find p_{ij} by integration of the component equations with the boundary condition $\mathbf{P}(t_f) = \mathbf{0}$.

Alternatively, we could define $\tau = t - t_f$ and obtain a reverse Riccati equation in τ with the boundary condition now transformed to $\mathbf{P}(0) = \mathbf{0}$.

If $t_f = \infty$ then, as expected, $\frac{d\mathbf{P}}{dt} = \mathbf{0}$ and we leave you to solve the resulting algebraic equations and show that

$$u_{\text{opt}} = \frac{1}{k}\left[(x_1 - x_3) + \sqrt{(0.4k)}\, x_2\right]$$

In general, to see whether steady-state has been reached, it is customary to integrate over a time span two to three times the length of the longest time constant in the system.

Problem

The system $\dot{x}_1 = -x_1 + u$, $\dot{x}_2 = x_1$ is to be controlled so as to minimise

$$\int_0^T (x_2^2 + 0.1u^2)\,dt$$

Show that $u_{\text{opt}} = -1.7x_1 - 3.155x_2$.

Chapter Five

Random Processes

5.1 INTRODUCTION

There are many phenomena which produce data that are deterministic; that is they can be represented by explicit mathematical relationships. There are many others which produce non-deterministic data. Examples are the noise voltage at the terminals of a resistor, the oscillations of a high-speed train, the amount of water in a reservoir, the behaviour of queueing systems, the height of waves in the ocean, the output of a fading signal. It is impossible in each case to predict an exact value of the variable at a future time. Since the data are random in nature, they must be described by statistical averages and probability models.

The purpose of this chapter is to study methods of describing these **random processes**, also referred to as **stochastic processes**. Attention will be paid to the important problem in communications engineering of designing a system to receive the required signal and reject the noise element of the input. The particular example of the Gaussian process will be considered. When an input to a system is random it is important to know how the output behaves and this will be studied.

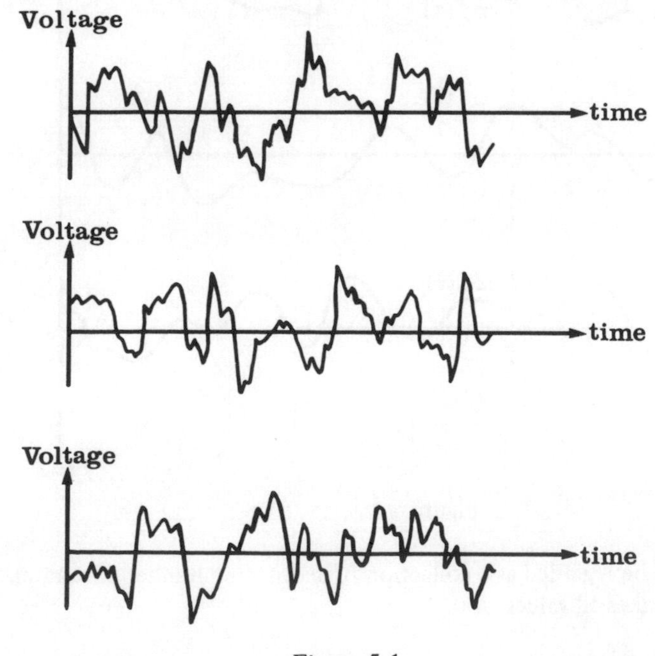

Figure 5.1

Basic Ideas

Figure 5.1 shows three records of the noise voltage at the terminals of a resistor. Each of these observations is but one of many possible records which could have occurred. When an observation is taken over a finite interval, it is referred to as a sample record. Each of the observations is called a **realization** of the random process; sometimes it is called a **sample function**. The set of all possible sample functions is called an **ensemble**. The process is random in the sense that a particular realization is a random selection from the ensemble of sample functions.

In addition to selecting one realization from the ensemble we may consider all possible results at time $t = \tau$. In Figure 5.2 is shown an ensemble for four functions, written $\{ x(t, \zeta_i) \}$ where ζ_i is the particular sample function. An alternative notation is $x_i(t)$. At time τ, the **sample value of the random process is a random variable** $x(\tau)$. In the figure, $x(\tau)$ takes possible values $\{ x_i(\tau) \}$. In a sense, then, the

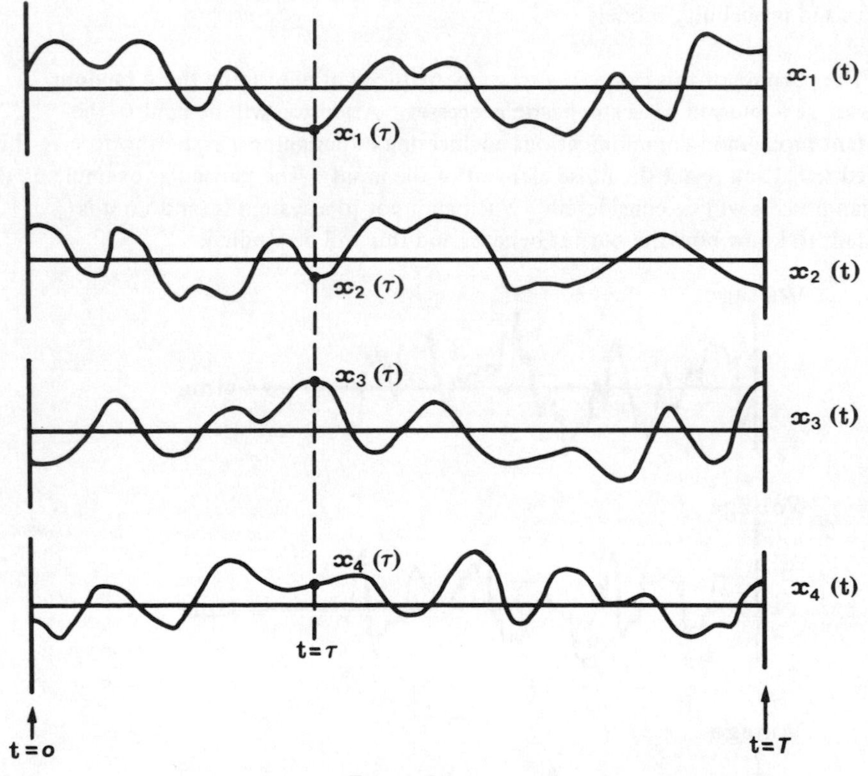

Figure 5.2

random process can be regarded as a collection of an infinite number of random variables where τ takes all values $\geqslant 0$.

In many situations the ensemble contains infinitely many members. It may be that there is but one device generating the 'signals' and each observation of the output represents a realization; it may be that there are several 'identical' devices. Suppose that we consider a signal transmitted from a station to a large number of receivers in the neighbourhood. Since each receiver is in a different location the phase of the signal received will be different. Also, the gain of each antenna will be different. Hence the ensemble of the waveforms received may be described as

$$x(t) = A \cos(\omega t + \theta) \tag{5.1}$$

where the amplitude A and the phase θ are random variables. It may be that θ is uniformly distributed over $[0, 2\pi]$ and A follows a Gaussian distribution.

The random process is defined by the equation (5.1) together with the probability density functions for amplitude and phase. It is left to you to generate some sample functions for this process. It can be done with the aid of a computer if there are routines available for selecting random numbers from uniform and Gaussian distributions.

5.2 DESCRIPTION OF A RANDOM PROCESS

To describe the fundamental properties of a random process, three types of statistical measures are used:

(i) **probability density function** which describes the properties of the process in the amplitude domain.

(ii) **autocorrelation function** which provides information on the process in the time domain, i.e. how rapidly the signal varies with time.

(iii) **power spectral density function** which provides information in the frequency domain.

In this section we shall concentrate mainly on the first of these measures. However, because of the importance of autocorrelation functions, there will be a brief introduction to them here. They and power spectral density functions will be treated at length in subsequent sections.

Probability density function

It is necessary to make some preliminary definitions. Let $x(\zeta)$ be a random variable where ζ are points in the sample space of outcomes. The **cumulative probability distribution function** $P(x)$ is defined as the probability that the random variable takes a value less than or equal to x. Then it follows that $P(a) \leqslant P(b)$ if

$a < b$ and $P(-\infty) = 0$, $P(\infty) = 1$. If the random variable assumes a continuous range of values then a (first-order) **probability density function** $p(x)$ is defined by the relationship

$$\frac{\mathrm{d}}{\mathrm{d}x} P(x) = p(x) \tag{5.2}$$

Note that

$$P(x) = \int_{-\infty}^{x} p(\zeta)\,\mathrm{d}\zeta \tag{5.3}$$

and that the probability that $x(\zeta)$ takes a value in the range $[a, b]$ is $\int_a^b p(\zeta)\,\mathrm{d}\zeta$. It follows that $p(x) \geqslant 0$.

From the density function we may determine the **mean value** of the random variable $x(\zeta)$; this is

$$E[x(\zeta)] = \int_{-\infty}^{\infty} x p(x)\,\mathrm{d}x = \mu_x \tag{5.4}$$

The **mean square value** of the random variable is

$$E[\{x(\zeta)\}^2] = \int_{-\infty}^{\infty} x^2 p(x)\,\mathrm{d}x = \psi_x^2 \tag{5.5}$$

The **variance** of the random variable can be found as

$$E[\{x(\zeta) - \mu_x\}^2] = \psi_x^2 - \mu_x^2 = \sigma_x^2 \tag{5.6}$$

Now let $g(x)$ be any real single-valued continuous function of x. Let its probability density function be $p(g)$. If $\frac{\mathrm{d}g}{\mathrm{d}x}$ exists and is non-zero then it may be shown that

$$p(g) = \frac{n p(x)}{|\mathrm{d}g/\mathrm{d}x|} \tag{5.7}$$

where n is the multiplicity of the inverse function $x(g)$. For example, the function $\sin^{-1} x$ is double-valued in $[0, 2\pi]$.

xample 1

Let a random process be given by

$$x(t) = at + b \qquad (5.8)$$

here a is normally distributed $N(0, 1^2)$ and b is constant.

Now $a = \frac{1}{t}(x - b)$ and $\frac{dx}{da} = t$. The density function of a is

$(a) = \frac{1}{\sqrt{2\pi}} e^{-\frac{1}{2}a^2}$, therefore using (5.7) in the form $p(x) = \frac{p(a)}{|dx/da|}$ since a is a

ngle-valued function of x, we obtain

$$p(x) = \frac{1}{t\sqrt{2\pi}} e^{-(x-b)^2/t^2} \qquad (5.9)$$

his varies with t and we might well use the notation $p(x; t)$. For example, when

$= 2$, $p(x) = \frac{1}{2\sqrt{2\pi}} e^{-(x-b)^2/4}$ whilst when $t = 4$, $p(x) = \frac{1}{4\sqrt{2\pi}} e^{-(x-b)^2/16}$.

hese density functions are graphed in Figure 5.3.

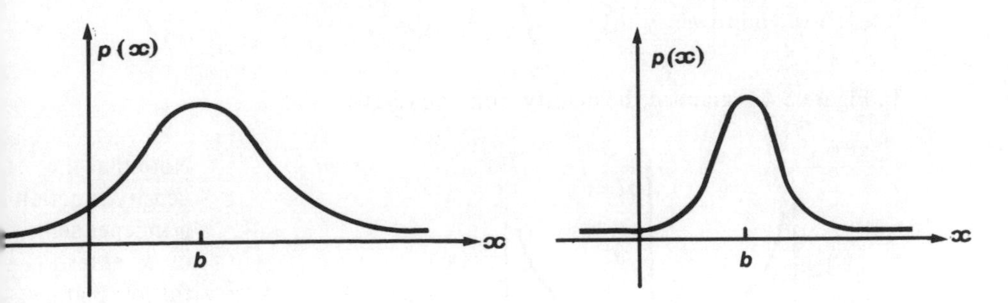

Figure 5.3

Note that using (5.4) to (5.6) it is possible to show that $\mu_x = b$, $\sigma_x = t$.

xample 2

The random process

$$x(t) = A \cos(\omega t + \theta) \qquad (5.1)$$

is such that θ is a random variable uniformly distributed in $[0, 2\pi]$, A and ω are constants. First we note that

$$p(\theta) = \begin{cases} \dfrac{1}{2\pi}, & 0 \leqslant \theta \leqslant 2\pi \\ 0, & \text{elsewhere} \end{cases}$$

Remember that t is effectively a constant and (5.1) gives x as a function of θ. Using (5.7) with $n = 2$†, $p(x) = \dfrac{2p(\theta)}{|\mathrm{d}x/\mathrm{d}\theta|}$ and noting that

$$\frac{\mathrm{d}x}{\mathrm{d}\theta} = A \sin(\omega t + \theta) = \pm A\sqrt{1 - \frac{x^2}{A^2}}$$

we see that

$$p(x) = \frac{1}{\pi\sqrt{A^2 - x^2}}, \quad -A \leqslant x \leqslant A \tag{5.10}$$

We need to add the requirement that $p(x) = 0$ for $|x| > A$.

It is left to you to verify that $\displaystyle\int_{-\infty}^{\infty} p(\zeta)\mathrm{d}\zeta = 1.$

In Figure 5.4 is graphed the density function (5.10).

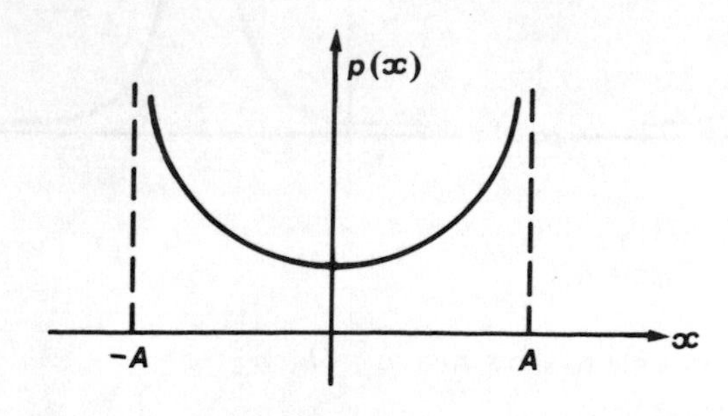

Figure 5.4

Note that the density function is independent of t; note further that $\mu_x = 0$, $\sigma_x^2 = A^2/2$.

This density function reflects the fact that a cosine wave spends proportionally more of its time near its crests and troughs. If the cosine wave is contaminated by

† Draw a graph of x against θ, for $0 < \theta < 2\pi$ to see why.

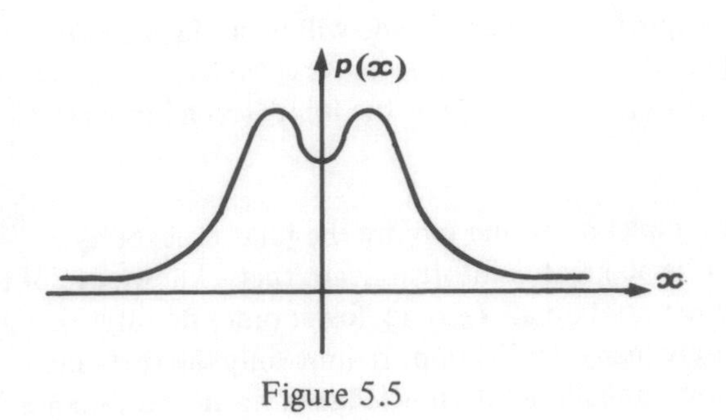

Figure 5.5

random noise, which may be assumed to be distributed about zero mean, then the density function will have the characteristics of both components and a typical case is shown in Figure 5.5.

To summarise, the first-order density function is the probability density of the amplitude of the sample functions at time t. Hence $p(x)\delta x$ is the probability that the amplitude of a sample function will lie in the range $(x, x + \delta x)$ at time t. If there are M sample functions observed and m satisfy the requirement then $p(x) \simeq \dfrac{m}{M\delta x}$.

Second-Order statistics

The first-order statistics provide information about the distribution of amplitudes. Is this sufficient? Figure 5.6 shows two random processes with the same first-order density. They are clearly quite different processes.

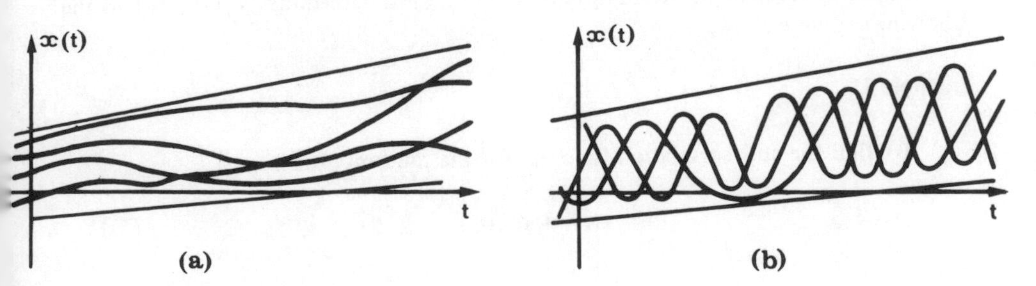

Figure 5.6

The process in (a) shows slow variability with time whereas that of (b) shows a much more rapid variation. In the first case, the values of $x(t_1)$ and $x(t_2)$ are not much different and provided $(t_2 - t_1)$ is sufficiently small it is clear that $x(t_1)$ and $x(t_2)$ are not statistically independent (the 'signal' contains predominantly low frequency components). In the second case, where high frequency components dominate, values of $x(t)$ separated by the same interval $(t_2 - t_1)$ bear little relation to each other.

A measure of the relationship between two random variables is provided by their correlation. The values of the sample functions at times t_1 and t_2, viz. $x(t_1)$ and $x(t_2)$ will be denoted x_1 and x_2 respectively. Consider the joint probability density function. The probability that a sample function amplitude will be in the interval $(x_1, x_1 + dx_1)$ at time t_1 and $(x_2, x_2 + dx_2)$ at time t_2 is given by $p(x_1, x_2; t_1, t_2)dx_1 dx_2$ where $p(x_1, x_2; t_1, t_2)$ is the joint (second-order) probability density function.

To specify a random process completely would require the joint probability density functions of all orders n. It is no great comfort to learn that a knowledge of the nth order density function allows one to obtain all $(n - 1)$ lower order density functions by successive integration. Fortunately many applications require only the first- and second-order functions for satisfactory modelling (cf knowledge of mean and variance for a simple random variable).

A statistic of much value is the **autocorrelation function.** It is the correlation between two variables defined on the same process and is a measure of the dependence of the amplitudes of the sample functions at time t_1 on the amplitudes at time t_2. It is formally defined as the average over the ensemble of the product of two random variables , i.e.

$$R_x(t_1, t_2) = \int_{-\infty}^{\infty} \int_{-\infty}^{\infty} x_1 x_2 p(x_1, x_2; t_1, t_2) dx_1 \, dx_2 \qquad (5.11)$$

The autocorrelation function is written sometimes as $E(x_1 x_2)$. Fortunately we do not always need the formidable defining equation (5.11) to determine $R_x(t_1, t_2)$ as the following example shows.

Example 3

We find the autocorrelation function for the random process

$$x(t) = A \cos(\omega t + \theta) \qquad (5.1)$$

Now

$$\begin{aligned} R_x(t_1, t_2) &= E[x(t_1) . x(t_2)] \\ &= A^2 E[\cos(\omega t_1 + \theta) . \cos(\omega t_2 + \theta)] \\ &= \frac{A^2}{2} E[\cos \omega(t_1 - t_2) + \cos(\omega t_1 + \omega t_2 + 2\theta)] \\ &= \frac{A^2}{2} \left\{ E[\cos \omega(t_1 - t_2)] + E[\cos(\omega t_1 + \omega t_2 + 2\theta)] \right\} \end{aligned}$$

But $E[\cos \omega(t_1 - t_2)] = \cos \omega(t_1 - t_2)$ since θ, the only random variable on the right-hand-side of (5.1), is not involved. Further,

$$E[\cos(\omega t_1 + \omega t_2 + 2\theta)] = \int_{-\infty}^{\infty} \cos(\omega t_1 + \omega t_2 + 2\theta) \,.\, p(\theta)\,d\theta$$

$$= \int_0^{2\pi} \frac{1}{2\pi} \cos(\omega t_1 + \omega t_2 + 2\theta) \,.\, d\theta$$

$$= 0$$

Hence

$$R_x(t_1, t_2) = \frac{A^2}{2} \cos \omega (t_1 - t_2) = \frac{A^2}{2} \cos \omega (t_2 - t_1) \qquad (5.12)$$

A graph of (5.12) appears in Figure 5.7. We have used the notation $\tau = t_2 - t_1$. Such a graph is sometimes referred to as an **autocorrelogram**.

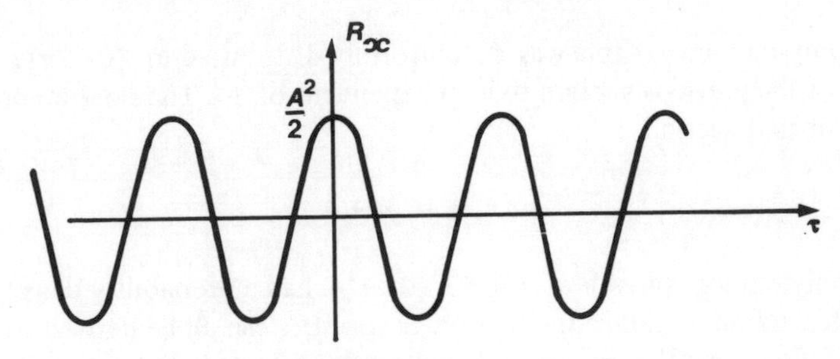

Figure 5.7

Note that $R_x(\tau) \equiv R_x(t_1, t_2)$ in this case depends only on the difference $(t_2 - t_1)$, not specifically on the individual values of t_1 and t_2. Note further that it is an even function of τ. We return to the study of autocorrelation functions in Section 5.5.

Problems

1. Sketch some sample functions for the random process $x(t) = A \cos(\omega t + \theta)$

   (i) when the random variable is A and is uniformly distributed in $[-1, 1]$

   (ii) when θ is the random variable, uniformly distributed in $[-\pi, \pi]$

   (iii) when the random variable is ω uniformly distributed in $[0, 5]$.

2. For each of the processes in Problem 1 find the first-order probability density function, the mean and mean-square values and the autocorrelation function.

3. For the random process of Example 3 in this section find the autocorrelation function. Comment.

4. The Poisson process $x(t)$ may be described by the formula for the probability of r occurrences in time interval t, which is $p(r, t) = \frac{(\mu t)^r}{r!} e^{-\mu t}$.

Show that the expected value of $x(t)$ is μt, that the expected value of $[x(t)]^2$ is $\mu^2 t^2 + \mu t$ and hence find the variance of the process. Show that the autocorrelation function is given by

$$R_x(t_1, t_2) = \begin{cases} \mu t_1 (1 + t_2), & t_1 \leqslant t_2 \\ \mu t_2 (1 + t_1), & t_1 \geqslant t_2 \end{cases}$$

5.3 STATIONARY AND ERGODIC PROCESSES

The probability density function of the process

$$x(t) = A \cos(\omega t + \theta) \tag{5.1}$$

where the only random variable was θ, uniformly distributed in $[0, 2\pi]$, was found in Example 2 of the previous section to be independent of t. The random process of Example 1 of that section,

$$x(t) = at + b \tag{5.8}$$

where the only random variable was $a \sim N(0, 1^2)$ had a probability density function which did depend on t. Since the first-order statistics can all be derived from the density function it follows that the process (5.1) will not be affected, at least as far as the first-order statistics are involved, by a change in the time origin and is in some sense **stationary**. The same cannot be said for the process (5.8).

A random process is said to be **stationary to order 1** if its (first-order) probability density function is independent in time. A process is said to be **stationary to order n** if its nth order density function depends only on time *differences* between the sample values. Note that since the $(n - 1)$ lower order density functions can be derived from the nth order one, then those functions will also depend only on time differences.

A process which is stationary to *all* orders n is called **strictly stationary**.

As an example we may note that a process for which

$$p(x_1, x_2; t_1, t_2) = p(x_1, x_2; \tau) \tag{5.13}$$

where $\tau = t_2 - t_1$ is stationary to second-order.

The most important second-order statistic is the autocorrelation function. A

›rocess stationary to second-order will have

$$R_x(t_1, t_2) = R_x(\tau)$$

Iowever, the converse is not necessarily true.

The actual determination of whether a process is stationary is not often traightforward. Some processes may be reasonably stationary once the generating levice reaches a steady state, but initially may be non-stationary. The analytical letermination is very difficult. However, a less severe requirement for stationarity is ›rovided by an examination of the mean and autocorrelation functions.

A process is said to be **stationary in the wide sense** (or **weakly stationary**) if

(i) μ_x is constant for all time (5.14a)

and (ii) $R_x(t_1, t_2) = R_x(\tau)$ (5.14b)

vhere $\tau = t_2 - t_1$.

Note that a process which is stationary to order 2 (and certainly one which is trictly stationary) is stationary in the wide sense. The converse is not true, since we ›nly require one second-order statistic to be independent of time origin. For the process

$$x(t) = A \cos(\omega t + \theta) \qquad (5.1)$$

t was shown that $\mu_x = 0$ and $R_x(t_1, t_2) = R_x(\tau)$. Hence the process is at least tationary in the wide sense.

Example 1

Let $x(t)$ be defined by the differential equation $\dot{x} = -\alpha x + u$ where α is a ›ositive constant and u is a random variable with mean zero and constant variance σ^2. Assume that u does not depend on the initial value $x(0) = 0$.

Now any sample function has the form

$$x(t) = \frac{u}{\alpha}(1 - e^{-\alpha t}), \quad t \geq 0$$

t is left to you to show that $\mu_x = 0$. (Note that the mean is taken over the ensemble.) Further

$$R_x(t_1, t_2) = E[\frac{u}{\alpha}(1 - e^{-\alpha t_1}).\frac{u}{\alpha}(1 - e^{-\alpha t_2})]$$

Since $E(u^2) = \sigma^2 + \mu_x^{\,2} = \sigma^2$, it follows that

$$R_x(t_1, t_2) = \sigma^2 \cdot \frac{1}{\alpha^2}(1 - e^{-\alpha t_1}) \cdot (1 - e^{-\alpha t_2})$$

and the process is seen to be non-stationary.

Note, however, that if t_1 and t_2 are both very large

$$R_x(t_1, t_2) \to \frac{\sigma^2}{\alpha^2}$$

and therefore in the steady state the process is weakly stationary.

Example 2

Consider the random process defined by

$$x(t) = A \cos \omega t \tag{5.15}$$

where ω is constant and A is a random variable uniformly distributed in $[0, A^*]$. Note that the density function of A is

$$p(A) = \begin{cases} 1/A^* & , \ 0 \leqslant A \leqslant A^* \\ 0 & , \ \text{elsewhere} \end{cases}$$

In Figure 5.8 are shown three sample functions of the ensemble.

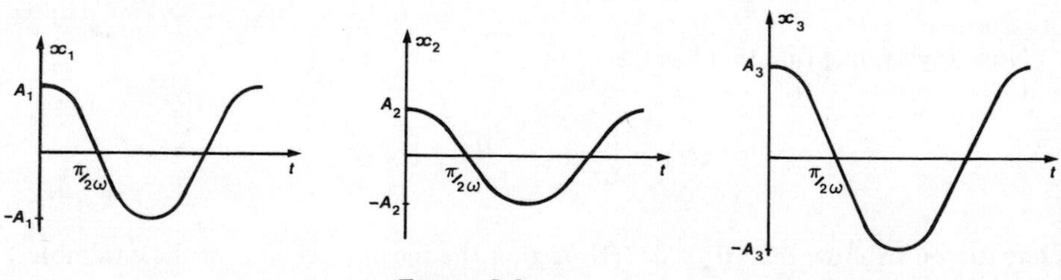

Figure 5.8

First consider the situation at $t = \dfrac{\pi}{2\omega}$. All the sample functions assume the value ero there. The density function at that time, $p(x, \pi/2\omega)$, must be given by $(x) = \delta(x)$.†

However, at $t = 0$, $x(t) = A$ and the density function is given by

$$p(x, 0) = \begin{cases} \dfrac{1}{A^*}, & 0 \leqslant x \leqslant A^* \\ 0, & \text{elsewhere} \end{cases}$$

ince $p(x, t)$ is not independent of time the process is not stationary to order 1, let lone stationary in the strict sense.

rgodic Processes

The trouble with all that has been deveolped so far is that in order to obtain the tatistical measures it is necessary to consider the whole ensemble of sample functions.

Certain processes are such that a single sample function carries the same statistical nformation as any other sample function and hence one sample function can provide all he statistics of the entire process. When a process possesses this property, which may be xpressed by saying that time averages (which are the only averages that may be obtained rom a single sample function) are identical to ensemble averages, it is said to be **ergodic**. here is no agreement about the precise definition of ergodic. The one we develop here s used by many authors.

Before we present a formal definition of ergodicity, note that an ergodic process nust necessarily be stationary, whereas the converse is not in general true. From a ingle sample function only one set of statistics can be obtained. If the process were on-stationary there would be an infinite number of probability density functions, epending on the time origin chosen. Hence an ergodic process cannot be non-stationary. n Figure 5.9 we depict the relationship between the various classes of random process.

If a single record of an ergodic process is obtained for $0 \leqslant t \leqslant T$ then it is possible o estimate the statistics of the process by examining this record. For example the first-order) probability density function may be determined as follows. If the total ime during which the variable $x(t)$ takes values between x_0 and $x_0 + \delta x_0$ is T_{x_0} hen $p(x_0) = T_{x_0}/(T\delta x_0)$.

It is implicit that in an ergodic process any sample function takes all the possible alues with the same relative frequency that an ensemble will assume at any instant of ime.

† $\delta(x)$ is the **Dirac** delta function with the properties that $\delta(x) = 0$ for $x \neq 0$ and

$$\int_{-\infty}^{\infty} \delta(x)\,dx = 1.$$

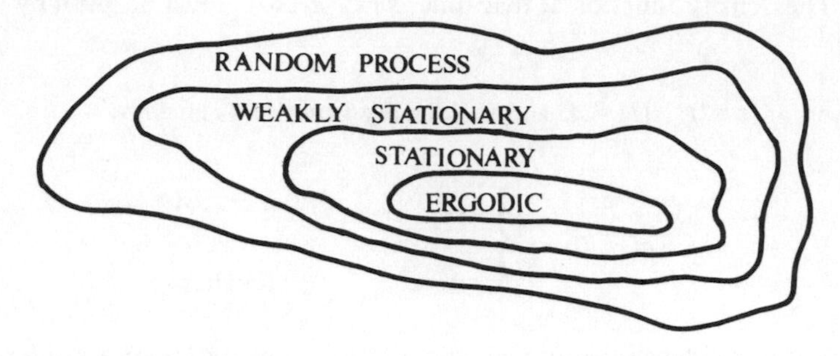

Figure 5.9

A process is **ergodic in the mean** if the ensemble mean

$$E(X) \equiv \mu_x = \int_{-\infty}^{\infty} x \,.\, p(x)\,\mathrm{d}x \tag{5.4}$$

and the time mean

$$\overline{x(t)} = \lim_{T\to\infty} \frac{1}{2T} \int_{-T}^{T} x(t)\,\mathrm{d}t \tag{5.16}$$

are identical.

A process is **ergodic to the first-order** if the first-order ensemble average and the first-order time average of any function $\phi\,[x(t)]$ of a sample function are equal , i.e.

$$E[\phi(x)] = \int_{-\infty}^{\infty} \phi(x)\,.\,p(x)\,\mathrm{d}x \tag{5.17a}$$

is identical to

$$\overline{\phi[x(t)]} = \lim_{T\to\infty} \frac{1}{2T} \int_{-T}^{T} \phi[x(t)]\,\mathrm{d}t \tag{5.17b}$$

A process is **ergodic to the second-order** if

$$E[\phi(x_1\,,\,x_2)] = \int_{-\infty}^{\infty}\int_{-\infty}^{\infty} \phi(x_1\,,x_2)\,.\,p(x_1\,,x_2\,;\,\tau)\mathrm{d}x_1\,\mathrm{d}x_2 \tag{5.18a}$$

and

$$\overline{\phi[x(t),\ x(t+\tau)]} = \lim_{T\to\infty} \frac{1}{2T} \int_{-T}^{T} \phi[x(t),\ x(t+\tau)]\, dt \qquad (5.18b)$$

are identical. Recall that an ergodic process is stationary; hence a process which is ergodic to second-order will be stationary to second-order.

A particular case to consider is the autocorrelation function which is a second-order statistic.

Since $R_x(t_1, t_2) = R_x(\tau)$ for a process stationary to order 2 we have the definition

$$R_x(\tau) = E[x_1(t).x_2(t+\tau)] \qquad (5.19a)$$

It follows from (5.18b) that for a process ergodic to the second-order,

$$R_x(\tau) = \overline{x(t)\,.\,x(t+\tau)} \qquad (5.19b)$$

It is possible to generalise equations (5.17) to the definition of a process ergodic to order n. A **strictly ergodic process** is one which is ergodic to any integral order n.

It is not easy to determine whether a process is ergodic or not. Certainly if a process is not stationary it cannot be ergodic and hence the process (5.15) of Example 2 in this section is not ergodic. In practice if the mean and mean-square values of a process are time-independent it is worthwhile calculating time averages (of say mean and mean-square values) for a selection of sample functions. Should these values be reasonably in agreement with each other, then the process may be assumed to be ergodic to the order of the averages calculated.

Example 3

The random process

$$x(t) = A\cos(\omega t + \phi) \qquad (5.1)$$

where ϕ is uniformly distributed in $[0,\ 2\pi]$ has a mean $E[x(t)] \equiv \mu_x = 0$. Now

$$\overline{x(t)} = \lim_{T\to\infty} \frac{1}{2T} \int_{-T}^{T} A\cos(\omega t + \phi)\, dt$$

Since

$$\int_{-T}^{T} A\cos(\omega t+\phi)\,dt = \left[\frac{A}{\omega}\sin(\omega t+\phi)\right]_{-T}^{T}$$

$$= \frac{A}{\omega}\left\{\sin(\omega T+\phi) - \sin(-\omega T+\phi)\right\}$$

$$= \frac{2A}{\omega}\sin\omega T\cos\omega\phi$$

then $\overline{x(t)} = 0$.

The process is therefore ergodic in the mean. Further, the time autocorrelation function is

$$\lim_{T\to\infty}\frac{1}{2T}\int_{-T}^{T} A\cos(\omega t+\phi)\,.\,A\cos(\omega t+\omega\tau+\phi)\,dt$$

The integral is equal to

$$\frac{A^2}{2}\int_{-T}^{T}[\cos(2\omega t+\omega\tau+2\phi)+\cos\omega\tau]\,dt$$

$$= \frac{A^2}{4\omega}\left[\sin(2\omega t+\omega\tau+2\phi)\right]_{-T}^{T} + \frac{A^2}{2}\left[t\cos\omega\tau\right]_{-T}^{T}$$

$$= \frac{A^2}{4\omega}\left[\sin(2\omega T+\omega\tau+2\phi) - \sin(-2\omega T+\omega\tau+2\phi)\right] + A^2T\cos\omega\tau$$

$$= \frac{A^2}{4\omega}\,.\,2\sin 2\omega T\,.\,\cos(\omega\tau+2\phi) + A^2T\cos\omega\tau$$

Hence the time autocorrelation is $\frac{A^2}{2}\cos\omega\tau$ which agrees with result (5.12). The process is therefore ergodic in the autocorrelation function.

vo random processes

The **cross-correlation function** for two random processes $x(t)$, $y(t)$ is defined to

$$R_{xy}(t_1, t_2) = E[x(t_1)y(t_2)] \tag{5.20}$$

or wide-sense jointly stationary processes,

$$R_{xy}(t_1, t_2) = R_{xy}(\tau) \tag{5.21}$$

hen $x(t)$ and $y(t)$ are jointly ergodic to the **second-order**

$$R_{xy}(\tau) = \lim_{T\to\infty} \frac{1}{2T} \int_{-T}^{T} x(t)y(t+\tau)\mathrm{d}t \tag{5.22}$$

ote Unless otherwise stated, the rest of this **chapter** will deal solely with ergodic processes.

oblems

Show that the random process (5.15) is not ergodic.

Show that the process $x(t) = A \sin(\omega t + \phi)$, where only A is a random variable, is not stationary. Sketch a few realisations.

The random process $x(t) = A \sin(\omega t + \phi)$ is such that A is a random variable with probability density function $p(a)$ and ϕ is uniformly distributed in $[0, 2\pi]$ and is independent of A. Show that the process is stationary in the wide sense and ergodic in the mean and autocorrelation function.

Show that no deterministic function which varies with time can be ergodic.

The process $x(t) = A \sin \omega t + B \cos \omega t$ is such that A and B are uncorrelated random variables, each with mean zero and variance σ^2. Show that the process is ergodic in the mean.

Consider this definition of ergodicity: A process is ergodic if

$$E[x(t) > x_0] = \int_{x_0}^{\infty} p(x)\mathrm{d}x \equiv \text{prob}\,[x(t) > x_0]$$

Show that, with this definition, the process (5.1) is ergodic.

(Note that $E[x(t) > x_0]$ is the fraction of one period during which $A \sin(\omega t + \phi)$ exceeds x_0.)

7. A random process is given by $x(t) = A \cos(t + \phi)$ where A is a normally distributed random variable with mean zero and variance 1 and ϕ is uniformly distributed in $[-\pi, \pi]$. Show that $x(t)$ is stationary in the wide sense. Is it ergodic in the mean and in the autocorrelation function?

5.4 POWER DENSITY SPECTRUM

There are many situations when it is advantageous to think of a random process in terms of the distribution of its frequency components.

For a deterministic signal it is useful to describe a linear system via the transfer function which measures the frequency response and to discuss the signals themselves via the relative amplitudes of their frequency content using Fourier Transforms.

Consider a particular realisation of a random process, $x_1(t)$ say. Then this is a completely determined function of time. Assuming that the Fourier Transform of $x_1(t)$ exists then its amplitude spectral density will be given by

$$A_1(\omega) = \int_{-\infty}^{\infty} x_1(t)\, e^{j\omega t}\, dt = \int_{0}^{\infty} x_1(t)\, e^{j\omega t}\, dt \qquad (5.23)$$

since the realisation exists for $t \geqslant 0$. Here, $\omega = 2\pi f$ is assumed to be the angular frequency in radians/second.

Certain properties are well-known.

(i) $A_1(\omega)$ is the amplitude spectral density in an infinitesimal interval $d\omega$

(ii) $A_1(\omega)A_1^*(\omega)$ is the energy density in the interval $d\omega$†

(iii) $|A_1(\omega)|^2 d\omega$ is the energy in the frequency interval $(\omega, \omega + d\omega)$

(iv) In a continuous spectrum, the power at a particular frequency, ω_1, is infinitesimally small

(v) In a discrete spectrum, the spectral density at a specific value will be infinite.

Suppose we regard $x_1(t)$ as a voltage across a one-ohm resistor, then

† $A^*(\omega)$ is the complex conjugate of $A(\omega)$.

$$\lim_{T\to\infty} \frac{1}{T}\left| A_1(\omega)\right|^2$$

the power density per radian per second. It represents the instantaneous power issipated by the resistance.

If, now, several realisations of $x(t)$ are examined, there will generally be different alues of $A(\omega)$ for each realisation. Suppose, then, we take an ensemble average of he power density. It can be shown by an argument which falls outside our scope that *the random process is stationary* then the power density spectrum, viz.

$$\Phi(\omega) = \lim_{T\to\infty} \frac{1}{T}\left\{ E[\,|A(\omega)|^2\,]\right\}$$

equal to the Fourier Transform of its autocorrelation function. Stated more formally: rovided the autocorrelation function of a stationary random process satisfies the equirement that

$$\int_{-\infty}^{\infty} \left| R_x(\tau)\right| \mathrm{d}\tau$$

xists then its Fourier Transform, namely

$$\Phi(\omega) = \int_{-\infty}^{\infty} R_x(\tau)\, \mathrm{e}^{-j\omega\tau}\, \mathrm{d}\tau \tag{5.24a}$$

the **power density spectrum** for the process.

Note that $E[\{x(t)\}^2]$ represents the power that would be dissipated by a one-ohm esistance if $x(t)$ represents voltage across the resistance.

Applying the inversion formula for Fourier Transforms, we obtain

$$R(\tau) = \frac{1}{2\pi} \int_{-\infty}^{\infty} \mathrm{e}^{j\omega\tau}\, \Phi(\omega)\, \mathrm{d}\omega \tag{5.24b}$$

Equations (5.24a) and (5.24b) are known as the **Wiener-Khinchin** relationships. ote that since $R(\tau)$ and $\Phi(\omega)$ are Fourier Transforms of each other then the wider e spectrum $\Phi(\omega)$ the shorter the correlation time, i.e. that value of τ above which (τ) falls below $\frac{1}{\mathrm{e}}$. That is, the more rapidly $x(t)$ varies, the higher the frequency components it must contain and hence the wider its spectrum.

Putting $\tau = 0$ in equation (5.24b) produces

$$R(0) = \frac{1}{2\pi} \int_{-\infty}^{\infty} 1 \,.\, \Phi(\omega)\, d\omega = \int_{-\infty}^{\infty} \Phi(2\pi f)\, df$$

Now $R(\tau) = E[x(t) \,.\, x(t+\tau)]$ so that

$$R(0) = E[x(t) \,.\, x(t)] = E[(x(t))^2] \tag{5.25}$$

Hence

$$E[(x(t))^2] = \int_{-\infty}^{\infty} \Phi(2\pi f)\, df$$

which justifies the use of the name ‘power density’.

It is worth remarking that if we agree that $\Phi(2\pi f)$ may be written simply as $\Phi(f)$ equations (5.24a) and (5.24b) become

$$\Phi(f) = \int_{-\infty}^{\infty} e^{-j2\pi f\tau} R(\tau)\, d\tau \tag{5.26a}$$

and

$$R(\tau) = \int_{-\infty}^{\infty} e^{+j2\pi f\tau} \Phi(f)\, df \tag{5.26b}$$

In the case of an ergodic process, the power spectral density of any sample function is given by the Fourier Transform of its autocorrelation function. Hence the power density spectra are identical for all sample functions. In this instance the quantity

$$E[(x(t))^2] = \lim_{T\to\infty} \frac{1}{2T} \int_{-T}^{T} [x(t)]^2\, dt$$

is the time average power dissipated in a one-ohm resistor.

roperties of $\Phi(\omega)$

It can be shown that

) $\Phi(\omega)$ is a real function and hence contains no information about phase

i) $\Phi(\omega)$ is an even function of ω

ii) $\Phi(\omega)$ is non-negative

v) If $\mu_x = 0$, $\quad \int_{-\infty}^{\infty} \Phi(\omega)\mathrm{d}\omega = \sigma_x^2$

Note that, being the FourierTransform of $R(\tau)$, the power density spectrum arries information about a second-order statistic of a stationary random process. onsequently, it does not *uniquely* describe a process.

xample 1 Random telegraph signal

This process is such that $x(t)$ takes values of either a or $-a$ and changes from ne value to the other in such a way that the time interval between changes follows a oisson distribution with an average of ν change-overs per second. Sketch a typical ealisation. It will be shown in the next section that the autocorrelation function is iven by

$$R_x(\tau) = a^2 \mathrm{e}^{-2\nu|\tau|}$$

low

$$\Phi_x(\omega) = \int_{-\infty}^{0} a^2 \mathrm{e}^{+2\nu\tau}\,\mathrm{e}^{-j\omega\tau}\,\mathrm{d}\tau + \int_{0}^{\infty} a^2 \mathrm{e}^{-2\nu\tau}\,\mathrm{e}^{-j\omega\tau}\,\mathrm{d}\tau$$

$$= \frac{a^2}{2\nu - j\omega}\left[\mathrm{e}^{(2\nu - j\omega)\tau}\right]_{-\infty}^{0} - \frac{a^2}{2\nu + j\omega}\left[\mathrm{e}^{-(2\nu + j\omega)\tau}\right]_{0}^{\infty}$$

$$= \frac{a^2}{2\nu - j\omega} + \frac{a^2}{2\nu + j\omega} = \frac{4\nu a^2}{4\nu^2 + \omega^2}$$

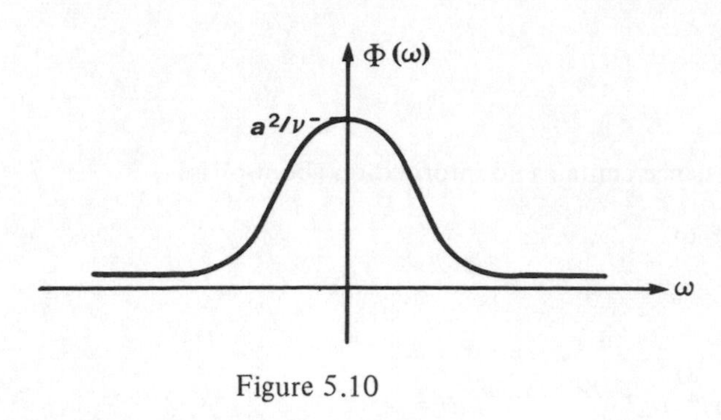

Figure 5.10

The graph of $\Phi(\omega)$ is shown as Figure 5.10.

This process can represent **thermal noise** in conductors.

Note that properties (i), (ii) and (iii) hold.

Example 2 White noise

White noise is defined as a stationary random process with zero mean whose power density spectrum is a constant value, say η, which is independent of frequency. The autocorrelation function is

$$R_x(\tau) = \frac{1}{2\pi} \int_{-\infty}^{\infty} \eta \, e^{j\omega\tau} \, d\omega = \eta \cdot \frac{1}{2\pi} \int_{-\infty}^{\infty} e^{j\omega\tau} \, d\omega$$

Now consider the Fourier Transform of the delta function $\delta(\tau)$. It is

$$\int_{-\infty}^{\infty} \delta(\tau) \, e^{-j\omega\tau} \, d\tau = \left[e^{-j\omega\tau} \right]_{\tau=0} = 1$$

Hence the inverse transform of $g(\omega) = 1$ is $f(\tau) = \delta(\tau)$. Therefore $R_x(\tau) = \eta \, . \, \delta(\tau)$.

This implies that the autocorrelation function is zero for $\tau \neq 0$ and therefore that $x(t)$ is uncorrelated with $x(t + \tau)$ if $\tau \neq 0$. Physically, white noise cannot exist since it requires infinite power; this follows because $\int_{-\infty}^{\infty} \eta \, . \, e^{j\omega\tau} \, d\omega$ is not finite.
It is customary, however, to refer to a random process as white if over the band of frequency of interest in a particular problem the power density function is constant.

Example 3

$\Phi(\omega) = \omega/(\omega^2 + 1)$ is not a valid power spectral density function since it is not an even function of ω.

Problems

1. Show that the power spectral density for the process (5.1), $x(t) = A\cos(\omega_0 t + \phi)$ is

$$\Phi(\omega) = \frac{\pi A^2}{2}\left[\delta(\omega + \omega_0) + \delta(\omega - \omega_0)\right].$$

Where is all the power concentrated? Is the result reasonable?

2. The Binary Transmission process is a sequence of pulses with the properties that (i) each pulse is of duration L, (ii) pulses are equally likely to be of values +1 or −1, (iii) pulses are statistically independent, (iv) the starting time of the first pulse after $t = 0$ can be any value between 0 and 1 with equal probability (pulse sequence not synchronised). The autocorrelation function is

$$R_x(\tau) = \begin{cases} 1 - \dfrac{|\tau|}{L}, & |\tau| < L \\ 0, & |\tau| > L \end{cases}$$

Show that $\Phi(\omega) = L\left(\dfrac{\sin\frac{\omega L}{2}}{\frac{\omega L}{2}}\right)^2$. Sketch it.

3. Which of the functions below might be a possible power density function?

(i) $\omega^2/(\omega^2 + 2)$ (ii) $(2 + j\omega)/(\omega^2 + 2)$

(iii) $\delta(\omega + 6) + \delta(\omega - 6)$ (iv) $\dfrac{1}{\omega^2 - 4}$

4. Using equation (5.25) find the mean-square values of

(i) $\dfrac{1}{(\omega^2 + 1)(\omega^2 + 9)}$ (ii) $\dfrac{\omega^2 + 3}{(\omega^2 + 4)(\omega^2 + 6)}$

(iii) $\dfrac{1}{\omega^4 + 4\omega^2 + 5}$

5. By obtaining $\Phi(\omega)$ explain why the proposed autocorrelation function

$$R_x(\tau) = \begin{cases} 1, & |\tau| < T \\ 0, & \text{otherwise} \end{cases}$$

is not realisable. Sketch both $R_x(\tau)$ and $\Phi(\omega)$.

6. Noting that

$$R(\tau) = \begin{cases} 2T^* \left[1 - \dfrac{|\tau|}{2T^*} \right], & |\tau| < 2T^* \\ 0 & , \text{ elsewhere} \end{cases}$$

is the convolution of the autocorrelation of Problem 5 with itself, and that this is equivalent to multiplication in the frequency domain, show that $\Phi(\omega) = 2T^* \left(\dfrac{\sin \omega T^*}{\omega T^*} \right)$. Sketch $R_x(\tau)$ and $\Phi(\omega)$. Is $R_x(\tau)$ realisable?

7. Band-limited white noise has a power density spectrum

$$\Phi(\omega) = \begin{cases} A, & \omega_1 < |\omega| < \omega_2 \\ 0, & \text{otherwise} \end{cases}$$

Show that

$$R_x(\tau) = \frac{2A}{\pi\tau} \cos \left[\frac{\omega_1 + \omega_2}{2} \right] \tau \sin \left[\frac{\omega_2 - \omega_1}{2} \right] \tau$$

If the band is narrow, i.e. $\Delta\omega = \omega_2 - \omega_1 << \dfrac{\omega_1 + \omega_2}{2} = \omega_0$ then $R_x(\tau)$ becomes $\dfrac{2A}{\pi\tau} \cos \omega_0 \tau \sin \left[\dfrac{\Delta\omega}{2} \right] \tau$. Sketch $R_x(\tau)$ in this case.

5.5 A FURTHER LOOK AT AUTOCORRELATION FUNCTIONS

We met in Section 5.2 the autocorrelation function as a descriptor of the behaviou of a process in the time domain and in Section 5.3 we saw how it helped to define a wide-sense stationary process. In the last section we found that via Fourier Transforms it could also provide information about the frequency content of a signal. It is a sufficiently important statistic to merit further attention.

Figure 5.11(a) shows a typical autocorrelation function for a slowly varying signal and Figure 5.11(b) shows an autocorrelation function for a rapidly varying function. For large values of τ, $R_x(\tau)$ dies out unless the signal is periodic or contains a dc – component (i.e. a constant). If the signal is periodic, $R_x(\tau)$ will be periodic (see Figure 5.7). If the signal contains a dc –component, then $R_x(\tau) \to E(x_1) \,.\, E(x_2)$ as $\tau \to \infty$. We have remarked earlier that for signals with high frequency content the autocorrelation function is compressed in the τ-direction; if the signal contains low frequencies then $x(\tau)$ is more spread out. (c.f. power density spectrum.)

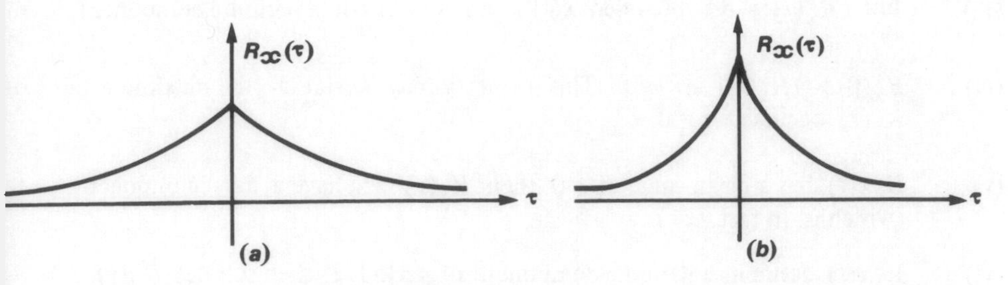

Figure 5.11

Figure 5.12 shows four autocorrelation functions. Diagram (a) depicts the utocorrelation function of a wide-band noise with mean value zero (note that as the and width $\to \infty$ the autocorrelation function tends to the Dirac delta function $\delta(\tau)$). Diagram (b) represents a narrow band noise: $R_x(\tau)$ will decay to zero as $|\tau| \to \infty$. Diagram (c) is for a sine wave with noise added while diagram (d) is the case of a purely

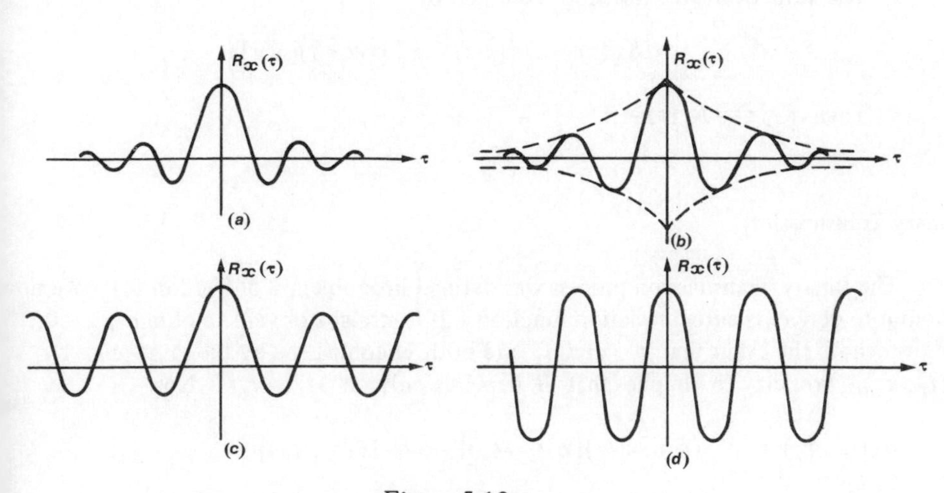

Figure 5.12

eterministic sine wave. It is therefore possible to deduce something about the nature f a random process by examining the features of its autocorrelation function.

roperties of the autocorrelation function for ergodic processes

) $\quad R_x(-\tau) = R_x(\tau)$ (5.27)

i) $\quad R_x(0) = \overline{x^2} = E[(x(t))^2]$

(iii) $\lim_{T \to \infty} R_x(\tau) = \bar{x}^2$ provided $x(t)$ does not contain a periodic component.

(iv) $R_x(0) \geqslant |R_x(\tau)|$, $\tau \neq 0$. This property may be stated: the maximum value of $R_x(\tau)$ occurs at $\tau = 0$.

(v) If $x(t)$ has a mean value $\bar{x} \neq 0$ then $R_x(\tau)$ will have a dc – component (which is, in fact, $\bar{x}^2$).

(vi) If $x(t)$ contains a periodic component of period T then so does $R_x(\tau)$.

(vii) If $z(t) = \dfrac{d}{dt} x(t)$, $R_z(\tau) = \dfrac{-d^2}{d\tau^2} R_x(\tau)$.

(viii) If $z(t) = x(t) + y(t)$ where $x(t)$, $y(t)$ are uncorrelated processes with zero mean then $R_z(\tau) = R_x(\tau) + R_y(\tau)$.

(ix) If $z(t) = x(t) \, . \, y(t)$ where $x(t)$, $y(t)$ are statistically independent then $R_z(\tau) = R_x(\tau) \, . \, R_y(\tau)$.

(x) The **autocovariance function** is defined by

$$K_x(\tau) = E\left\{[x(t) - \bar{x}]\,[x(t+\tau) - \bar{x}]\right\}$$

Then $K_x(\tau) = R_x(\tau) - \bar{x}^2$.

Binary Transmission

The Binary Transmission process was defined in Problem 2 of Section 5.4. We now attempt to derive its autocorrelation function. It is straightforward to obtain $\mu_x = 0$. Let us denote the event that t_1 and t_2 are both encompassed by the same pulse by $B(t_1, t_2)$. Property (iv) implies that B depends only on $|t_1 - t_2|$. Now

$$\begin{aligned} Rx(t_1, t_2) = {} & E[x(t_1) \, . \, x(t_2) | B(t_1, t_2)] \text{ prob } [B(t_1, t_2)] \\ & + E[x(t_1) \, . \, x(t_2) |^c B(t_1, t_2)] \left\{1 - \text{prob } [B(t_1, t_2)]\right\} \end{aligned}$$

where $E[A|B]$ is the expected value of the event A, given that event B has occurred, ${}^cB$ is the event 'not B' and prob $[B]$ is the probability of event B occurring. Then

$$E[x(t_1) \, . \, x(t_2) | B(t_1, t_2)] = \tfrac{1}{2} \, . \, 1^2 + \tfrac{1}{2}(-1)^2 = 1$$

$$E[x(t_1) \, . \, x(t_2) |^c B(t_1, t_2)] = E[x(t_1)] \, . \, E[x(t_2)] = 0$$

$$\text{prob } [B(t_1, t_2)] = 0, \quad \text{for} \quad |t_1 - t_2| > L$$

Now when $|t_1 - t_2| < L$, property (iv) leads to the result that
rob $[t_1$ and t_2 occur on different pulses$] = |t_1 - t_2|/L$. Hence prob $[B(t_1, t_2)]$
$1 - |t_1 - t_2|/L$.

Finally, it can be deduced that

$$R_x(t_1, t_2) = \begin{cases} 1 - |t_1 - t_2|/L, & |t_1 - t_2| < L \\ 0, & |t_1 - t_2| > L \end{cases}$$

hich is the result quoted in the problem if $t_1 - t_2 = \tau$.

oblems

The signal $x(t) = \begin{cases} A, & 0 \leqslant t \leqslant T \\ 0, & \text{elsewhere} \end{cases}$. Sketch the signal and show that

$$R_x(\tau) = \begin{cases} A^2(T - |\tau|), & |\tau| < T \\ 0, & \text{elsewhere} \end{cases}$$

Sketch $R_x(\tau)$ and check properties (i) to (iv).

A signal $x(t)$ is periodic with period $T_0 > 2T$; $x(t) = A$, $-T \leqslant t \leqslant T$ and $x(t) = 0$, $T < t < T_0 - T$. Repeat the strategy of Problem 1, showing that

$$R_x(\tau) = \frac{1}{T_0} \sum_{n=-\infty}^{\infty} R_x(\tau - nT_0)$$

. The random process $x(t)$ consists of an infinite sequence of identical rectangular pulses of duration T. The possible locations for the centres of the pulses are $\pm T_0$, $\pm 2T_0$, $\pm 3T_0$, ... where $T_0 > 2T$. The probability that a pulse does occur at a location is p and is independent of the absence or presence of other pulses. Sketch a sample function of $x(t)$ and the product $x(t) \,.\, x(t + \tau)$ for $0 \leqslant \tau \leqslant T$. The graph of $R_x(\tau)$ is shown below. Decompose $R_x(\tau)$ into a periodic and an aperiodic component.

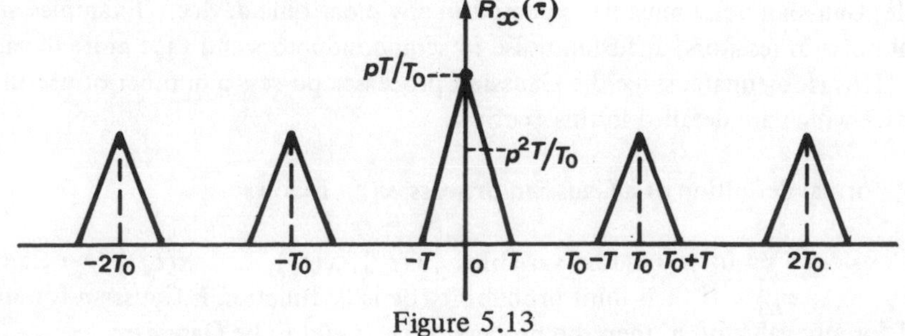

Figure 5.13

4. Prove properties (i), (ii), (iii), (vi), (x).

5. The cross-correlation function $R_{xy}(\tau)$ was defined in Section 5.3. Prove that

   (i) $R_{xy}(\tau) = R_{xy}(-\tau)$

   (ii) $[R_x(0)R_y(0)]^{1/2} > |R_{xy}(\tau)|$

   (iii) If $x(t)$ and $y(t)$ are independent, then $R_{xy}(\tau) = R_{yx}(\tau) = \bar{x}\,.\,\bar{y}$.

6. The **cross-covariance function** is defined by $C_{xy}(\tau) = E\left\{[x(t) - \mu_x]\,[y(t+\tau) - \mu_y]\right\}$. Show that $R_{xy}(\tau) = C_{xy}(\tau) + \mu_x \,.\, \mu_y$.

7. Show that $2\,|R_{xy}(\tau)| < R_x(0) + R_y(0)$.

8. The **cross-power density spectrum** is defined by

$$\Phi_{xy}(\omega) = \int_{-\infty}^{\infty} R_{xy}(\tau)\, e^{-j\omega\tau}\, d\tau$$

   Show that $\Phi_{xy}(\omega) = \Phi_{yx}(-\omega)$.

9. Two random processes are given by $x(t) = \cos(t + \theta)$, $y(t) = \cos(t + \phi)$ where θ and ϕ are independent and each is distributed uniformly in $[-\pi, \pi]$. Show that $R_{xy}(\tau) = 0$.

10. If the processes of Problem 9 are such that $\theta = \phi$ and θ is uniformly distributed in $[-\pi, \pi]$, show that $R_{xy}(\tau) = \frac{1}{2}\cos\tau$.

5.6 GAUSSIAN PROCESSES

A Gaussian, or normal, random process occurs as the idealisation of many natural phenomena which are concerned with the superposition of several small effects: for example, Gaussian noise must be expected in any electronic device. Examples are thermal noise in resistors, diffusion noise in semi-conductors and shot noise in vacuum tubes. This is fortunate, since the Gaussian processes possess a number of useful properties which are detailed in this section.

A formal definition of a **Gaussian process** $x(t)$ follows:

Consider a set of n random variables $\{x(t_1), x(t_2), \ldots\ldots, x(t_n)\}$, written as $\{x_1, x_2, \ldots\ldots, x_n\}$. If their joint probability density function is Gaussian for any suc' set and for any value of n then the process $x(t)$ is said to be Gaussian.

As an illustration consider $n = 2$ and the set of functions $\{x_1, x_2\}$. The joint nd-order)probability function is required to be

$$p(x_1, x_2) = \frac{1}{\sqrt{2\pi}} \frac{1}{|K|} \exp\left\{-\frac{1}{2|K|} \sum_{i=1}^{2} \sum_{j=1}^{2} \Delta_{ij} (x_i - \mu_{x_i})(x_j - \mu_{x_j})\right\} \tag{5.28}$$

ıere $|K|$ is the determinant of the covariant matrix $\mathbf{K} = [K_{ij}]$ and $_{j} = K_x(t_i, t_j) = E[(x_i - \mu_{x_i})(x_j - \mu_{x_j})]$. Hence

$$\begin{aligned} K_{ij} &= E[x_i x_j] - \mu_{x_i} E(x_j) - \mu_{x_j} E(x_i) + \mu_{x_i} \mu_{x_j} \\ &= E[x_i x_j] - E(x_i) E(x_j) - E(x_i) E(x_j) + E(x_i) E(x_j) \\ &= E(x_i x_j) - E(x_i) E(x_j) \\ &= R_x(t_i, t_j) - \mu_{x_i} \mu_{x_j} \end{aligned}$$

so Δ_{ij} is the co-factor of K_{ij}.

The discussion can be generalised to any n and we deduce the result that:

a Gaussian process is completely specified by its autocorrelation function and its mean.

If, therefore, the process is stationary in the wide sense both the mean and the ıtocorrelation depend only on time differences. Then all the statistics of the process e dependent only on time differences. Hence a Gaussian process which is stationary in e wide sense is also strictly stationary.

Another virtue of a Gaussian process is that when it is the input to a linear system, e output of the system is also a Gaussian process. A knowledge of the mean and ıtocorrelation function of the input process is sufficient to determine the autocorrelation nction and mean of the output or response process (as will be shown in the next ction). Hence the complete statistics of the output process can be calculated.

In white noise any two variables $x(t_1)$ and $x(t_2)$ are uncorrelated no matter how ose are t_1 and t_2. If the process is also Gaussian then zero correlation implies that ıe two variables are independent. Therefore, with Gaussian white noise any two ımples are independent irrespective of how close they may be in time. Clearly this is a ıathematical idealisation.

With any ergodic process it is necessary only to take one sample function in order to estimate the autocorrelation function. However, if the sample is for a time duration T then a random error is introduced into the estimation. It can be shown that for Gaussian white noise band-limited to $B\,Hz$ (i.e. power density spectrum uniform in $(-B/2,\ B/2)$ and zero outside) the mean-square error in estimating $R_x(\tau)$ is given by

$$\overline{\epsilon^2} \doteqdot \frac{1}{2TB}\left\{[R_x(\tau)]^2 + [R_x(0)]^2\right\} \tag{5.29}$$

It is left as an exercise to show that to reduce the mean-squared error to 1% of $[R_x(\tau)$ requires a sample time T of at least 0.13 sec for that τ such that $R_x(\tau) = 0.2\,R_x(0)$. It can also be shown that to reduce the mean-squared error to below 1% of $[R_x(0)]^2$ requires $T > 100/B$. Taking the effective width of the autocorrelation function to be $1/2B$ suggests that the length of the sample record T should be at least 200 times the width of the autocorrelation function. (The averaging required to estimate $R_x(\tau)$ can be achieved using a finite time-averaging filter.)

5.7 TRANSMISSION OF A RANDOM PROCESS THROUGH A LINEAR SYSTEM

When a random process is transmitted through a linear system it can be shown tha the power density spectrum of the output is related to that of the input via the transfer function of the system.

It may be shown that the relationship is

$$\Phi_y(\omega) = |H(\omega)|^2\,\Phi_x(\omega) \tag{5.30}$$

where $\Phi_x(\omega)$ is the power density spectrum of the input, $\Phi_y(\omega)$ is that of the output and $H(\omega)$ is the transfer function of the system. In the τ-domain we have the corresponding relationship

$$R_y(\tau) = h(-\tau) * h(\tau) * R_x(\tau) \tag{5.31}$$

where $R_x(\tau)$ is the autocorrelation of the input and $h(\tau)$ is the impulse response of th system. The operation $*$ represents the convolution operation; for two functions

$g(t)$ and $f(t)$, $g(t) * f(t) = \int_{-\infty}^{\infty} g(u)\,f(t-u)\,du$. Hence we may rewrite (5.31) as

$$R_y(\tau) = \int_{-\infty}^{\infty}\int_{-\infty}^{\infty} h(u)\,h(v)\,R_x(\tau - v + u)\,du\,dv \tag{5.32}$$

Note that an impulse response $h(t)$ is physically realisable only if $h(t) = 0$ for $t < 0$. It is left as an exercise to show that in such a case (5.31) becomes

$$R_y(\tau) = \int_0^\infty \int_0^\infty h(u)\,h(v)\,R_x(\tau - v + u)\,du\,dv \qquad (5.33)$$

Example 1

Let the input to a linear system be the random telegraph signal (Example 1, Section 5.4). The impulse response of the system is

$$h(t) = \begin{cases} Ae^{-\alpha t}, & t \geqslant 0 \\ 0, & t < 0 \end{cases} \qquad (5.34)$$

It is required to find the power density spectrum of the output. First, we find the Fourier Transform of $h(t)$, which is

$$H(\omega) = \int_{-\infty}^{\infty} Ae^{-\alpha t}\,e^{j\omega t}\,dt = \frac{A}{j\omega + \alpha}$$

From Example 1 of Section 5.4 we find that $\Phi_x(\omega) = \dfrac{4\nu a^2}{4\nu^2 + \omega^2}$ and using (5.30) we obtain the result

$$\Phi_y(\omega) = \frac{4A^2\nu a^2}{(4\nu^2 + \omega^2)(\alpha^2 + \omega^2)}$$

Example 2

The input to a linear system is a random process which has autocorrelation function

$$R_x(\tau) = \frac{A\omega_1}{\pi}\;\frac{\sin \omega_1 \tau}{\omega_1 \tau} \qquad (5.35)$$

The impulse response of the system is $h(t) = \dfrac{\omega_0}{\pi}\;\dfrac{\sin \omega_0 t}{\omega_0 t}$. Find the autocorrelation function of the output. It can be shown that

$$\Phi_x(\omega) = \begin{cases} A, & |\omega| < \omega_1 \\ 0, & \text{elsewhere} \end{cases}$$

Since $h(t)$ is of a similar form to $R_x(\tau)$ it follows that

$$H(\omega) = \begin{cases} 1, & |\omega| < \omega_0 \\ 0, & \text{elsewhere} \end{cases}$$

Then

$$|H(\omega)|^2 = \begin{cases} 1, & |\omega| < \omega_0 \\ 0, & \text{elsewhere} \end{cases}$$

There are two cases to consider

(i) $\omega_1 \geqslant \omega_0$. Here, $\Phi_y(\omega) = A|H(\omega)|^2$ (Why?)

Hence $R_y(\tau) = A \dfrac{\omega_0}{\pi} \dfrac{\sin \omega_0 \tau}{\omega_0 \tau}$

(ii) $\omega_1 < \omega_0$. Here $\Phi_y(\omega) = \Phi_x(\omega)$ (Why?)

Hence $R_y(\tau) = R_x(\tau) = A \dfrac{\omega_1}{\pi} \dfrac{\sin \omega_1 \tau}{\omega_1 \tau}$

Example 3

White noise is input to the linear system whose impulse response is $Be^{-\alpha t}$. The noise bandwidth for a linear system is defined to be $\omega_n = \int_0^\infty \left| \dfrac{H(\omega)}{H(0)} \right|^2 d\omega$. Find ω_n for the system.

Now $H(\omega) = \dfrac{B}{j\omega + \alpha}$ so that $|H(\omega)|^2 = \dfrac{B^2}{\omega^2 + \alpha^2}$. Then $|H(0)|^2 = \dfrac{B^2}{\alpha^2}$ and hence

$$\omega_n = \int_0^\infty \frac{\alpha^2}{\omega^2 + \alpha^2} d\omega = \alpha^2 \left[\frac{1}{\alpha} \tan^{-1} \left(\frac{\omega}{\alpha} \right) \right]_0^\infty = \frac{\pi\alpha}{2}$$

Note that if $\omega = \alpha$

$$|H(\alpha)|^2 = \frac{B^2}{2\alpha^2} = \tfrac{1}{2}|H(0)|^2$$

'iltering

If a signal is contaminated with noise it is necessary to design a filter which will emove as much of the noise as possible. The design of the optimum filter implies nding its transfer function. One criterion is that of minimising the mean-square error etween actual and desired outputs. The **Wiener-Hopf condition** requires that for such n optimum filter the cross-correlation of the desired output with the input shall equal he cross-correlation of the actual output with the input.

The condition may be written as

$$R_{xy}(\tau) = \int_{-\infty}^{\infty} h(u)\, R_x(\tau - u)\, du \tag{5.36}$$

f it is required that the filter be **physically realisable**, then

$$R_{xy}(\tau) = \int_{0}^{\infty} h(u)\, R_x(\tau - u)\, du \tag{5.37}$$

t can further be shown that the **mean-square error** for a physically realisable system is iven by

$$\overline{\epsilon^2(t)} = R_y(0) - \int_{0}^{\infty} R_{xy}(\tau)\, h(\tau)\, d\tau \tag{5.38}$$

roblems

. Show that if white noise of power density B is input to the system of Example 1, the autocorrelation function of the output is $\dfrac{BA^2}{2\alpha}\, e^{-\alpha|\tau|}$.

. A linear system has the transfer function

$$H(\omega) = \begin{cases} H, & \omega_1 < |\omega| < \omega_2 \\ 0, & \text{elsewhere} \end{cases}$$

The input is a random process with $\Phi_x(\omega) = \dfrac{A\omega_0^2}{\omega_0^2 + \omega^2}$. Show that if $\omega_2, \omega_1 >> \omega_0$

and $(\omega_2 - \omega_1) << \omega_2$ then

$$R_y(\tau) \simeq \frac{2A\omega_0^2 H^2}{\pi\tau\omega_1^2} [\cos \tfrac{1}{2}(\omega_1 + \omega_2)\tau . \sin \tfrac{1}{2}(\omega_2 - \omega_1)\tau]$$

(Hint: First show that $\Phi_x(\omega_1) \simeq A\omega_0^2/\omega_1^2$ and similarly for $\Phi_x(\omega_2)$. Then show that the ratio of the two functions $\simeq 1 - 2 \dfrac{(\omega_2 - \omega_1)}{\omega_2} \simeq 1$. Hence show that

$$\Phi_y(\omega) = |H(\omega)|^2 \frac{A\omega_0^2}{\omega_1^2} .)$$

3. A random voltage signal $x(t)$ for which $R_x(\tau) = K\delta(\tau) + K^2$, is input to an RC circuit for which $H(\omega) = \dfrac{1}{1 + j\omega}$. Show that $R_y(\tau) = \dfrac{K}{2} e^{-|\tau|} + K^2$ and that $\overline{y} = K$, $\overline{y^2} = \dfrac{K}{2} + K^2$

5.8 DISCRETE PARAMETER PROCESSES

Finally in this chapter we mention processes in which the independent variable t assumes only a discrete set of values. A continuous random process is effectively rendered discrete if it is sampled at specified values of t.

It is customary to define $n = t/T$ as the normalised variable for a periodically-sampled process. If the process is sampled at times ..... $-2T, -T, 0, T, 2T, 3T,$ ..... the sample values are denoted ....., $f(-2), f(-1), f(0), f(1), f(2), f(3),$ .....

The time average of the process $f(n)$ is

$$\overline{f(n)} = \lim_{N\to\infty} \frac{1}{2N+1} \sum_{r=-N}^{N} f(r) \tag{5.39}$$

and the ensemble average is given by

$$E[f(n)] = \lim_{N\to\infty} \frac{1}{N+1} \sum_{i=0}^{N} f_i(n) \tag{5.40}$$

where $f_i(n)$ are sample functions of the process.

If the basic continuous process is stationary then so will be the derived discrete process. If the continuous process is ergodic and is sampled periodically then the derived discrete process is ergodic so that $E[f(n)] = \overline{f(n)}$.

An example of a discrete process would be if the air temperature in a laboratory was measured every morning at 9 a.m.

For stationary and ergodic processes $f(n)$ and $g(n)$, the autocorrelation function $R_f(k)$, $k = 0, 1, 2, ...$ is defined as

$$R_f(k) = \overline{f(n)f(n+k)} \tag{5.41}$$

and the cross-correlation function of $f(n)$ and $g(n)$ is given by

$$R_{fg}(k) = \overline{f(n)g(n+k)} \tag{5.42}$$

Note that

$$R_{fg}(k) = R_{gf}(-k) \tag{5.43}$$

If $f(n)$ and $g(n)$ are independent then the autocorrelation function of the product $h(n) = f(n)g(n)$ is given by

$$R_h(k) = R_f(k) \,.\, R_g(k) \tag{5.44}$$

A system is called **time-invariant** if when an input $f(n)$ produces an output $g(n)$, the shifted input $f(n+i)$ produces the output $g(n+i)$, where i is an integer.

The **unit sample response** $h(n)$ is the output of a time-invariant system for an input given by

$$f_\circ(n) = \begin{cases} 1, & n = 0 \\ 0, & n \neq 0 \end{cases} \tag{5.45}$$

(c.f. unit impulse response for a continuous system.)

The output from an input $f(n)$ to a discrete linear system is

$$\left.\begin{aligned} g(n) &= \sum_{k=-\infty}^{\infty} f(k)\,h(n-k) = \sum_{k=-\infty}^{\infty} h(k)\,f(n-k) \\ \text{or} \quad g(n) &= f(n) * h(n) \qquad = h(n) * f(n) \end{aligned}\right\} \tag{5.46}$$

Example

The input to a discrete system is given by

$$f(n) = \begin{cases} a_1 , & n = 2 \\ a_2 , & n = 4 \\ 0 , & \text{otherwise} \end{cases}$$

Find the output $g(n)$.

Now using (5.44), and ignoring terms which vanish,

$$\begin{aligned} g(n) = \sum_{k=-\infty}^{\infty} f(k)\, h(n-k) &= f(2)\, h(n-2) + f(4)\, h(n-4) \\ &= a_1 h(n-2) \quad + \; a_2 h(n-4) \end{aligned}$$

In the place of the Fourier Transform for a continuous system we use the z-**transform**.

The z-transform of a discrete process $f(n)$ is given by

$$F(z) = \sum_{n=-\infty}^{\infty} f(n)\, z^{-n} \tag{5.47}$$

The transfer function $H(z)$ of a discrete system is the z-transform of its unit sample response. With the obvious notation,

$$G(z) = F(z)\, H(z) \tag{5.48}$$

The **spectrum** of a discrete process $f(n)$ is the z-transform where $z = e^{j\omega T}$. It is periodic in frequency with period $2\pi/T$. The spectrum can be used to study the behaviour of the processes passing through linear systems.

There are many other parallels between discrete and continuous processes which are beyond the scope of this book. Reference can be made to appropriate books in the bibliography.

Problems

1. Derive result (5.44).

2. A discrete process is obtained by successively throwing a fair coin. A head is denoted by 1, a tail by -1. Show that the autocorrelation function of the process is given by

$$R_f(k) = \begin{cases} 1, & k = 0 \\ 0, & k \neq 0 \end{cases}$$

3. The input to a linear system is $f(n) = \begin{cases} \alpha^n, & n \geqslant 0 \\ 0, & n < 0 \end{cases}$ and the unit sample response of the system is $h(n) = \begin{cases} \beta^n, & n \geqslant 0 \\ 0, & n < 0 \end{cases}$ where $|\alpha|, |\beta| < 1$. Show that the output response is given by

$$g(n) = \begin{cases} \dfrac{\beta^{n+1} - \alpha^{n+1}}{\beta - \alpha}, & n \geqslant 0 \\ 0, & n < 0 \end{cases}$$

4. Show that the z-transform of a delayed unit sample, i.e. $f(n) = \begin{cases} 1, & n = m \\ 0, & n \neq m \end{cases}$ is given by $F(z) = z^{-m}$.

5. The z-transform of a process is given by $F(z) = \dfrac{z^{-1}(1 + z^{-1})}{(1 - z^{-1})^3}$. Show that $f(3) = 9$.

(Hint: note that $f(3)$ is the coefficient of z^{-3} in the expansion of $F(z)$.)

6. Show that the transfer function of the system with unit sample response $h(n) = \alpha^{|n|}$, $|\alpha| < 1$ is given by

$$H(z) = \frac{1}{1 - \alpha z} + \frac{2z^{-1}}{1 - \alpha z^{-1}}$$

Chapter Six

Cartesian Tensors

6.1 INTRODUCTION

The Navier-Stokes equations of motion (in rectangular Cartesian co-ordinates) for a viscous, Newtonian fluid are

$$\left.\begin{aligned} F_x + \frac{1}{\rho}\left(\frac{\partial p_{xx}}{\partial x} + \frac{\partial p_{yx}}{\partial y} + \frac{\partial p_{zx}}{\partial z}\right) &= \frac{d}{dt} u_x \\ F_y + \frac{1}{\rho}\left(\frac{\partial p_{xy}}{\partial x} + \frac{\partial p_{yy}}{\partial y} + \frac{\partial p_{zy}}{\partial z}\right) &= \frac{d}{dt} u_y \\ F_z + \frac{1}{\rho}\left(\frac{\partial p_{xz}}{\partial x} + \frac{\partial p_{yz}}{\partial y} + \frac{\partial p_{zz}}{\partial z}\right) &= \frac{d}{dt} u_z \end{aligned}\right\} \tag{6.1}$$

where F_x, F_y and F_z are the components of the body forces/unit mass, ρ is the densit of the fluid, p_{xy} etc. are the pressure force components and u_x, u_y, u_z are the velocity components.

In vector notation the equations become

$$\mathbf{F} + \frac{1}{\rho}\boldsymbol{\nabla} p = \frac{d}{dt}\mathbf{u} \tag{6.2}$$

The tensor notation for the equations is

$$F_i + \frac{1}{\rho}\frac{\partial p_{ki}}{\partial x_k} = \frac{du_i}{dt} \tag{6.3}$$

Consider the components e_{ij} of strain in the deformation of an elastic solid. If the solid is at rest and body forces are negligible, the following equations hold:

$$\left.\begin{aligned} \frac{\partial e_{xx}}{\partial x} + \frac{\partial e_{xy}}{\partial y} + \frac{\partial e_{xz}}{\partial z} &= 0 \\ \frac{\partial e_{yx}}{\partial x} + \frac{\partial e_{yy}}{\partial y} + \frac{\partial e_{yz}}{\partial z} = 0 \qquad \frac{\partial e_{zx}}{\partial x} + \frac{\partial e_{zy}}{\partial y} + \frac{\partial e_{zz}}{\partial z} &= 0 \end{aligned}\right\} \tag{6.4}$$

In tensor notation the equations may be expressed as

$$\frac{\partial e_{ik}}{\partial x_k} = 0 \tag{6.5}$$

Tensor analysis is of value to the engineer in two ways. First, it allows complex physical relationships to be expressed in a compact way and simplifies the mechanics of the development of theory. Second, it enables the obtaining of relationships which do not depend directly on the coordinate axes chosen but which highlight the effects of the parameters and properties of the material or system in which the processes under study are taking place.

At first sight the notation of tensor analysis is somewhat complicated; the intention of this chapter is to provide a familiarity and from there a confidence with the notation which will enable the reader to study other texts and applications without trepidation.

General and Cartesian Tensors

In dealing with general co-ordinate transformations between arbitrary curvilinear co-ordinate systems (examples of curvilinear co-ordinate systems are cylindrical and spherical polar systems), the tensor quantities employed are known as **general tensors.** In this chapter we shall be concerned with only a small subset of such tensors, namely **Cartesian Tensors.** The transformations for such tensors are restricted to those transformations between two Cartesian frames of reference. Although this appears at first sight to be a major restriction, many areas of application can be examined by such tensors. Examples are stress and strain problems, transport processes in fluids, heat and mass transfer, electrical phenomena and field theory. We shall examine in the next section transformations between Cartesian co-ordinate frames of reference.

6.2 CHANGE OF AXES IN CARTESIAN CO-ORDINATES AND THE TRANSFORMATION MATRIX

It is essential to adopt a suitable notation in the development of the properties of transformations. Hence we use (x_1, x_2, x_3) for the co-ordinates in a Cartesian system of axes Ox_1, Ox_2, Ox_3 rather than (x, y, z) and Ox, Oy, Oz.

Consider a change of axes from Ox_1, Ox_2, Ox_3 to Ox_1', Ox_2', Ox_3' as shown in Figure 6.1.

The positions of the dashed axes relative to the undashed axes are established if we know the angles made by each dashed axis with each of the undashed axes. (These nine angles are not independent as will be shown later.) The cosines of these angles are denoted by λ_{11}, λ_{12}, ..., λ_{33} where λ_{pq} is the cosine of the angle between the axis Ox_p' and the axis Ox_q. (Notice that the order is dashed to undashed. For example

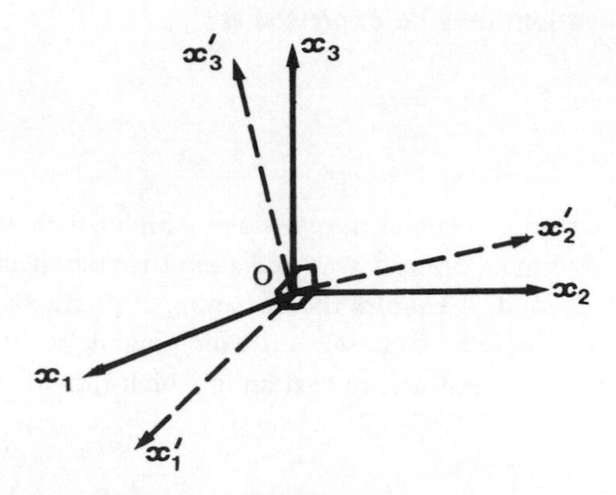

Figure 6.1

λ_{23} is the cosine of the angle between $0x_2'$ and $0x_3$ and λ_{31} is the cosine of the angle between $0x_3'$ and $0x_1$.) It is convenient to display these direction cosines in the form of a matrix

$$[\lambda] = \begin{bmatrix} \lambda_{11} & \lambda_{12} & \lambda_{13} \\ \lambda_{21} & \lambda_{22} & \lambda_{23} \\ \lambda_{31} & \lambda_{32} & \lambda_{33} \end{bmatrix} \tag{6.6}$$

$[\lambda]$ is called the **Transformation Matrix.**

As a simple example consider rotation about the $0x_3$ axis; see Figure 6.2. $0x_3'$ coincides with $0x_3$ in this case.

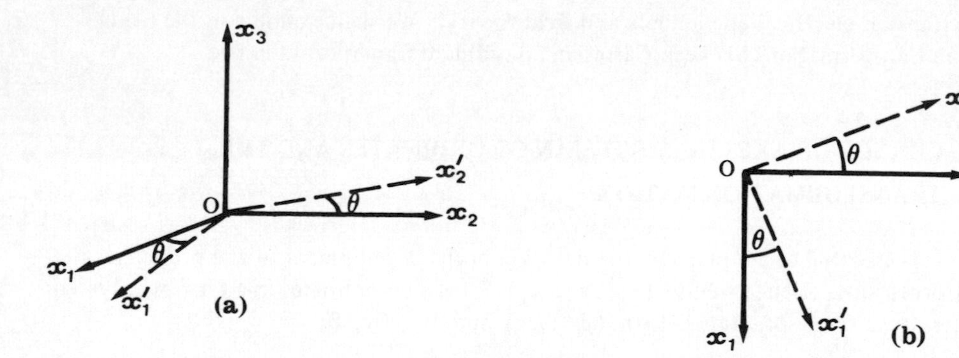

Figure 6.2

The angle $x_1' 0x_1 = x_2' 0x_2 = \theta$ say, defines the transformation. Then

$$\lambda_{11} = \cos(x_1' 0x_1) = \cos\theta,\ \lambda_{12} = \cos(x_1' 0x_2) = \cos(90^\circ - \theta) = \sin\theta,$$

$$\lambda_{13} = \cos 90^\circ = 0$$

Similarly

$$\lambda_{21} = \cos(x_2'0x_1) = \cos(90^\circ + \theta) = -\sin\theta,\ \lambda_{22} = \cos\theta,$$

$$\lambda_{23} = 0,\quad \lambda_{31} = 0,\quad \lambda_{32} = 0,\quad \lambda_{33} = 1$$

The Transformation Matrix is given by

$$[\lambda] = \begin{bmatrix} \cos\theta & \sin\theta & 0 \\ -\sin\theta & \cos\theta & 0 \\ 0 & 0 & 1 \end{bmatrix} \tag{6.7}$$

Effectively we are dealing with a transformation of axes in two dimensions only and we could express the rotation matrix simply as

$$[\lambda] = \begin{bmatrix} \cos\theta & \sin\theta \\ -\sin\theta & \cos\theta \end{bmatrix}$$

Problems

1. The angles between a dashed and undashed set of co-ordinates are given in the table below. Find the transformation matrix.

	x_1	x_2	x_3
x_1'	135°	60°	120°
x_2'	90°	45°	45°
x_3'	45°	60°	120°

2. Find the determinant of the transformation matrix in Problem 1.

3. If the transformation matrix in Problem 1 is denoted by $[\lambda]$, show that $[\lambda][\lambda]^T = I$. Deduce the value of $[\lambda]^{-1}$ and hence write down the transformation matrix for the transformation dashed to undashed axes.

4. A table of direction cosines is given by

	x_1	x_2	x_3
x_1'	3/5	–4/5	0
x_2'	0	0	1
x_3'	–4/5	–3/5	0

Show that the associated transformation has determinant 1 and is **orthogonal**, that is $[\lambda]^{-1} = [\lambda]^T$.

5. The transformation between co-ordinates x_1, x_2, x_3 and x_1', x_2', x_3' is given by

$$x_1' = \frac{1}{3}(2x_1 + 2x_2 - x_3)$$

$$x_2' = \frac{1}{3}(2x_1 - x_2 + 2x_3)$$

$$x_3' = \frac{1}{3}(-x_1 + 2x_2 + 2x_3)$$

Write down the transformation matrix and verify that this is orthogonal.

6. A transformation matrix is given by

$$[\lambda] = \begin{bmatrix} 1/\sqrt{3} & 1/\sqrt{3} & 1/\sqrt{3} \\ 0 & 1/\sqrt{2} & -1/\sqrt{2} \\ -2/\sqrt{6} & 1/\sqrt{6} & 1/\sqrt{6} \end{bmatrix}$$

Verify that $[\lambda]^T = [\lambda]^{-1}$.

7. Find the transformation matrix for the transformation from axes $0x_1$, $0x_2$, $0x_3$ to $0x_1'$, $0x_2'$, $0x_3'$ shown below.

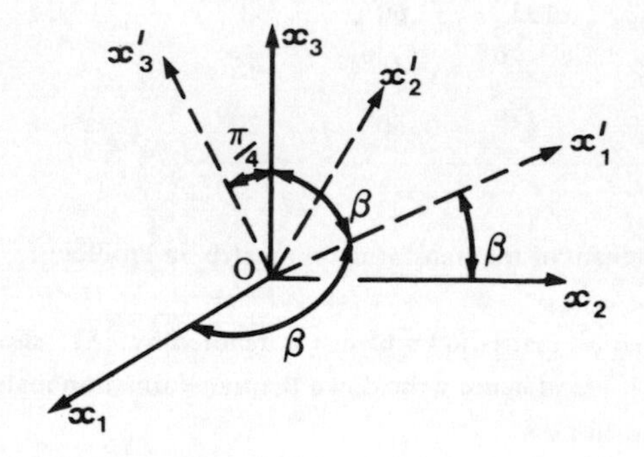

.3 SCALAR AND VECTOR QUANTITIES

A **scalar** quantity is one which can be specified in any co-ordinate system in three-imensional space by **one** component. The size of this component is quite independent f any system of axes taken and so we say that a scalar is **invariant** under a transformation f axes.

It is commonplace to describe a **vector** quantity as being one which possesses both agnitude and direction and which therefore may be represented by a directed line-gment in three-dimensional space. Let us assume that with respect to the system of xes Ox_1, Ox_2, Ox_3 the vector **A** has components A_1, A_2, A_3. We ask what will e form of **A** be when referred to different axes Ox_1', Ox_2', Ox_3'. It will almost rtainly have different components in the new system of axes but it must represent the me vector quantity **A**.

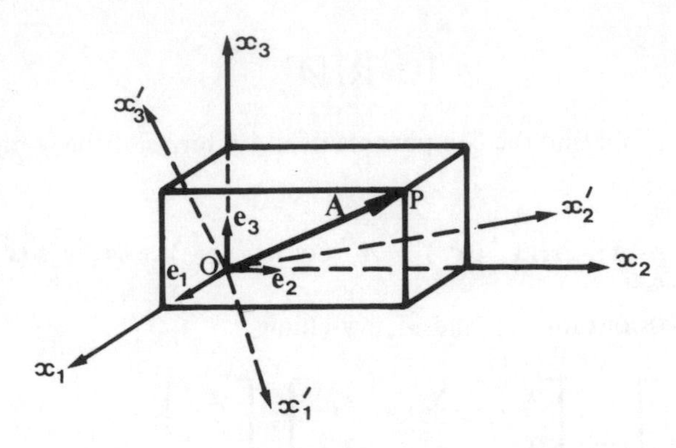

Figure 6.3

In Figure 6.3, OP represents the vector **A** and if $\mathbf{e}_1$, $\mathbf{e}_2$, $\mathbf{e}_3$ are unit vectors in e directions of the co-ordinate axes Ox_1, Ox_2, Ox_3 then

$$\mathbf{A} = A_1\mathbf{e}_1 + A_2\mathbf{e}_2 + A_3\mathbf{e}_3 \tag{6.8}$$

nd similarly, for axes Ox_1', Ox_2', Ox_3' we may write

$$\mathbf{A} = A_1'\mathbf{e}_1' + A_2'\mathbf{e}_2' + A_3'\mathbf{e}_3' \tag{6.9}$$

$'$, $\mathbf{e}_2'$, $\mathbf{e}_3'$ being unit vectors in the directions of the new axes.

The component A_1' will be made up of the resolved contributions from the omponents A_1, A_2, A_3 giving

$$A_1' = A_1 \cos(x_1'Ox_1) + A_2 \cos(x_1'Ox_2) + A_3 \cos(x_1'Ox_3)$$

hich may be written

$$A_1' = \lambda_{11}A_1 + \lambda_{12}A_2 + \lambda_{13}A_3$$

Similarly we find that

$$A_2' = \lambda_{21}A_1 + \lambda_{22}A_2 + \lambda_{23}A_3$$
$$A_3' = \lambda_{31}A_1 + \lambda_{32}A_2 + \lambda_{33}A_3$$

The three equations may be written in matrix form either as

$$\begin{bmatrix} A_1' \\ A_2' \\ A_3' \end{bmatrix} = \begin{bmatrix} \lambda_{11} & \lambda_{12} & \lambda_{13} \\ \lambda_{21} & \lambda_{22} & \lambda_{23} \\ \lambda_{31} & \lambda_{32} & \lambda_{33} \end{bmatrix} \begin{bmatrix} A_1 \\ A_2 \\ A_3 \end{bmatrix} \tag{6.10}$$

or as

$$[A'] = [\lambda][A] \tag{6.11}$$

Equally well we could find the components of **A** in terms of the components of **A**′. W then obtain

$$A_1 = A_1' \cos(x_1' 0x_1) + A_2' \cos(x_2' 0x_1) + A_3' \cos(x_3' 0x_1)$$

and similar expressions for A_2 and A_3, yielding

$$\begin{bmatrix} A_1 \\ A_2 \\ A_3 \end{bmatrix} = \begin{bmatrix} \lambda_{11} & \lambda_{21} & \lambda_{31} \\ \lambda_{12} & \lambda_{22} & \lambda_{32} \\ \lambda_{13} & \lambda_{23} & \lambda_{33} \end{bmatrix} \begin{bmatrix} A_1' \\ A_2' \\ A_3' \end{bmatrix} \tag{6.12}$$

or

$$[A] = [\lambda]^{\mathrm{T}}[A'] \tag{6.13}$$

Equations (6.10) or (6.11) give the **transformation law** for vector components. Furthermore any quantity having components which satisfy equation (6.10) *must* be a vector quantity. Equations (6.12) or (6.13) give the **inverse transformation law.** From equations (6.11) and (6.13) we can discover some further properties of the transformatio (or λ) matrix. Substituting for $[A]$ from (6.13) into (6.11),

$$[A'] = [\lambda][\lambda]^{\mathrm{T}}[A']$$

Therefore

$$[\lambda][\lambda]^{\mathrm{T}} = \mathbf{I} \tag{6.14}$$

or

$$\begin{bmatrix} \lambda_{11} & \lambda_{12} & \lambda_{13} \\ \lambda_{21} & \lambda_{22} & \lambda_{23} \\ \lambda_{31} & \lambda_{32} & \lambda_{33} \end{bmatrix} \begin{bmatrix} \lambda_{11} & \lambda_{21} & \lambda_{31} \\ \lambda_{12} & \lambda_{22} & \lambda_{32} \\ \lambda_{13} & \lambda_{23} & \lambda_{33} \end{bmatrix} = \begin{bmatrix} 1 & 0 & 0 \\ 0 & 1 & 0 \\ 0 & 0 & 1 \end{bmatrix}$$

› that

$$\left.\begin{aligned} \lambda_{11}^2 + \lambda_{12}^2 + \lambda_{13}^2 &= 1 \\ \lambda_{21}^2 + \lambda_{22}^2 + \lambda_{23}^2 &= 1 \\ \lambda_{31}^2 + \lambda_{32}^2 + \lambda_{33}^2 &= 1 \\ \lambda_{11}\lambda_{21} + \lambda_{12}\lambda_{22} + \lambda_{13}\lambda_{23} &= 0 \\ \lambda_{21}\lambda_{31} + \lambda_{22}\lambda_{32} + \lambda_{23}\lambda_{33} &= 0 \\ \lambda_{31}\lambda_{11} + \lambda_{32}\lambda_{12} + \lambda_{33}\lambda_{13} &= 0 \end{aligned}\right\} \qquad (6.15)$$

'e see therefore that the 9 quantities λ_{pq} are not independent and that only 3 f them are independent since they have to satisfy the six equations (6.15). It is worth oting that the first three equations of (6.15) express the fact that the λ's are direction osines *not* direction ratios. Notice further that since

$$[\lambda]\,[\lambda]^T = \mathbf{I}$$

ıen

$$[\lambda]^T = [\lambda]^{-1} \qquad (6.16)$$

'he $[\lambda]$ matrix is an *orthogonal* matrix. Its determinant is equal to 1. (Why?)

he Summation Convention

In order to explain the summation convention and illustrate its usefulness let us xamine some of the equations already obtained. For example, returning to (6.8):

$$\mathbf{A} = A_1\mathbf{e}_1 + A_2\mathbf{e}_2 + A_3\mathbf{e}_3 \qquad (6.8)$$

'e could equally well write

$$\mathbf{A} = \sum_{i=1}^{3} A_i\,\mathbf{e}_i$$

'he summation convention goes further than this and omits the Σ sign, it being assumed ıat if two suffices are *repeated* in a term then summation is implied. Thus we write 6.8) in the very economical form

$$\mathbf{A} = A_i \mathbf{e}_i$$

Consider now equations (6.10) which were written in the form

$$A_1' = \lambda_{11} A_1 + \lambda_{12} A_2 + \lambda_{13} A_3$$

$$A_2' = \lambda_{21} A_1 + \lambda_{22} A_2 + \lambda_{23} A_3$$

$$A_3' = \lambda_{31} A_1 + \lambda_{32} A_2 + \lambda_{33} A_3$$

The left-hand side terms are A_i' where i can take the values 1, 2 and 3; similar considerations apply to the right-hand sides and we could write the set of equations as

$$A_i' = \lambda_{i1} A_1 + \lambda_{i2} A_2 + \lambda_{i3} A_3 \tag{6.17}$$

If i has the value 1 we recover the first equation of the set while if i has the values 2 and 3 we recover the 2nd and 3rd equations. Equation (6.17) therefore represents eac of the 3 equations (6.10). We can take one step further and write this in the form

$$A_i' = \sum_{j=1}^{3} \lambda_{ij} A_j$$

Remembering the summation convention that summation is implied if a suffix is repeate in one term we see that this becomes

$$A_i' = \lambda_{ij} A_j \tag{6.18}$$

it being understood that j takes the values 1, 2, 3 and that the ensuing terms are summed.

Finally compare the forms of equations (6.11) and (6.18) and note the similar economy of writing. Equations (6.13) would now become

$$A_i = \lambda_{ji} A_j' \tag{6.19}$$

Kronecker Delta

We see even more the effectiveness of the notation if equations (6.15) are written in summation convention form. All six equations can be written

$$\lambda_{ik} \lambda_{jk} = 1 \text{ if } i \text{ and } j \text{ are the same}$$

$$\lambda_{ik} \lambda_{jk} = 0 \text{ if } i \text{ and } j \text{ are different}$$

We now introduce the **Kronecker Delta** δ_{ij} which is defined by

$$\delta_{ij} = 1 \quad \text{if} \quad i \text{ and } j \text{ are the same}$$
$$\delta_{ij} = 0 \quad \text{if} \quad i \text{ and } j \text{ are different}$$

Then equations (6.15) take the form

$$\lambda_{ik}\lambda_{jk} = \delta_{ij} \tag{6.20}$$

You should now take $i = 1$, $j = 1$ and sum the terms for $k = 1, 2$ and 3 to recover the first of equations (6.15). Then take $i = 2$, $j = 3$ and sum the terms for $k = 1, 2$ and 3 to recover the fifth of those equations. Convince yourself that equations (6.15) and (6.20) are really the same.

Many other examples will occur in the following pages of the value of the summation convention. To emphasise the meaning and reinforce understanding two examples are given now.

i) The equations of motion for an inviscid fluid are

$$\frac{\partial u}{\partial t} + u\frac{\partial u}{\partial x} + v\frac{\partial u}{\partial y} + w\frac{\partial u}{\partial z} = -\frac{1}{\rho}\frac{\partial p}{\partial x}$$

$$\frac{\partial v}{\partial t} + u\frac{\partial v}{\partial x} + v\frac{\partial v}{\partial y} + w\frac{\partial v}{\partial z} = -\frac{1}{\rho}\frac{\partial p}{\partial y}$$

$$\frac{\partial w}{\partial t} + u\frac{\partial w}{\partial x} + v\frac{\partial w}{\partial y} + w\frac{\partial w}{\partial z} = -\frac{1}{\rho}\frac{\partial p}{\partial z}$$

where (u, v, w) are the velocity components, p is the pressure in the fluid and ρ is the density.

Remember that we now use x_1, x_2, x_3 instead of co-ordinates x, y, z and (u_1, u_2, u_3) will represent the velocity components. The three equations of motion can all be represented by the one equation

$$\frac{\partial u_i}{\partial t} + u_j\frac{\partial u_i}{\partial x_j} = -\frac{1}{\rho}\frac{\partial p}{\partial x_i} \tag{6.21}$$

Again, take $i = 1, 2$ and 3 in turn and sum the terms for the repeated suffix $j = 1$, 2 and 3 to recover the full equations.

ii) Newton's law of viscosity (or Stokes' relationship) for a viscous fluid is

$$\sigma_{ij} = -p\delta_{ij} + \lambda\delta_{ij}\frac{\partial u_k}{\partial x_k} + \mu\left(\frac{\partial u_i}{\partial x_j} + \frac{\partial u_j}{\partial x_i}\right)$$

where σ_{ij} are shear stress components and λ, μ are the bulk and shear coefficients of viscosity. Take $i = 1, 2, 3$ in turn and values of $j = 1, 2, 3$ in turn, summing over the repeated suffix to obtain the nine equations involved.

A repeated suffix is often called a **dummy suffix** since it can be replaced by any other symbol (except those occurring in the same equation) and the equation will not be altered in any way.

Further results on vectors

We have seen that a vector **A** with components A_i transforms under a change of axes according to the law (6.18).

Moreover, if a quantity obeys the above law under transformation then it is a vector.

The inverse transformation law is given by (6.19).

Many results already known for vectors follow from these definitions of vector quantities. It is left to you to prove that :

(i) If a vector is multiplied by a scalar m, then the result is a vector quantity.

(ii) The sum of two vectors is also a vector.

The **gradient** of a scalar Φ, defined by $\left(\frac{\partial\Phi}{\partial x_1}, \frac{\partial\Phi}{\partial x_2}, \frac{\partial\Phi}{\partial x_3}\right) \equiv \frac{\partial\Phi}{\partial x_i}$, is a valid vector quantity which obeys the transformation law (6.18). To see this let $B_i = \frac{\partial\Phi}{\partial x_i}$

then $B_i = \frac{\partial\Phi}{\partial x_j'}\frac{\partial x_j'}{\partial x_i}$. (Notice the implied summation.)

But the co-ordinates x_i are transformed by the law $x_j' = \lambda_{ji}\, x_i$ and so $\frac{\partial x_j'}{\partial x_i} = \lambda_{ji}$. Thus $B_i = \lambda_{ji}\frac{\partial\Phi}{\partial x_j'}$.

Now $\frac{\partial\Phi}{\partial x_j'}$ represents the gradient in the dashed axes and is therefore B_j'. Finally we obtain $B_i = \lambda_{ji}\, B_j'$ which is of exactly the same form as equation (6.19).

B_i is a vector quantity; that is, gradient is a vector quantity. This means, of course, that the gradient represents the same physical quantity whatever the system of axes chosen.

oblems

Write each of the following in summation convention form:

(a) $d\phi = \frac{\partial \phi}{\partial x_1} dx_1 + \frac{\partial \phi}{\partial x_2} dx_2 + \frac{\partial \phi}{\partial x_3} dx_3$

(b) $a_1 x_1 x_3 + a_2 x_2 x_3 + a_3 x_3 x_3$

(c) $\sigma'_{11} = \lambda_{11}\lambda_{11}\sigma_{11} + \lambda_{11}\lambda_{12}\sigma_{12} + \lambda_{11}\lambda_{13}\sigma_{13} + \lambda_{12}\lambda_{11}\sigma_{21} + \lambda_{12}\lambda_{12}\sigma_{22}$
$+ \lambda_{12}\lambda_{13}\sigma_{23} + \lambda_{13}\lambda_{11}\sigma_{31} + \lambda_{13}\lambda_{12}\sigma_{32} + \lambda_{13}\lambda_{13}\sigma_{33}$

(d)

$$\begin{cases} \lambda_{11}\lambda_{11} + \lambda_{12}\lambda_{21} + \lambda_{13}\lambda_{31} = 1, & \lambda_{21}\lambda_{11} + \lambda_{22}\lambda_{21} + \lambda_{23}\lambda_{31} = 0 \\ \lambda_{11}\lambda_{12} + \lambda_{12}\lambda_{22} + \lambda_{13}\lambda_{32} = 0, & \lambda_{21}\lambda_{12} + \lambda_{22}\lambda_{22} + \lambda_{23}\lambda_{32} = 1 \\ \lambda_{11}\lambda_{13} + \lambda_{12}\lambda_{23} + \lambda_{13}\lambda_{33} = 0, & \lambda_{21}\lambda_{13} + \lambda_{22}\lambda_{23} + \lambda_{23}\lambda_{33} = 0 \\ \lambda_{31}\lambda_{11} + \lambda_{32}\lambda_{21} + \lambda_{33}\lambda_{31} = 0 & \\ \lambda_{31}\lambda_{12} + \lambda_{32}\lambda_{22} + \lambda_{33}\lambda_{32} = 0 & \\ \lambda_{31}\lambda_{13} + \lambda_{32}\lambda_{23} + \lambda_{33}\lambda_{33} = 1 & \end{cases}$$

Write out the following in full :

(a) $a_i x_i x_j$ (b) $a_{pq} x_q = b_p$ (c) $u'_{ik} = \lambda_{ij} \lambda_{kl} u_{jl}$ (d) $\lambda_{ij} \lambda_{jk} = \delta_{ik}$

(e) $A_{im} = B_{kk} \delta_{im} - B_{im}$ (f) $\frac{1}{2}\left(\frac{\partial u_i}{\partial x_k} + \frac{\partial u_k}{\partial x_i}\right) = e_{ik}$

(g) $a_{ik} = \frac{1}{2}\left(\frac{\partial u_i}{\partial x_k} - \frac{\partial u_k}{\partial x_i}\right)$

Let $P^o(x^o)$ and $P(x)$ be two points in the x-frame determining the vector **A** with components $A_i = x_i - x_i^o$. Under the transformation $x_i' = \alpha_{ij} x_j$ from the x- to the x'-frame, **A** transforms into the vector **A'** with components $A_i' = x_i' - x_i^{o\prime}$. Prove that $A_i' = \alpha_{ij} A_j$. Hence prove that

(i) Two vectors **A, B** with equal components transform into two vectors **A', B'** also having equal components;

(ii) Two parallel vectors transform into two parallel vectors.

A vector referred to axes $0x_1, 0x_2, 0x_3$ has components $(-1, 0, 3)$. Find its components referred to the axes $0x_1', 0x_2', 0x_3'$ given in Problem 7 on page 206.
A vector A referred to axes $0x_1, 0x_2, 0x_3$ has components $[2, 1, -2]$. Find its components when referred to axes $0x_1', 0x_2', 0x_3'$ of Problem 5on page 206.

6.4 THE SECOND-ORDER TENSOR

Scalars and vectors are both special cases of a more general object, the **tensor** of order n whose specification needs 3^n numbers called **components**. In fact a scalar is a tensor of order zero having $3^0 = 1$ component and a vector is a tensor of order one having $3^1 = 3$ components.

Now consider the tensor of order two which has $3^2 = 9$ components and examine how it is changed under a transformation of axes. Many quantities do require 9 components for their specifications; the examples examined in this section are stress and strain tensors.

Since a second-order tensor has 9 components, it can be written in the form of a 3 x 3 matrix $\begin{bmatrix} A_{11} & A_{12} & A_{13} \\ A_{21} & A_{22} & A_{23} \\ A_{31} & A_{32} & A_{33} \end{bmatrix}$ which in suffix notation is represented by A_{ij} (i and j both taking the values 1, 2, 3).

A second-order tensor is **defined** as a quantity which transforms from one set of co-ordinate axes to another according to the law

$$A'_{kl} = \lambda_{ki}\lambda_{lj} A_{ij} \tag{6.22}$$

Written in matrix form this law is

$$\begin{bmatrix} A'_{11} & A'_{12} & A'_{13} \\ A'_{21} & A'_{22} & A'_{23} \\ A'_{31} & A'_{32} & A'_{33} \end{bmatrix} = \begin{bmatrix} \lambda_{11} & \lambda_{12} & \lambda_{13} \\ \lambda_{21} & \lambda_{22} & \lambda_{23} \\ \lambda_{31} & \lambda_{32} & \lambda_{33} \end{bmatrix} \begin{bmatrix} A_{11} & A_{12} & A_{13} \\ A_{21} & A_{22} & A_{23} \\ A_{31} & A_{32} & A_{33} \end{bmatrix} \begin{bmatrix} \lambda_{11} & \lambda_{21} & \lambda_{31} \\ \lambda_{12} & \lambda_{22} & \lambda_{32} \\ \lambda_{13} & \lambda_{23} & \lambda_{33} \end{bmatrix}$$

or

$$[A'] = [\lambda]\,[A]\,[\lambda]^{\mathrm{T}} \tag{6.23}$$

The inverse transformation is probably easiest seen at this stage from this last equation for, remembering that $[\lambda]\,[\lambda]^{\mathrm{T}} = \mathrm{I} = [\lambda]^{\mathrm{T}}[\lambda]$, then we obtain $[\lambda]^{\mathrm{T}}[A']\,[\lambda] = [\lambda]^{\mathrm{T}}[\lambda]\,[A]\,[\lambda]^{\mathrm{T}}[\lambda] = \mathrm{I}[A]\mathrm{I} = [A]$. Therefore

$$[A] = [\lambda]^{\mathrm{T}}[A']\,[\lambda] \tag{6.24}$$

which can be written in the form

$$A_{ij} = \lambda_{ki}\,\lambda_{lj}\,A'_{kl} \tag{6.25}$$

Notice the corresponding positions of the suffices in equation (6.22) $A'_{kl} = \lambda_{ki}\lambda_{lj}A_{ij}$ and equation (6.25) $A_{ij} = \lambda_{ki}\lambda_{lj}A'_{kl}$.

Before considering examples of such quantities it is necessary to develop their algebra. The following properties hold.

(a) *If two second-order tensors are added the result is a second-order tensor.* Suppose the tensors are A_{ij} and B_{ij}, then $A'_{kl} = \lambda_{ki}\lambda_{lj}A_{ij}$ and $B'_{kl} = \lambda_{ki}\lambda_{lj}B_{ij}$. Therefore $A'_{kl} + B'_{kl} = \lambda_{ki}\lambda_{lj}A_{ij} + \lambda_{ki}\lambda_{lj}B_{ij} = \lambda_{ki}\lambda_{lj}(A_{ij} + B_{ij})$ and it is seen that the sum transforms according to the rule for second-order tensors. Hence it is a second-order tensor.

(b) *If m is a scalar and A_{ij} is a second-order tensor then $m \times A_{ij}$ is a second-order tensor.* (Show this.)

(c) *$A_{ij} + A_{ji}$ is a second-order tensor which is* **symmetric**. Reverting to matrix form, $A_{ij} + A_{ji}$ is

$$\begin{bmatrix} A_{11} & A_{12} & A_{13} \\ A_{21} & A_{22} & A_{23} \\ A_{31} & A_{32} & A_{33} \end{bmatrix} + \begin{bmatrix} A_{11} & A_{21} & A_{31} \\ A_{12} & A_{22} & A_{32} \\ A_{13} & A_{23} & A_{33} \end{bmatrix} = \begin{bmatrix} 2A_{11} & A_{12}+A_{21} & A_{13}+A_{31} \\ A_{21}+A_{12} & 2A_{22} & A_{23}+A_{32} \\ A_{31}+A_{13} & A_{32}+A_{23} & 2A_{33} \end{bmatrix}$$

and it is seen that this is a **symmetric** matrix. We define a **symmetric tensor** as one for which $B_{ij} = B_{ji}$.

(d) *$A_{ij} - A_{ji}$ is a second-order tensor which is* **antisymmetric**. Now

$$A_{ij} - A_{ji} = \begin{bmatrix} 0 & A_{12} - A_{21} & A_{13} - A_{31} \\ -(A_{12} - A_{21}) & 0 & A_{23} - A_{32} \\ -(A_{13} - A_{31}) & -(A_{23} - A_{32}) & 0 \end{bmatrix}$$

which is an antisymmetric matrix. We define an **antisymmetric tensor** as one for which $B_{ij} = -B_{ji}$.

(e) *Any second-order tensor can be expressed as the sum of a symmetric and an antisymmetric tensor.*

It is clear that $A_{ij} = \frac{1}{2}(A_{ij} + A_{ji}) + \frac{1}{2}(A_{ij} - A_{ji})$ which gives the result.

(f) *The unit matrix* $\begin{bmatrix} 1 & 0 & 0 \\ 0 & 1 & 0 \\ 0 & 0 & 1 \end{bmatrix}$ *can be written as δ_{ij}; it is a second order tensor.*

This can easily be shown using equations (6.23). (It would be instructive for you to do this.) It transforms into itself and we show this from equation (6.22).

The R.H.S. of (6.22) becomes $\lambda_{ki}\lambda_{lj}\delta_{ij}$. But $\delta_{ij} = 0$ if $i \neq j$ and $\delta_{ij} = 1$ if $i = j$. Thus the R.H.S. is $\lambda_{ki}\lambda_{li}$. Using equation (6.20) this is seen to be simply δ_{kl}. Therefore we have the result $\delta'_{kl} = \delta_{kl} = \lambda_{ki}\lambda_{lj}\delta_{ij}$.

The unit matrix is defined as the **unit tensor** of order 2 and has the same form no matter what axes are chosen.

Examples of second-order tensors

(1) **Stress Tensor:** σ_{ij}

Stress is defined as the force per unit area acting on a surface. The notation and sign conventions are illustrated in Figure 6.4.

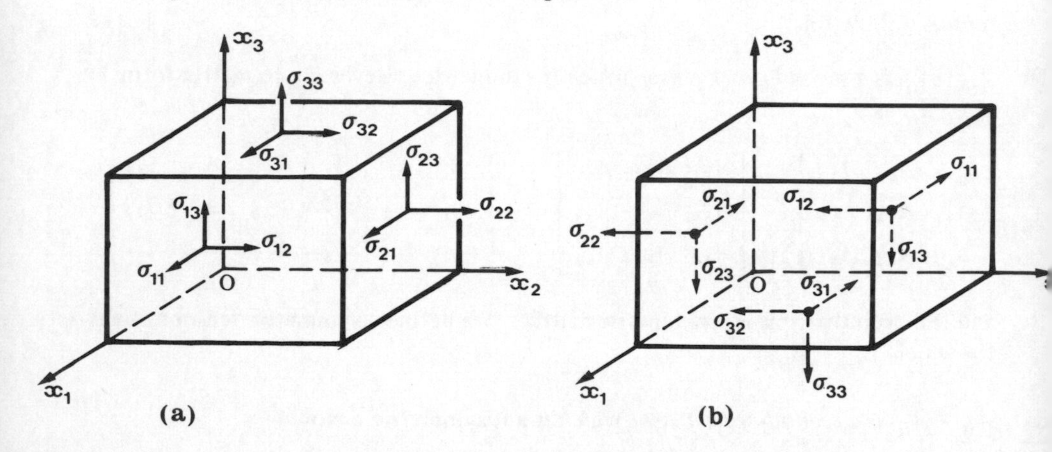

Figure 6.4

Figure 6.4(a) shows the stresses σ_{ij} on the *positive faces* (i.e. those with an emergent normal in the direction of the co-ordinate axes). The component σ_{ij} acts in the direction of the jth co-ordinate axis and on the plane whose outward normal is parallel to the ith co-ordinate axis. Thus, for example, σ_{12} acts on the face normal to x_1 in the direction of x_2. Figure 6.4(b) shows the stresses on the negative faces (i.e. those with an emergent normal in the opposite direction to the co-ordinate axes) and here σ_{ij} acts in the negative j direction. The components σ_{11}, σ_{22}, σ_{33} which are **perpendicular** to the faces are called **norma** or **direct stresses** while those acting tangential to the **plane faces**, viz., σ_{12}, σ_{13}, σ_{23}, σ_{21}, σ_{31}, σ_{32} are called **shear stresses.**

How do the stress components change under **transformation** of axes? The *continuum* means a volume of space which is filled by the material being studied (solid, liquid or gas) without reference to the molecular structure. Consider a small tetrahedron of the continuum, shown in Figure 6.5(a) with its slant face perpendicular to x'_1.

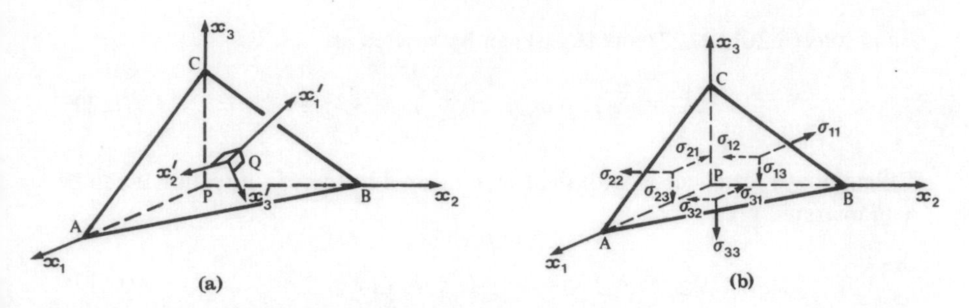

Figure 6.5

Let the area of the face ABC be S. The areas of the faces APB, PBC and PCA are then $\lambda_{13} S$, $\lambda_{11} S$ and $\lambda_{12} S$ respectively, since the cosines of the angles between x_1' and the normals to the faces are λ_{13}, λ_{11} and λ_{12} respectively.

There will be a normal stress σ'_{11} acting on the slant face and hence the force at Q in the direction of x_1' is $\sigma'_{11} S$. If the tetrahedron is in equilibrium, this force must equal the aggregate of resolved forces acting in the opposite direction on the faces ABP, BPC, CPA, provided by the stresses shown in Figure 6.5(b).

Take for example the stress σ_{11}. This provides a force $\sigma_{11}(\lambda_{11} S)$ in the negative x_1 direction which is resolved in the direction of x_1' to give $\sigma_{11}(\lambda_{11} S)\lambda_{11}$ and similarly the stress σ_{12} gives rise to a resolved force $\sigma_{12}(\lambda_{11} S)\lambda_{12}$ in the x_1' direction. Repeating this procedure for all the stresses we obtain the result

$$\begin{aligned} S\sigma'_{11} = {} & \sigma_{11}(\lambda_{11} S)\lambda_{11} + \sigma_{12}(\lambda_{11} S)\lambda_{12} + \sigma_{13}(\lambda_{11} S)\lambda_{13} \\ & + \sigma_{21}(\lambda_{12} S)\lambda_{11} + \sigma_{22}(\lambda_{12} S)\lambda_{12} + \sigma_{23}(\lambda_{12} S)\lambda_{13} \\ & + \sigma_{31}(\lambda_{13} S)\lambda_{11} + \sigma_{32}(\lambda_{13} S)\lambda_{12} + \sigma_{33}(\lambda_{13} S)\lambda_{13} \end{aligned}$$

Cancelling S throughout gives

$$\begin{aligned} \sigma'_{11} = {} & \lambda_{11}\lambda_{11}\sigma_{11} + \lambda_{11}\lambda_{12}\sigma_{12} + \lambda_{11}\lambda_{13}\sigma_{13} \\ & + \lambda_{12}\lambda_{11}\sigma_{21} + \lambda_{12}\lambda_{12}\sigma_{22} + \lambda_{12}\lambda_{13}\sigma_{23} \\ & + \lambda_{13}\lambda_{11}\sigma_{31} + \lambda_{13}\lambda_{12}\sigma_{32} + \lambda_{13}\lambda_{13}\sigma_{33} \end{aligned}$$

which in compressed notation is simply

$$\sigma'_{11} = \lambda_{1i}\lambda_{1j}\sigma_{ij} \tag{6.26}$$

Similarly, resolving in the x_2' direction and x_3' direction respectively we find

$$\sigma'_{12} = \lambda_{1i}\lambda_{2j}\,\sigma_{ij} \tag{6.27}$$

$$\sigma'_{13} = \lambda_{1i}\,\lambda_{3j}\,\sigma_{ij} \tag{6.28}$$

Equations (6.26), (6.27) and (6.28) can be written as

$$\sigma'_{1l} = \lambda_{1i}\,\lambda_{lj}\,\sigma_{ij}, \qquad l = 1, 2, 3 \tag{6.29}$$

Taking now a tetrahedron with slant face normal to the x'_2 direction it can be demonstrated that

$$\sigma'_{2l} = \lambda_{2i}\,\lambda_{lj}\,\sigma_{ij}, \qquad l = 1, 2, 3 \tag{6.30}$$

and similarly we could obtain

$$\sigma'_{3l} = \lambda_{3i}\,\lambda_{lj}\,\sigma_{ij}, \qquad l = 1, 2, 3 \tag{6.31}$$

Equations (6.29), (6.30) and (6.31) give *nine* equations which may all be represented by the *one* relationship

$$\sigma'_{kl} = \lambda_{ki}\,\lambda_{lj}\,\sigma_{ij} \tag{6.32}$$

However, this is identical to the relationship (6.22) which defines a second-order tensor quantity and so the stress-tensor is a second-order tensor.
The inverse relationship is of course

$$\sigma_{ij} = \lambda_{ki}\,\lambda_{lj}\,\sigma'_{kl} \tag{6.33}$$

The equivalent matrix representation of equation (6.32) is

$$[\sigma'] = [\lambda]\,[\sigma]\,[\lambda]^{\mathrm{T}} \tag{6.34}$$

(2) Plane Stress

The special case of two dimensions is worth considering since some of the results are probably already known to you.

Consider now plane stress as depicted in Figure 6.6. There the four components o the stress are shown. It is easily verified by taking moments that $\sigma_{12} = \sigma_{21}$. The stress tensor can be represented by the 2 x 2 matrix $\begin{bmatrix} \sigma_{11} & \sigma_{12} \\ \sigma_{21} & \sigma_{22} \end{bmatrix}$ which, sinc $\sigma_{12} = \sigma_{21}$, is symmetric. (This is easily shown to be equally true in 3 dimensior and the general result is that $\sigma_{ij} = \sigma_{ji}$ for the stress tensor.)

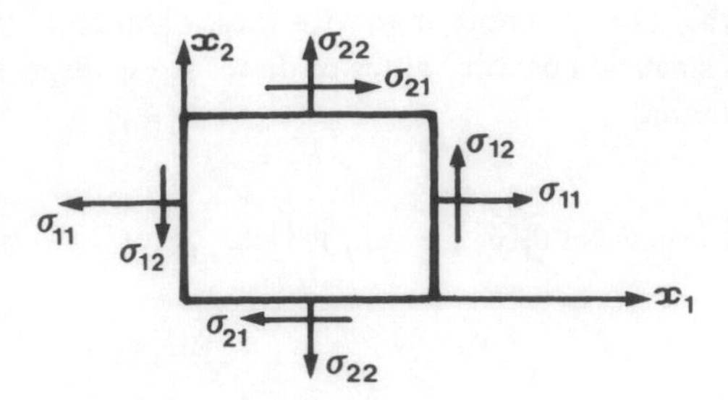

Figure 6.6

In the two-dimensional case we can define the transformation of axes in terms of θ, the angle of rotation shown in Figure 6.7.

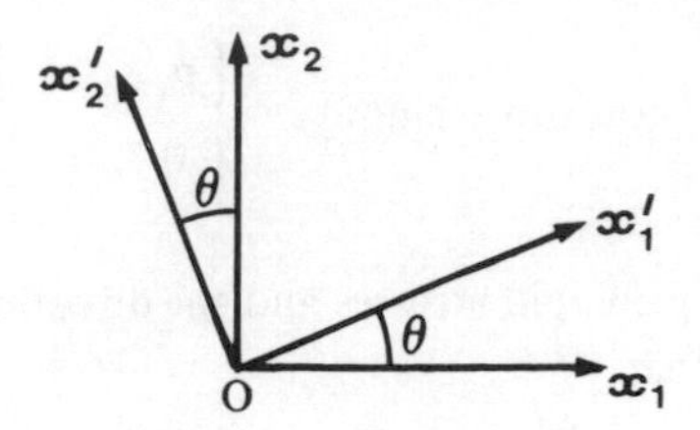

Figure 6.7

The matrix $[\lambda]$ reduces to $\begin{bmatrix} \lambda_{11} & \lambda_{12} \\ \lambda_{21} & \lambda_{22} \end{bmatrix}$, which is $\begin{bmatrix} \cos x_1' Ox_1 & \cos x_1' Ox_2 \\ \cos x_2' Ox_1 & \cos x_2' Ox_2 \end{bmatrix}$

giving $[\lambda] = \begin{bmatrix} \cos\theta & \sin\theta \\ -\sin\theta & \cos\theta \end{bmatrix}$. Hence equations (6.34) become

$$\begin{bmatrix} \sigma'_{11} & \sigma'_{12} \\ \sigma'_{21} & \sigma'_{22} \end{bmatrix} = \begin{bmatrix} \cos\theta & \sin\theta \\ -\sin\theta & \cos\theta \end{bmatrix} \begin{bmatrix} \sigma_{11} & \sigma_{12} \\ \sigma_{21} & \sigma_{22} \end{bmatrix} \begin{bmatrix} \cos\theta & -\sin\theta \\ \sin\theta & \cos\theta \end{bmatrix}$$

and the right-hand side when multiplied out becomes

$$\begin{bmatrix} \sigma_{11}\cos^2\theta + 2\sigma_{12}\cos\theta\sin\theta + \sigma_{22}\sin^2\theta & \sigma_{12}(\cos^2\theta - \sin^2\theta) - (\sigma_{11} - \sigma_{22})\sin\theta\cos\theta \\ \sigma_{12}(\cos^2\theta - \sin^2\theta) - (\sigma_{11} - \sigma_{22})\sin\theta\cos\theta & \sigma_{11}\sin^2\theta - 2\sigma_{12}\sin\theta\cos\theta + \sigma_{22}\cos^2\theta \end{bmatrix}$$

Notice that $\sigma'_{12} = \sigma'_{21}$ which is to be expected. We can choose θ so that $\sigma'_{12} = \sigma'_{21} = 0$, by requiring that

$$\sigma_{12}(\cos^2\theta - \sin^2\theta) - (\sigma_{11} - \sigma_{22})\sin\theta\cos\theta = 0$$

That is, we choose $\tan 2\theta = 2\sigma_{12}/(\sigma_{11} - \sigma_{22})$ giving two values of θ, differin by 90°. Furthermore, we note that with these choices of θ, σ'_{11} and σ'_{22} are the largest and smallest possible values of direct stress respectively (or vice-versa). To see this note that

$$\frac{d\sigma'_{11}}{d\theta} = [2 \sin\theta \cos\theta](\sigma_{22} - \sigma_{11}) + 2\sigma_{12}(\cos^2\theta - \sin^2\theta) = 0$$

and

$$\frac{d\sigma'_{22}}{d\theta} = [2 \sin\theta \cos\theta](\sigma_{11} - \sigma_{22}) - 2\sigma_{12}(\cos^2\theta - \sin^2\theta) = 0$$

giving the maximum and minimum values of σ'_{11}, σ'_{22}.

For these values of θ, then, we know that the stress tensor becomes

$$\begin{bmatrix} \sigma'_{11} & 0 \\ 0 & \sigma'_{22} \end{bmatrix} \quad \text{or, more simply,} \quad \begin{bmatrix} \sigma_1 & 0 \\ 0 & \sigma_2 \end{bmatrix}$$

σ_1 and σ_2 are called the **principal stresses** and the **directions in which they act are called principal directions.**

If we start with these directions and rotate the axes through an angle Φ then

$$\begin{bmatrix} \sigma''_{11} & \sigma''_{12} \\ \sigma''_{21} & \sigma''_{22} \end{bmatrix} = \begin{bmatrix} \cos\Phi & \sin\Phi \\ -\sin\Phi & \cos\Phi \end{bmatrix} \begin{bmatrix} \sigma_1 & 0 \\ 0 & \sigma_2 \end{bmatrix} \begin{bmatrix} \cos\Phi & -\sin\Phi \\ \sin\Phi & \cos\Phi \end{bmatrix}$$

which gives

$$\begin{bmatrix} \sigma''_{11} & \sigma''_{12} \\ \sigma''_{21} & \sigma''_{22} \end{bmatrix} = \begin{bmatrix} \sigma_1 \cos^2\Phi + \sigma_2 \sin^2\Phi & -(\sigma_1 - \sigma_2)\sin\Phi\cos\Phi \\ -(\sigma_1 - \sigma_2)\sin\Phi\cos\Phi & \sigma_1 \sin^2\Phi + \sigma_2 \cos^2\Phi \end{bmatrix}$$

i.e.

$$\begin{bmatrix} \sigma''_{11} & \sigma''_{12} \\ \sigma''_{21} & \sigma''_{22} \end{bmatrix} = \begin{bmatrix} \frac{(\sigma_1 + \sigma_2)}{2} + \frac{(\sigma_1 - \sigma_2)}{2}\cos 2\Phi & -\frac{(\sigma_1 - \sigma_2)}{2}\sin 2\Phi \\ -\frac{(\sigma_1 - \sigma_2)}{2}\sin 2\Phi & \frac{(\sigma_1 + \sigma_2)}{2} - \frac{(\sigma_1 - \sigma_2)}{2}\cos \end{bmatrix} \tag{6.35}$$

The results can conveniently be represented by **Mohr's circle** shown in Figure 6.8

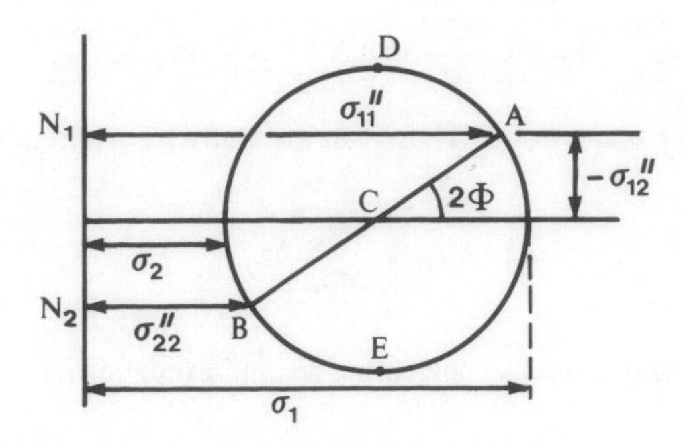

Figure 6.8

The principal stresses σ_1 and σ_2 are plotted on the horizontal axis and a circle is drawn as depicted. A diameter making an angle 2Φ with the horizontal axis is drawn to cut the circle at A and B. N_1A then gives the direct stress σ''_{11} and N_2B gives the other direct stress σ''_{22}. The shear stress σ''_{12} is determined from the vertical distance of A above the horizontal axis but note that this distance is $-\sigma''_{12}$. The points D and E represent maximum shear stress values attained when the axes are rotated by $\pm 45°$ from the principal directions.

) **Strain Tensor** e_{ij}

Consider the relative movement of two points close to each other in a body which is subjected to a general deformation. Let us denote these two points as P_1 with co-ordinates (x_1, x_2, x_3) and P_2 with co-ordinates $(x_1 + \delta x_1, x_2 + \delta x_2, x_3 + \delta x_3)$ as shown in Figure 6.9. We consider only infinitesimal strains or so-called **small deformations.**

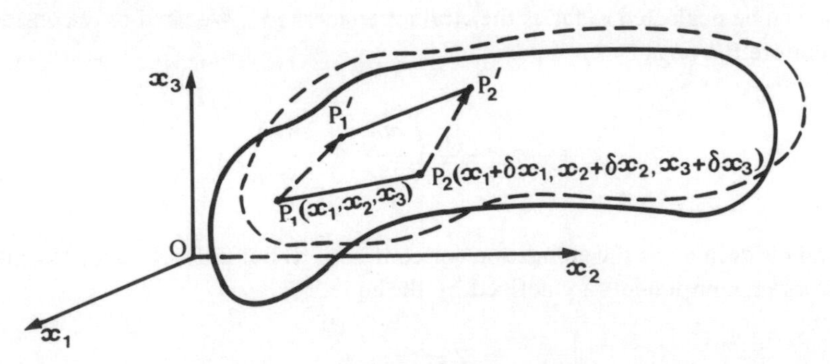

Figure 6.9

Suppose that in the deformation P_1 undergoes a movement (u_1, u_2, u_3) and that P_2 undergoes a movement $(u_1 + \delta u_1, u_2 + \delta u_2, u_3 + \delta u_3)$ to reach the points P_1' and P_2' respectively. The relative movement of the points P_1 and P_2

is given by $(\delta u_1, \ \delta u_2, \ \delta u_3)$.

Now $\delta u_1 \simeq \dfrac{\partial u_1}{\partial x_1} \delta x_1 + \dfrac{\partial u_1}{\partial x_2} \delta x_2 + \dfrac{\partial u_1}{\partial x_3} \delta x_3 = \dfrac{\partial u_1}{\partial x_j} \delta x_j$ with similar expressions for $\delta u_2, \ \delta u_3$. These can generally be written in the form

$$\delta u_i \simeq \frac{\partial u_i}{\partial x_j} \delta x_j \tag{6.36}$$

which can be expressed as (neglecting second-order terms)

$$\delta u_i = \frac{1}{2}\left(\frac{\partial u_i}{\partial x_j} + \frac{\partial u_j}{\partial x_i}\right) \delta x_j + \frac{1}{2}\left(\frac{\partial u_i}{\partial x_j} - \frac{\partial u_j}{\partial x_i}\right) \delta x_j$$

i.e. $\delta u_i = e_{ij}\, \delta x_j + \xi_{ij}\, \delta x_j$ where $e_{ij} = \dfrac{1}{2}\left(\dfrac{\partial u_i}{\partial x_j} + \dfrac{\partial u_j}{\partial x_i}\right)$, **a symmetric term**

and $\xi_{ij} = \dfrac{1}{2}\left(\dfrac{\partial u_i}{\partial x_j} - \dfrac{\partial u_j}{\partial x_i}\right)$ an **antisymmetric term.**

The relative movement between the two points P_1 and P_2 will be composed of that arising from a "stretching" of the material together with that arising from a rigid body motion in which the distance between P_1 and P_2 stays the same. It can be shown that the *antisymmetric* part $\dfrac{1}{2}\left(\dfrac{\partial u_i}{\partial x_j} - \dfrac{\partial u_j}{\partial x_i}\right) \delta x_j$ comes solely from a rigid body rotation of the elements without any shearing of the elements and can be neglected as far as the strain is concerned. We need only consider therefore the term

$$e_{ij}\, \delta x_j = \frac{1}{2}\left(\frac{\partial u_i}{\partial x_j} + \frac{\partial u_j}{\partial x_i}\right) \delta x_j$$

Strain is defined as the elongation (or contraction) per unit length of the material; the **strain components** are defined by the equation

$$e_{ij} = \frac{1}{2}\left(\frac{\partial u_i}{\partial x_j} + \frac{\partial u_j}{\partial x_i}\right) \tag{6.37}$$

It will help perhaps to describe in words the nature of these components.

In the unstrained state, let a line of unit length be drawn in the body in the direction of x_i. Then in the absence of rigid-body rotation the relative movements of the ends of this line will be e_{ij} in the direction of x_j. That is, if the line is drawn in the x_1 direction, the relative movements of the ends of this line will be e_{11} in the direction of x_1, e_{12} in the direction of x_2 and e_{13} in the direction of x_3. Therefore if one end P_1 of the line is held fixed, the other end P_2 would move to P_2' as shown in Figure 6.10.

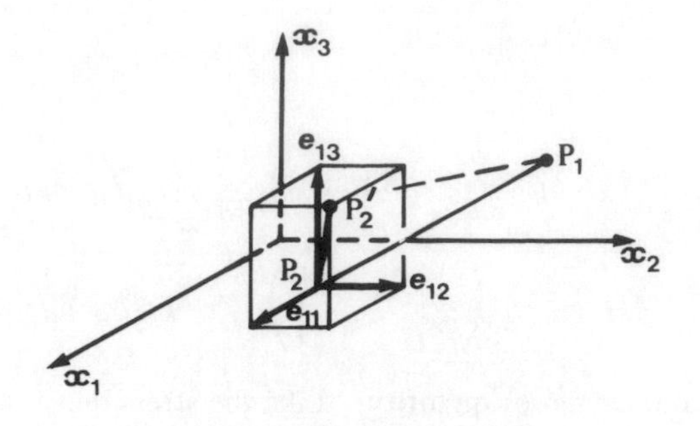

Figure 6.10

Similar results hold for lines of unit length drawn in the directions of x_2 and x_3.

The components of e_{ij} can be displayed in a matrix as

$$\begin{bmatrix} e_{11} & e_{12} & e_{13} \\ e_{21} & e_{22} & e_{23} \\ e_{31} & e_{32} & e_{33} \end{bmatrix} = \begin{bmatrix} \dfrac{\partial u_1}{\partial x_1} & \dfrac{1}{2}\left(\dfrac{\partial u_1}{\partial x_2} + \dfrac{\partial u_2}{\partial x_1}\right) & \dfrac{1}{2}\left(\dfrac{\partial u_1}{\partial x_3} + \dfrac{\partial u_3}{\partial x_1}\right) \\ \dfrac{1}{2}\left(\dfrac{\partial u_2}{\partial x_1} + \dfrac{\partial u_1}{\partial x_2}\right) & \dfrac{\partial u_2}{\partial x_2} & \dfrac{1}{2}\left(\dfrac{\partial u_2}{\partial x_3} + \dfrac{\partial u_3}{\partial x_2}\right) \\ \dfrac{1}{2}\left(\dfrac{\partial u_3}{\partial x_1} + \dfrac{\partial u_1}{\partial x_3}\right) & \dfrac{1}{2}\left(\dfrac{\partial u_3}{\partial x_2} + \dfrac{\partial u_2}{\partial x_3}\right) & \dfrac{\partial u_3}{\partial x_3} \end{bmatrix}$$

and it is seen to be symmetric.

We can show that the quantity e_{ij} is a tensor. Now $e'_{kl} = \dfrac{1}{2}\left(\dfrac{\partial u'_k}{\partial x'_l} + \dfrac{\partial u'_l}{\partial x'_k}\right)$ and

$$\frac{\partial u'_k}{\partial x'_l} = \frac{\partial}{\partial x'_l}(\lambda_{ki}\, u_i) \qquad (u_i \text{ is a vector quantity})$$

$$= \lambda_{ki} \frac{\partial u_i}{\partial x_l'}$$

$$= \lambda_{ki} \frac{\partial u_i}{\partial x_j} \frac{\partial x_j}{\partial x_l'} \qquad \text{(Note the implied summation)}$$

$$= \lambda_{ki} \frac{\partial u_i}{\partial x_j} \lambda_{lj}$$

That is, $\dfrac{\partial u_k'}{\partial x_l'} = \lambda_{ki}\, \lambda_{lj} \dfrac{\partial u_i}{\partial x_j}$. Similarly, $\dfrac{\partial u_l'}{\partial x_k'} = \lambda_{lj}\, \lambda_{ki} \dfrac{\partial u_j}{\partial x_i}$. Thus

$$e_{kl}' = \lambda_{ki}\, \lambda_{lj}\, \frac{1}{2}\left(\frac{\partial u_i}{\partial x_j} + \frac{\partial u_j}{\partial x_i}\right) = \lambda_{ki}\, \lambda_{lj}\, e_{ij} \tag{6.38}$$

and e_{ij} is a 2nd-order tensor quantity. Like the stress tensor at a particular point it is a symmetric second-order tensor and consequently it is possible to find 3 mutually perpendicular directions with respect to which the shear strains are zero. A small cube drawn in the body at this point with sides along these directions would suffer only direct straining and would therefore remain cube-shaped.

It is important to notice that there are differences here from the shear strain components normally used in engineering. The direct strain components are e_{11} e_{22}, e_{33} as given above, but the **shear strain components** γ_{12}, γ_{13}, γ_{23} are in fact given by

$$\left.\begin{aligned} \gamma_{12} = 2e_{12} &= \frac{\partial u_1}{\partial x_2} + \frac{\partial u_2}{\partial x_1} \qquad \gamma_{13} = 2e_{13} = \frac{\partial u_1}{\partial x_3} + \frac{\partial u_3}{\partial x_1} \\ \gamma_{23} = 2e_{23} &= \frac{\partial u_2}{\partial x_3} + \frac{\partial u_3}{\partial x_2} \end{aligned}\right\} \tag{6.39}$$

However, these components, viz. $\begin{bmatrix} e_{11} & \gamma_{12} & \gamma_{13} \\ \gamma_{12} & e_{22} & \gamma_{23} \\ \gamma_{13} & \gamma_{23} & e_{33} \end{bmatrix}$, do not form a tensor

Further properties of second-order tensors

(1) *A second-order symmetric tensor remains a second-order symmetric tensor under transformation.*

This has already been seen in the examples of tensors given. The proof is very similar to that given below for antisymmetric tensors and it is left to you to perform it.

) *A second-order antisymmetric tensor remains a second-order antisymmetric tensor under transformation.*

Let the antisymmetric tensor be a_{ij}, so that $a_{ji} = -a_{ij}$. Under transformation $a'_{kl} = \lambda_{ki}\,\lambda_{lj}\,a_{ij}$. But $a_{ij} = -a_{ji}$. Therefore $a'_{kl} = \lambda_{ki}\,\lambda_{lj}(-a_{ji}) = -\lambda_{lj}\,\lambda_{ki}\,a_{ji}$. However, $\lambda_{lj}\,\lambda_{ki}\,a_{ji} = a'_{lk}$ and so $a'_{kl} = -a'_{lk}$, proving the result.

) *A second-order tensor* a_{ij} *has the following invariants under transformation.*

(i) a_{ii}, i.e., $a_{11} + a_{22} + a_{33}$, called the **trace**.

(ii) $\frac{1}{2}(a_{ii}\,a_{jj} - a_{ij}\,a_{ij})$, i.e. $a_{11}a_{22} + a_{22}a_{33} + a_{33}a_{11} - a_{12}a_{21} - a_{23}a_{32} - a_{31}a_{13}$

(iii) $|a_{ij}|$, i.e. the *determinant* of the tensor.

The first result given by (i) is very useful for it tells us for example that *the sum of* *e direct stresses* $\sigma_{11} + \sigma_{22} + \sigma_{33}$ *is invariant under transformation of axes* and that *the* *m of the direct strain components* $e_{11} + e_{22} + e_{33}$ *is also invariant.*

The second result can be written in the form

$$\begin{vmatrix} a_{11} & a_{12} \\ a_{21} & a_{22} \end{vmatrix} + \begin{vmatrix} a_{22} & a_{23} \\ a_{32} & a_{33} \end{vmatrix} + \begin{vmatrix} a_{33} & a_{31} \\ a_{13} & a_{11} \end{vmatrix} \quad \text{is invariant.}$$

he reader may have already noticed the similarity of results obtained when considering e eigenvalues of a 3 x 3 matrix. (See AEM.)

The above results can be proved by the following method which exploits such milarities.

Consider $a_{ij} - \alpha\delta_{ij}$ where α is a scalar constant. Now a_{ij} is a second-order nsor and so is δ_{ij}. Therefore $a_{ij} - \alpha\delta_{ij}$ is also a second-order tensor. Changing axes e obtain $a'_{kl} - \alpha\delta'_{kl} = \lambda_{ki}\,\lambda_{lj}\,(a_{ij} - \alpha\delta_{ij})$.

It was shown that $\delta'_{kl} = \delta_{kl}$ (the unit matrix) and also that $|\lambda_{ki}| = |\lambda_{lj}| = 1$, so at $|(a'_{kl} - \alpha\delta_{kl})| = |(a_{ij} - \alpha\delta_{ij})|$, i.e.

$$\begin{vmatrix} a'_{11} - \alpha & a'_{12} & a'_{13} \\ a'_{21} & a'_{22} - \alpha & a'_{23} \\ a'_{31} & a'_{32} & a'_{33} - \alpha \end{vmatrix} = \begin{vmatrix} a_{11} - \alpha & a_{12} & a_{13} \\ a_{21} & a_{22} - \alpha & a_{23} \\ a_{31} & a_{32} & a_{33} - \alpha \end{vmatrix}$$

Each side of this equation is a cubic in α and since it must hold for any value of α then the coefficients of each cubic must be the same.

The coefficient of α^2 gives

$$a'_{11} + a'_{22} + a'_{33} = a_{11} + a_{22} + a_{33} \qquad \text{which is result (i)}$$

The coefficient of α gives

$$\begin{vmatrix} a'_{11} & a'_{12} \\ a'_{21} & a'_{22} \end{vmatrix} + \begin{vmatrix} a'_{22} & a'_{23} \\ a'_{32} & a'_{33} \end{vmatrix} + \begin{vmatrix} a'_{33} & a'_{31} \\ a'_{13} & a'_{11} \end{vmatrix}$$

$$= \begin{vmatrix} a_{11} & a_{12} \\ a_{21} & a_{22} \end{vmatrix} + \begin{vmatrix} a_{22} & a_{23} \\ a_{32} & a_{33} \end{vmatrix} + \begin{vmatrix} a_{33} & a_{31} \\ a_{13} & a_{11} \end{vmatrix} \text{; result (ii)}$$

From the constant term we find

$$\begin{vmatrix} a'_{11} & a'_{12} & a'_{13} \\ a'_{21} & a'_{22} & a'_{23} \\ a'_{31} & a'_{32} & a'_{33} \end{vmatrix} = \begin{vmatrix} a_{11} & a_{12} & a_{13} \\ a_{21} & a_{22} & a_{23} \\ a_{31} & a_{32} & a_{33} \end{vmatrix} \text{; result (iii)}$$

Limitations of the use of matrix notation to represent tensor quantities

For second-order tensors the results and expressions can be represented in matrix form using 3×3 matrices and indeed we can regard a scalar as being represented by a 1×1 matrix (one component) and a vector by a 3×1 matrix (3 components). However, higher order tensors exist; the third-order tensor, having 3^3 components, would require a $3 \times 3 \times 3$ cubical array for its display; the fourth-order tensor has 3^4 components, and so on. For tensors of higher order than the second we need to use the suffix notation and there is no directly applicable matrix analogue in terms of matrices having scalar elements. We consider briefly higher order tensors in the next section.

Problems

1. The components of a second-order tensor are given by

$$\begin{bmatrix} 10 & 8 & -1 \\ 2 & 7 & 5 \\ 6 & -9 & 3 \end{bmatrix}$$

Find its symmetric and antisymmetric parts.

2. A symmetric second-order tensor is given by

$$\mathbf{B} = \begin{bmatrix} 1 & -6 & 3 \\ -6 & 7 & -2 \\ 3 & -2 & -5 \end{bmatrix}$$

and two vectors are given by $U_i = [3,\ 6,\ 9]$ and $V_i = [-2,\ 5,\ -7]$. Find values for the expressions $B_{ij}\, U_i\, V_j$ and $B_{ij}\, V_i\, U_j$. What do the expressions represent and why are they equal?

3. A plane area in two dimensions is referred to axes $0x_1$, $0x_2$ (mutually perpendicular). A quantity $\mathbf{A}$ is defined by

$$A_{11} = \iint x_2^2\, dx_1\, dx_2, \quad A_{12} = A_{21} = -\iint x_1 x_2\, dx_1\, dx_2, \quad A_{22} = \iint x_1^2\, dx_1\, dx_2$$

each integral being taken over the plane area. Verify that under planar rotation of axes $\mathbf{A}$ behaves as a second-order tensor.

. For a set of undashed axes a stress tensor is given by

$$\sigma_{ij} = \begin{bmatrix} \tau & 0 & 0 \\ 0 & \tau & 0 \\ 0 & 0 & \tau \end{bmatrix}$$

Determine the stress tensor for the dashed axes defined in Problem 7 on page 206. You should obtain a surprising result - all axes are principal axes, which is always true for isotropic tensors of the form $\tau\delta_{ij}$. Show this is true in general by using the transformation law for second-order tensors.

. A cube of metal with faces perpendicular to the axes $0x_1$, $0x_2$, $0x_3$ is subjected to direct stresses $\sigma_{11} = 600\ \text{kg/cm}^2$, $\sigma_{22} = 400\ \text{kg/cm}^2$, $\sigma_{33} = 100\ \text{kg/cm}^2$. The shear stresses on its faces are zero. Write down the stress tensor. A rotation of axes to $0x_1'$, $0x_2'$, $0x_3'$ is defined by means of the direction cosines array

$$\begin{bmatrix} 1/\sqrt{3} & 1/\sqrt{3} & 1/\sqrt{3} \\ 0 & 1/\sqrt{2} & -1/\sqrt{2} \\ -2/\sqrt{6} & 1/\sqrt{6} & 1/\sqrt{6} \end{bmatrix}$$

Find the stress tensor components referred to these axes.

The principal axes of a second-order tensor can be found from the eigenvalues and corresponding eigenvectors of the tensor given by $a_{ij}\, x_j = \lambda x_i$ where λ is an eigenvalue and x_i is the corresponding eigenvector. Establish the following results:

(a) $(a_{ij} - \lambda\delta_{ij})\, x_j = 0$ (b) $\det\,(a_{ij} - \lambda\delta_{ij}) = 0$

It is a well-known result in matrix algebra that if a matrix is real and symmetric then the eigenvalues are real and the eigenvectors are mutually orthogonal. If a matrix $\mathbf{P}$ is made up of the **normalised** eigenvectors (made into unit vectors) of a symmetric matrix $\mathbf{A}$ then $\mathbf{P}^{-1} = \mathbf{P}^T$ and $\mathbf{P}^T \mathbf{AP} = \mathbf{D}$, a diagonal matrix having the eigenvalues as its diagonal matrix. Thus the principal stresses are the eigenvalues and the corresponding eigenvectors give the direction of the principal axes. Use this result to find the principal stress values of the stress tensor

$$\sigma_{ij} = \begin{bmatrix} 3 & 1 & 1 \\ 1 & 0 & 2 \\ 1 & 2 & 0 \end{bmatrix}$$

and the corresponding directions of the principal axes.

7. Determine the principal stress values of the tensor

$$\begin{bmatrix} \tau & \tau & \tau \\ \tau & \tau & \tau \\ \tau & \tau & \tau \end{bmatrix}$$

and the directions of the principal axes.

8. The stress tensor for a plane stress problem is given relative to the axes $0x_1$, $0x_2$, $0x_3$ as

$$\sigma_{ij} = \begin{bmatrix} 5 & \sqrt{3} \\ \sqrt{3} & 3 \end{bmatrix}$$

Find the principal stress values and the directions of the principal axes. Hence using Mohr's circle, find the maximum shear stress.

9. A displacement u_i is defined by

$$[x_1 - 3x_2 + \sqrt{2}x_3, \quad -3x_1 + x_2 - \sqrt{2}x_3, \quad \sqrt{2}x_1 - \sqrt{2}x_2 + 4x_3]$$

Find the strain tensor e_{ij}. Show that the principal strains are 6, 2 and -2.

10. A displacement u_i is given by $u_1 = 3x_1x_2^2$, $u_2 = 2x_3x_1$, $u_3 = x_3^2 - x_1x_2$. Find the strain tensor e_{ij}. The compatibility conditions (which ensure that the displacement components u_i are single-valued and continuous) are given by

$$\frac{\partial^2 e_{ij}}{\partial x_k \partial x_m} + \frac{\partial^2 e_{km}}{\partial x_i \partial x_j} - \frac{\partial^2 e_{ik}}{\partial x_j \partial x_m} - \frac{\partial^2 e_{jm}}{\partial x_i \partial x_k} = 0$$

Show that this represents 6 distinct equations and verify that they are satisfied for the displacement given.

11. For the displacement given in Problem 10 find the strain tensor at the points (0, 2, 2) and (0, 0, ½), and determine the principal strains at the latter point and the directions of the principal axes.

2. The invariants of a second-order tensor take a particularly simple form if they are related to the principal components. Write these out.

3. A symmetric tensor has components $\begin{bmatrix} 5 & 2 & 5 \\ 2 & 2 & 2 \\ 5 & 2 & 5 \end{bmatrix}$. Using the invariants of a second-order tensor, find the principal components.

.5 HIGHER ORDER TENSORS

hird-order tensor The third-order tensor T_{ijk} obeys the transformation law

$$T'_{lmn} = \lambda_{li}\lambda_{mj}\lambda_{nk}T_{ijk} \tag{6.40}$$

he inverse transformation law being

$$T_{ijk} = \lambda_{li}\lambda_{mj}\lambda_{nk}T'_{lmn} \tag{6.41}$$

t has 27 components of the form T_{111}, T_{112}, T_{113}, T_{123}, etc.

ourth-order tensor The fourth-order tensor similarly obeys the transformation law

$$T'_{lmnp} = \lambda_{li}\lambda_{mj}\lambda_{nk}\lambda_{pt}T_{ijkt} \tag{6.42}$$

nd inverse transformation law

$$T_{ijkt} = \lambda_{li}\lambda_{mj}\lambda_{nk}\lambda_{pt}T_{lmnp} \tag{6.43}$$

t has $3^4 = 81$ components of the form T_{1111}, T_{1112}, T_{1113}, etc.

You can probably now see the pattern for tensors of order five, six etc. Try vriting down their transformation laws.

evi-Civita tensor

As a special case of a third-order tensor we now examine the properties of the **evi-Civita tensor** (or **alternating symbol**) ϵ_{ijk}. This is defined as follows:-

$\epsilon_{ijk} = 1$ if i, j, k are all different and in cyclic order, that is the numbers progress in the order obtained by going clockwise round the circle

$$\epsilon_{ijk} = -1 \quad \text{if } i, j, k \text{ are all different and in anticyclic order.}$$

$$\epsilon_{ijk} = 0 \quad \text{if the } i, j, k \text{ are not all different.}$$

For example, $\epsilon_{132} = -1$, $\epsilon_{312} = 1$, $\epsilon_{211} = 0$. It is a third-order tensor but the proof is beyond our scope here.

Example 1 Show that $\epsilon_{ijk}\,\epsilon_{kij} = 6$

First sum over i

$$\epsilon_{ijk}\,\epsilon_{kij} = \epsilon_{1jk}\,\epsilon_{k1j} + \epsilon_{2jk}\,\epsilon_{k2j} + \epsilon_{3jk}\,\epsilon_{k3j}$$

Next sum over j

$$\epsilon_{ijk}\,\epsilon_{kij} = \cancel{\epsilon_{11k}\,\epsilon_{k11}} + \epsilon_{12k}\,\epsilon_{k12} + \epsilon_{13k}\,\epsilon_{k13} + \epsilon_{21k}\,\epsilon_{k21} + \cancel{\epsilon_{22k}\,\epsilon_{k22}}$$
$$+ \epsilon_{23k}\,\epsilon_{k23} + \epsilon_{31k}\,\epsilon_{k31} + \epsilon_{32k}\,\epsilon_{k32} + \cancel{\epsilon_{33k}\,\epsilon_{k33}}$$

the terms which are zero being crossed out.

Finally we sum over k and write down only non-zero terms

$$\epsilon_{ijk}\,\epsilon_{kij} = \epsilon_{123}\,\epsilon_{312} + \epsilon_{132}\,\epsilon_{213} + \epsilon_{213}\,\epsilon_{321} + \epsilon_{231}\,\epsilon_{123} + \epsilon_{312}\,\epsilon_{231}$$
$$+ \epsilon_{321}\,\epsilon_{132}$$
$$= (1)(1) + (-1)(-1) + (-1)(-1) + (1)(1) + (1)(1) + (-1)(-1) = 6$$

Example 2 Show that $\delta_{ij}\,\epsilon_{ijk} = 0$

Summing first over i we obtain

$$\delta_{ij}\,\epsilon_{ijk} = \delta_{1j}\,\epsilon_{1jk} + \delta_{2j}\,\epsilon_{2jk} + \delta_{3j}\,\epsilon_{3jk}$$

Summing now over j gives

$$\delta_{ij}\,\epsilon_{ijk} = \delta_{11}\,\epsilon_{11k} + \delta_{12}\,\epsilon_{12k} + \delta_{13}\,\epsilon_{13k} + \delta_{21}\,\epsilon_{21k} + \delta_{22}\,\epsilon_{22k}$$
$$+ \delta_{23}\,\epsilon_{23k} + \delta_{31}\,\epsilon_{31k} + \delta_{32}\,\epsilon_{32k} + \delta_{33}\,\epsilon_{33k}$$
$$= 0 \text{ (remember } \delta_{ij} = 0 \text{ unless } i = j \text{ and } \epsilon_{ijk} = 0 \text{ if } i = j)$$

ensor Multiplication

The two previous examples just considered were particular cases of the ıultiplication of two tensors. Consider first the multiplication of two vector quantities $4_1, A_2, A_3)$ and (B_1, B_2, B_3). The possible terms obtainable by multiplying the omponents of these vector quantities are $A_i B_j$; in matrix form these may be isplayed as:

$$\begin{bmatrix} A_1 B_1 & A_1 B_2 & A_1 B_3 \\ A_2 B_1 & A_2 B_2 & A_2 B_3 \\ A_3 B_1 & A_3 B_2 & A_3 B_3 \end{bmatrix}$$

his is a second-order tensor T_{ij}, since

$$T'_{kl} = A'_k B'_l = \lambda_{ki} A_i \lambda_{lj} B_j = \lambda_{ki} \lambda_{lj} A_i B_j = \lambda_{ki} \lambda_{lj} T_{ij}$$

hich is the transformation law for second-order tensors.

Notice that the process has produced a tensor of an order which is the *sum* of the rders of the factor tensors. This is true in general and, for example, $A_i B_{jk}$ is a third-rder tensor C_{ijk} while $D_{ij} E_{kl} = F_{ijkl}$, a fourth-order tensor. This is also true with ıultiplication by a scalar (which is a zero-order tensor) which produces a tensor of the ame order. The above type of multiplication is called the **outer product** of two tensors.

roducts involving the Kronecker Delta δ_{ij}

Consider the product $\delta_{ij} T_{ij}$ where T_{ij} is a second-order tensor. Now

$$\delta_{ij} T_{ij} = T_{11} + T_{22} + T_{33} = T_{ii}$$

ence the multiplication by δ_{ij} has produced a scalar. (Multiplication by δ_{ij} ffectively says put $i = j$ everywhere.)

Similarly $\delta_{ij} T_{ijk} = T_{iik} = T_{11k} + T_{22k} + T_{33k}$ which gives the 3 omponents $(T_{111} + T_{221} + T_{331}, T_{112} + T_{222} + T_{332}, T_{113} + T_{223} + T_{333})$.

Notice then that the multiplication of a second-order tensor by δ_{ij} has produced a ensor of order zero (a scalar) and multiplication of a third-order tensor by δ_{ij} has roduced a tensor of order one (a vector). Thus in each case the order of the tensor has een reduced by 2. The process of putting two suffices the same in a tensor quantity is alled **contraction** and reduces the order of the tensor by 2. Another example is $_{ij} A_i B_j$ where A_i and B_j are vectors and $A_i B_j$ is a second-order tensor. Then $_{ij} A_i B_j = A_i B_i = A_1 B_1 + A_2 B_2 + A_3 B_3$ which we readily recognise as being the calar product of the two vectors. That is

$$\mathbf{A}\,.\,\mathbf{B} = \delta_{ij} A_i B_j \qquad (6.44)$$

Referring back to Example 2 it is easily seen that $\delta_{ij}\,\epsilon_{ijk}$ must equal ϵ_{iik} which has the value zero (from the definition of ϵ_{ijk}).

The Levi-Civita tensor and vector cross products

Consider the expression $\epsilon_{ijk} A_i B_j$. Summing first over i we obtain

$$\epsilon_{ijk} A_i B_j = \epsilon_{1jk} A_1 B_j + \epsilon_{2jk} A_2 B_j + \epsilon_{3jk} A_3 B_j$$

Summing over j, the non-zero terms are

$$\epsilon_{12k} A_1 B_2 + \epsilon_{13k} A_1 B_3 + \epsilon_{21k} A_2 B_1 + \epsilon_{23k} A_2 B_3 + \epsilon_{31k} A_3 B_1 + \epsilon_{32k} A_3 B_2$$

Now k can take the values 1, 2 and 3 giving three components; using the definition of ϵ_{ijk} we find that these are $[A_3 B_1 - A_1 B_3,\ A_1 B_2 - A_2 B_1,\ A_2 B_3 - A_3 B_2]$. These may be recognised as the components of the vector cross product $\mathbf{A} \times \mathbf{B}$ so that $\epsilon_{ijk} A_i B_j = C_k\,\mathbf{i}_k$ where $\mathbf{i}_k$ is the unit vector in the x_k direction. The scalar triple vector product $\mathbf{A}\,.\,(\mathbf{B} \times \mathbf{C})$ becomes in suffix notation $A_k\,\mathbf{i}_k\,.\,(\epsilon_{ijk} B_i C_j\,\mathbf{i}_k) = A_k B_i C_j\,\epsilon_{ijk}$ and it can be easily shown (you should try this) that

$$\mathbf{A}\,.\,(\mathbf{B} \times \mathbf{C}) = (\mathbf{A} \times \mathbf{B})\,.\,\mathbf{C} = \mathbf{B}\,.\,(\mathbf{C} \times \mathbf{A}) = [\mathbf{A},\ \mathbf{B},\ \mathbf{C}]$$

The vector triple product $\mathbf{A} \times (\mathbf{B} \times \mathbf{C})$ can also be written in suffix notation. Since $\mathbf{B} \times \mathbf{C} = \epsilon_{ijk} B_i C_j\,\mathbf{i}_k$. Then

$$\begin{aligned} \mathbf{A} \times (\mathbf{B} \times \mathbf{C}) &= \epsilon_{lkm} A_l (\epsilon_{ijk} B_i C_j)\,\mathbf{i}_m \\ &= \epsilon_{lkm}\,\epsilon_{ijk} A_l B_i C_j\,\mathbf{i}_m \end{aligned}$$

Using this we can prove the result $\mathbf{A} \times (\mathbf{B} \times \mathbf{C}) = (\mathbf{A}\,.\,\mathbf{C})\,\mathbf{B} - (\mathbf{A}\,.\,\mathbf{B})\,\mathbf{C}$

Problems

1. Prove that $\epsilon_{ijk}\,\epsilon_{ilm} = \delta_{jl}\,\delta_{km} - \delta_{jm}\,\delta_{kl}$.

2. Show that the tensor $A_{ik} = \epsilon_{ijk}\,a_j$ is antisymmetric.

3. Write down the vector $\mathbf{v} = \mathbf{b} \times \mathbf{c}$ in index notation and similarly write down the vector $\mathbf{w} = \mathbf{a}$ ×
Hence using the identity given in Problem 1 show that $\mathbf{a} \times (\mathbf{b} \times \mathbf{c}) = (\mathbf{a}\,.\,\mathbf{c})\,\mathbf{b} - (\mathbf{a}\,.\,\mathbf{b})\,\mathbf{c}$

Show that the determinant of a second-order tensor can be written in the form $\epsilon_{ijk} A_{1i} A_{2j} A_{3k}$.

.6 CARTESIAN TENSOR CALCULUS

We shall not delve very deeply into the calculus of tensors, but rather concentrate n becoming familiar with standard vector field theory results written in suffix form.

Recall that the gradient of $\phi = \left(\dfrac{\partial \phi}{\partial x_1}, \dfrac{\partial \phi}{\partial x_2}, \dfrac{\partial \phi}{\partial x_3} \right) \equiv \mathbf{i}_k \dfrac{\partial \phi}{\partial x_k}$: a vector uantity. It is written grad ϕ or $\nabla \phi$ where ∇ is the vector operator

$$\left(\mathbf{i}_1 \frac{\partial}{\partial x_1} + \mathbf{i}_2 \frac{\partial}{\partial x_2} + \mathbf{i}_3 \frac{\partial}{\partial x_3} \right) \equiv \mathbf{i}_k \frac{\partial}{\partial x_k} \tag{6.45}$$

he **divergence** of a vector $\mathbf{A}$ is defined by $\text{div}\,\mathbf{A} = \nabla . \mathbf{A} = \dfrac{\partial A_1}{\partial x_1} + \dfrac{\partial A_2}{\partial x_2} + \dfrac{\partial A_3}{\partial x_3}$.

hus in suffix notation,

$$\text{div}\ \mathbf{A} = \frac{\partial A_k}{\partial x_k} \tag{6.46}$$

he **curl** of a vector is written

$$\text{curl}\ \mathbf{A} = \nabla \times \mathbf{A} = \epsilon_{ijk} \frac{\partial A_j}{\partial x_i} \mathbf{i}_k \tag{6.47}$$

.aplace's Operator

$$\nabla^2 \phi = \text{div grad}\ \phi = \frac{\partial^2 \phi}{\partial x_1^2} + \frac{\partial^2 \phi}{\partial x_2^2} + \frac{\partial^2 \phi}{\partial x_3^2} = \frac{\partial^2 \phi}{\partial x_i\, \partial x_i} \tag{6.48}$$

ote An alternative notation is often used for partial differentiation especially when ealing with general rather than Cartesian Tensors. This takes the form of a comma otation as follows

$$\phi_{,i} = \frac{\partial \phi}{\partial x_i} \qquad u_{i,j} = \frac{\partial u_i}{\partial x_j} \qquad \phi_{,ij} = \frac{\partial^2 \phi}{\partial x_i\, \partial x_j}$$

o that

$$\text{grad}\ \phi = \mathbf{i}_k\, \phi_{,k} \qquad \text{div}\ \mathbf{A} = A_{i,i} \qquad \text{curl}\ \mathbf{A} = \epsilon_{ijk} A_{j,i}\, \mathbf{i}_k$$

Try to show that the Integral Theorems take the following form when using suffix notation.

Gauss Divergence Theorem For a closed volume V with surface S

$$\iiint_V \frac{\partial A_i}{\partial x_i}\, dV = \iint_S A_i l_i \, dS \tag{6.49}$$

(Here the outward normal to the surface is $\hat{\mathbf{n}} = l_1\mathbf{i}_1 + l_2\mathbf{i}_2 + l_3\mathbf{i}_3$.)

Stokes' Theorem For an open surface S with perimeter C

$$\oint_C A_i \, dx_i = \iint_S \epsilon_{pqr} \frac{\partial A_q}{\partial x_p} l_r \, dS \tag{6.50}$$

Problems

1. Using index notation prove the vector identities curl grad $\phi = 0$, div curl $\mathbf{A} = 0$.

2. By means of Gauss Divergence theorem and the identity given in Problem 1 on page 232, show that $\iint_S d\mathbf{S} \times (\mathbf{a} \times \mathbf{r}) = 2\mathbf{a}V$ where V is the volume enclosed by the surface S, $\mathbf{r}$ is the position vector of any point in V and $\mathbf{a}$ is a constant vector.

3. Show that $\dfrac{\partial A_{ij}}{\partial x_k}$ is a third-order tensor.

4. A second-order tensor is given by $A_{ij} = \begin{bmatrix} 2 & 4 & 0 \\ 6 & 8 & 0 \\ 0 & 0 & 1 \end{bmatrix}$.

   (a) find the components of $C_{ij} = \delta_{ij} A_{jk}$

   (b) find $\delta_{ij} A_{ij}$.

6.7 APPLICATIONS OF CARTESIAN TENSORS

This section examines two applications: elasticity and viscous fluid dynamics.

(1) Elasticity

When discussing the stress tensor and the strain tensor no mention was made of aterial properties and in fact these two tensors do not depend for their validity on ıy particular material; the stress tensor is symmetric because considerations of ıuilibrium make it thus and the strain tensor is symmetric because considerations ˙ geometry require it to be.

ı order to establish the link between the stress and strain components it is ɛcessary to consider the properties of the material. A material is considered to be **astic** if the specification of the state of stress determines the strain state. Thus rains depend only on the stresses that exist and not on stress history (Hysteresis ·ops are impossible). We say a material is **linear** if the "effect" (change of rain) is proportional to the "cause" (change of stress). The material is assumed ı this subsection to be **homogeneous**; this means that one point in the body is ıdistinguishable from another point so far as the material properties are concerned. urther, we consider **isotropic** material; that is, one direction through a given point indistinguishable from another direction through that point as far as material roperties are concerned.

or a general linear elastic solid, a single stress component must be envisaged as eing capable of producing contributions to all 9 strain components in the strain ɛnsor. Thus $\epsilon_{ij} = C_{ijkl}\ \sigma_{kl}$ where there are 81 coefficients C_{ijkl} which epend on the material. This is called the **generalised Hooke's law.**

ince the stress and strain tensors are both symmetric and therefore each contain nly six distinct components, then C_{ijkl} has only 36 distinct components. By nergy considerations for "small strain theory" it can be shown that the number of ıdependent elastic constants C_{ijkl} is at most 21. If the material is restricted to eing isotropic it can be shown that the number of independent elastic constants ɛduces to 2.

'herefore two elastic constants of the material need to be specified. These are btained from the **Young's modulus** (E) and **Poisson's ratio** (ν). (We could equally vell specify the **Bulk modulus** (K) and the **shear modulus** (G) in which case 'oung's modulus and Poisson's ratio could be derived from these.)

'he relationship between ϵ_{ij} and σ_{ij} is derived as follows. Consider a rod of quare section and suppose there is a stress σ_{11} acting in the longitudinal direction s shown in Figure 6.11.

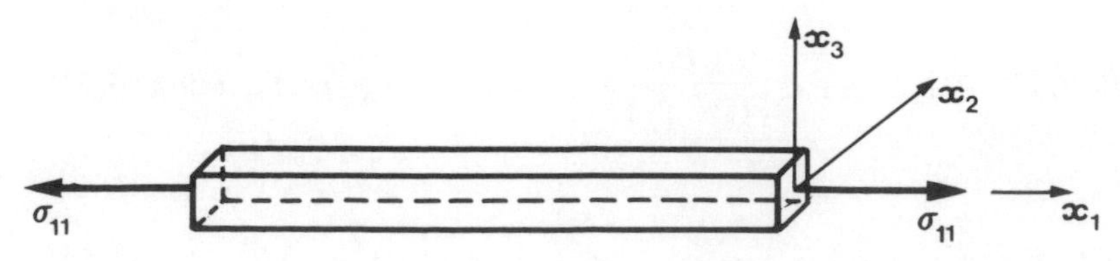

Figure 6.11

This will produce a stretching in the longitudinal direction and a contraction in the x_2 and x_3 directions giving strains e_{11}, e_{22}, e_{33}. Then we know that

$$Ee_{11} = \sigma_{11} \qquad Ee_{22} = -\nu\sigma_{11} \qquad Ee_{33} = -\nu\sigma_{11}$$

If we superimpose stresses σ_{22} and σ_{33} the following relationships are obtained

$$\left.\begin{aligned} Ee_{11} &= \sigma_{11} - \nu\sigma_{22} - \nu\sigma_{33} = (1+\nu)\,\sigma_{11} - \nu\,(\sigma_{11} + \sigma_{22} + \sigma_{33}) \\ Ee_{22} &= -\nu\sigma_{11} + \sigma_{22} - \nu\sigma_{33} = (1+\nu)\,\sigma_{22} - \nu\,(\sigma_{11} + \sigma_{22} + \sigma_{33}) \\ Ee_{33} &= -\nu\sigma_{11} - \nu\sigma_{22} + \sigma_{33} = (1+\nu)\,\sigma_{33} - \nu(\sigma_{11} + \sigma_{22} + \sigma_{33}) \end{aligned}\right\} \quad (6.51)$$

Here of course e_{12}, e_{13} and e_{23} are all zero, as are σ_{12}, σ_{13} and σ_{23}.

For principal directions therefore

$$Ee_{ij} = (1+\nu)\,\sigma_{ij} - \nu\,(\sigma_{11} + \sigma_{22} + \sigma_{33})\,\delta_{ij}$$

where e_{ij} and σ_{ij} are zero if $i \neq j$.

Suppose we change to a new set of axes. Assuming that e_{ij} and σ_{ij} are tensor quantities we obtain, remembering that $(\sigma_{11} + \sigma_{22} + \sigma_{33})$ is invariant,

$$\begin{aligned} Ee'_{kl} &= \lambda_{ki}\,\lambda_{lj}\,Ee_{ij} \\ &= (1+\nu)\,\lambda_{ki}\,\lambda_{lj}\,\sigma_{ij} - \nu\,(\sigma_{11} + \sigma_{22} + \sigma_{33})\,\lambda_{ki}\,\lambda_{lj}\,\delta_{ij} \end{aligned}$$

That is $Ee'_{kl} = (1+\nu)\,\sigma'_{kl} - \nu\,(\sigma'_{11} + \sigma'_{22} + \sigma'_{33})\,\delta'_{kl}$. Since this is true for any set of axes $(0x'_1, 0x'_2, 0x'_3)$, the result must be true with e'_{kl} and σ'_{kl} non-vanishing when $k \neq l$. Hence in general

$$Ee_{ij} = (1+\nu)\,\sigma_{ij} - \nu\,(\sigma_{11} + \sigma_{22} + \sigma_{33})\,\delta_{ij} \qquad (6.52)$$

The stresses can be obtained in terms of the strains by putting $i = j$ and summing to give

$$E(e_{11} + e_{22} + e_{33}) = (1+\nu)(\sigma_{11} + \sigma_{22} + \sigma_{33}) - 3\nu\,(\sigma_{11} + \sigma_{22} + \sigma_{33})$$

so that $\sigma_{11} + \sigma_{22} + \sigma_{33} = \dfrac{E}{(1-2\nu)}(e_{11} + e_{22} + e_{33})$. Equation (6.52) then gives

$$\sigma_{ij} = \frac{E}{(1+\nu)}\,e_{ij} + \frac{\nu E}{(1+\nu)(1-2\nu)}\,(e_{11} + e_{22} + e_{33})\,\delta_{ij} \qquad (6.53)$$

he quantities $\dfrac{E}{2(1+\nu)} = \mu$ and $\dfrac{\nu E}{(1+\nu)(1-2\nu)} = \lambda$ are called **Lamé constants**
nd equation (6.53) assumes the form

$$\sigma_{ij} = 2\mu e_{ij} + \lambda(e_{11} + e_{22} + e_{33})\delta_{ij} \tag{6.54}$$

quations of Motion

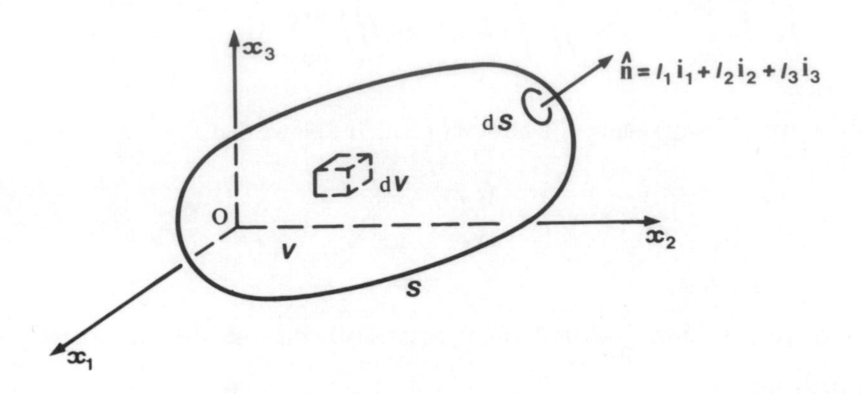

Figure 6.12

Consider a body of general shape with surface S enclosing a volume V as shown in Figure 6.12.

Suppose that there are body forces (such as gravity, inertia forces) of magnitude F_i/unit mass and surface forces X_i /unit area.

The total body forces are then $\iiint_V \rho F_i \, dV$ and the total surface forces are

$\iint_S \mathbf{X} \,.\, d\mathbf{S}$ where $d\mathbf{S} = \hat{\mathbf{n}} dS$. The surface forces will produce stresses e_{ij}

so that $\iint_S \mathbf{X} \,.\, d\mathbf{S} = \iiint_V e_{ij} \, l_j \, dS$.

Using the Gauss Divergence theorem, we obtain

$$\iint_S \mathbf{X} \,.\, d\mathbf{S} = \iint_S X_i \, dS = \iint_S e_{ij} \, l_j \, dS = \iiint_V \frac{\partial e_{ij}}{\partial x_j} \, dV$$

If the volume dV undergoes a displacement $u_i = (u_1, u_2, u_3)$ then its acceleration is $\dfrac{\partial^2 u_i}{\partial t^2}$. Newton's second law provides the equation

$$\iiint_V \rho \frac{\partial^2 u_i}{\partial t^2} \, dV = \iiint_V \rho F_i \, dV + \iint_S X_i \, dS$$

giving

$$\iiint_V \rho \frac{\partial^2 u_i}{\partial t^2} \, dV = \iiint_V \rho F_i \, dV + \iiint_V \frac{\partial e_{ij}}{\partial x_j} \, dV$$

Since this is true for any volume V, however small, it follows that

$$\rho \frac{\partial^2 u_i}{\partial t^2} = \rho F_i + \frac{\partial e_{ij}}{\partial x_j}$$

If the body is at rest, then $\dfrac{\partial^2 u_i}{\partial t^2} = 0$ and if, as is usually the case, the body forces are negligible then

$$\frac{\partial e_{ij}}{\partial x_j} = 0 \tag{6.55}$$

In full this gives three equations, which are often referred to as the boundary conditions. The first is

$$\frac{\partial e_{11}}{\partial x_1} + \frac{\partial e_{12}}{\partial x_2} + \frac{\partial e_{13}}{\partial x_3} = 0$$

Write down the other two.

Strain Energy

If a load is applied to a system then the work done by this load produces

(i) an increase in Kinetic Energy

(ii) an increase in the recoverable Potential Energy

(iii) non-measurable work done due to permanent deformation which will appear in the form of heat.

We deal only within the elastic limit of small deformations and therefore (iii) does not enter into the discussion. Also the kinetic energy may be taken as zero since we are not interested in the overall motion of the body. Hence we are left with the consideration of (ii). This comprises (a) the potential energy of the body as a whole and (b) the potential energy coming from the relative displacement of points of the body. The latter is called the **strain energy**.

We can find an expression for the strain energy. Consider the relative movement between the faces of a cube depicted in Figure 6.4. For example the relative movement between the horizontal faces in the x_1 direction is e_{31} and so the work done by the stress component σ_{31} which builds up from zero to its full value whilst the movement is taking place is $\frac{1}{2}\sigma_{31}e_{31}$. The other stress components give similar contributions so that the total strain energy per unit volume is

$$U = \tfrac{1}{2}\sigma_{ij}\, e_{ij} \tag{6.56}$$

The principal directions are those where the stress components are simply $\begin{bmatrix} \sigma_1 & 0 & 0 \\ 0 & \sigma_2 & 0 \\ 0 & 0 & \sigma_3 \end{bmatrix}$ and the strain components are $\begin{bmatrix} e_1 & 0 & 0 \\ 0 & e_2 & 0 \\ 0 & 0 & e_3 \end{bmatrix}$. (Notice that the principal axes of the stress tensor coincide with those of the strain tensor.) Then $U = \frac{1}{2}\sigma_k e_k$ and using equations (6.51) we find that

$$U = \tfrac{1}{2}E\left\{\sigma_1^2 + \sigma_2^2 + \sigma_3^2 - 2\nu(\sigma_1\sigma_2 + \sigma_2\sigma_3 + \sigma_3\sigma_1)\right\} \tag{6.57}$$

or

$$U = \tfrac{1}{2}(\lambda + 2\mu)(e_1^2 + e_2^2 + e_3^2) - 2\mu(e_1e_2 + e_2e_3 + e_3e_1) \tag{6.58}$$

) **Viscous Fluid Dynamics**

The theory of elastic solids can be used to obtain results in fluid dynamics. Observe the following:

(a) When a fluid is at rest the pressure at a point P in the fluid is the same in all directions and the assumption is now made that the pressure p in a viscous fluid may be introduced by defining it as the mean of the normal stresses p_{11}, p_{22}, p_{33} acting at P. That is,

$$p = -\frac{1}{3}(p_{11} + p_{22} + p_{33})$$

Since p, the hydrostatic pressure, will have no influence upon the shear of the viscous fluid it is necessary to replace the σ_{ij} in the elastic solid theory by the departure from its hydrostatic value. Hence we replace σ_{ij} by $p_{ij} + p\delta_{ij}$.

(b) In an elastic solid we are concerned with the rate of change of lengths in finding the strains. However in a fluid, the stresses will produce changes in the velocity rather than length. Hence the quantity employed is the rate of strain, that is $e_{ij} = \frac{1}{2}\left(\frac{\partial u_i}{\partial x_j} + \frac{\partial u_j}{\partial x_i}\right)$ where u_i now represents *velocit*

(c) From experimental results, the relationship between stress and velocity gradient for flow parallel to the x_1 axis is found to be of the form shown in Figure 6.13.

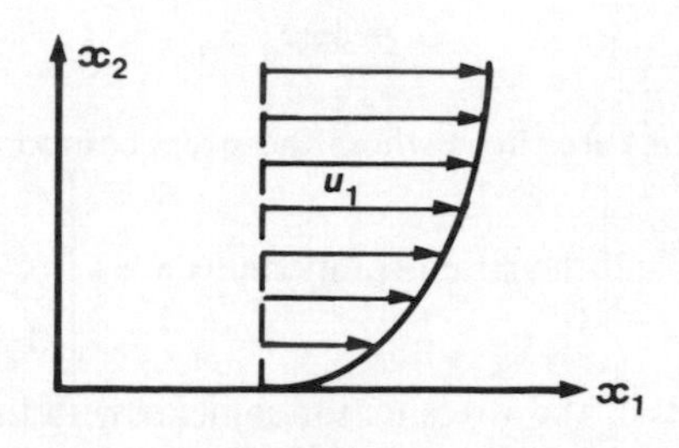

Figure 6.13

The pressure component $p_{21} = \mu \frac{\partial u_1}{\partial x_2}$ where μ is the coefficient of viscosity an

$u_2 = 0$. Had we studied flow parallel to the x_2 axis then we would find that

$p_{12} = \mu \frac{\partial u_2}{\partial x_1}$ $(u_1 = 0)$. Hence we can take in general

$$p_{12} = p_{21} = \mu\left(\frac{\partial u_1}{\partial x_2} + \frac{\partial u_2}{\partial x_1}\right) = 2\mu \,.\, ½\left(\frac{\partial u_1}{\partial x_2} + \frac{\partial u_2}{\partial x_1}\right) = 2\mu e_{12}$$

with similar expressions for p_{23} and p_{13}.

These expressions satisfy equation (6.54) with σ_{ij} replaced by $p_{ij} + p\delta_{ij}$. The connection between p_{ik} and e_{ik} is of the form

$$p_{ik} + p\delta_{ik} = 2\mu e_{ik} + \lambda(e_{11} + e_{22} + e_{33})\,\delta_{ik} \tag{6.59}$$

where λ is a constant to be determined.

To find λ, put $i = k$ in equation (6.59) and add, giving

$$p_{11}+p_{22}+p_{33}+3p = 2\mu(e_{11}+e_{22}+e_{33}) + 3\lambda(e_{11}+e_{22}+e_{33})$$

and, remembering that p has been taken as $-(p_{11}+p_{22}+p_{33})/3$, then $2\mu + 3\lambda = 0$. Hence

$$\lambda = -\frac{2\mu}{3} \tag{6.60}$$

We finally obtain for viscous flow

$$p_{ik} + p\delta_{ik} = 2\mu e_{ik} - \frac{2\mu}{3}(e_{11} + e_{22} + e_{33})\,\delta_{ik} \tag{6.61}$$

and it is seen that the elastic theory, with suitable interpretation has supplied the corresponding relations between stress and strain for a viscous fluid. Having established equation (6.61) we can now find the equations of motion of a viscous fluid.

Equations of Motion of a Viscous Fluid

We again consider a volume of fluid with body forces F_i/unit mass. The acceleration term is $\iiint_V \rho \dfrac{du_i}{dt}\, dV$ since u_i now represents the velocity; e_{ik} is of course replaced by p_{ik} so that

$$\rho \frac{du_i}{dt} = \rho F_i + \frac{\partial p_{ik}}{\partial x_k} \tag{6.62}$$

It is to be noted that the "total acceleration" is

$$\frac{du_i}{dt} = \frac{\partial u_i}{\partial t} + u_1 \frac{\partial u_i}{\partial x_1} + u_2 \frac{\partial u_i}{\partial x_2} + u_3 \frac{\partial u_i}{\partial x_3}$$

Now (6.62) becomes , using (6.61)

$$\rho \frac{du_i}{dt} = \rho F_i + \frac{\partial}{\partial x_k}\left\{-p\delta_{ik} + 2\mu e_{ik} - \frac{2}{3}\mu(e_{11} + e_{22} + e_{33})\,\delta_{ik}\right\}$$

giving

$$\rho \frac{du_i}{dt} = \rho F_i - \frac{\partial p}{\partial x_i} + 2\mu \frac{\partial e_{ik}}{\partial x_k} - \frac{2\mu}{3}\frac{\partial}{\partial x_i}(e_{11}+e_{22}+e_{33})$$

where μ is assumed to be constant.

Inserting expressions for $e_{ik} = \frac{1}{2}\left(\frac{\partial u_i}{\partial x_k} + \frac{\partial u_k}{\partial x_i}\right)$ yields

$$\rho \frac{du_i}{dt} = \rho F_i - \frac{\partial p}{\partial x_i} + \mu \frac{\partial}{\partial x_k}\left(\frac{\partial u_i}{\partial x_k} + \frac{\partial u_k}{\partial x_i}\right)$$

$$- \frac{2}{3}\mu \frac{\partial}{\partial x_i}\left(\frac{\partial u_1}{\partial x_1} + \frac{\partial u_2}{\partial x_2} + \frac{\partial u_3}{\partial x_3}\right)$$

As an exercise you should now write out in full the last two terms and show that

$$\rho \frac{du_i}{dt} = \rho F_i - \frac{\partial p}{\partial x_i} + \mu\left(\frac{\partial^2 u_i}{\partial x_1^2} + \frac{\partial^2 u_i}{\partial x_2^2} + \frac{\partial^2 u_i}{\partial x_3^2}\right)$$

$$+ \frac{\mu}{3} \frac{\partial}{\partial x_i}\left(\frac{\partial u_1}{\partial x_1} + \frac{\partial u_2}{\partial x_2} + \frac{\partial u_3}{\partial x_3}\right)$$

which is best written

$$\rho \frac{du_i}{dt} = \rho F_i - \frac{\partial p}{\partial x_i} + \mu \nabla^2 u_i + \frac{\mu}{3} \frac{\partial}{\partial x_i} (\text{div } \mathbf{V}) \tag{6.63}$$

where $\mathbf{V}$ is the velocity vector of the fluid. In vector form these equations are

$$\rho \frac{d\mathbf{V}}{dt} = \rho \mathbf{F} - \text{grad } p + \mu \nabla^2 \mathbf{V} + \frac{\mu}{3} \text{grad (div } \mathbf{V}) \tag{6.64}$$

which are the **Navier-Stokes** equations of motion of the fluid. There are only a exact solutions of these equations.

Equation of Continuity

Consider an arbitrary closed volume V with surface S at rest within the fluid. The rate of change of the mass of fluid inside this volume must be equal to the net rate at which fluid is crossing the surface. Therefore

$$- \frac{\partial}{\partial t} \iiint_V \rho \, dV = \iint_S \rho \mathbf{V} \, . \, d\mathbf{S}$$

which gives, using the Divergence Theorem,

$$\iiint_V \left(\frac{\partial \rho}{\partial t} + \operatorname{div}(\rho \mathbf{V}) \right) dV = 0$$

and since this is true for an arbitrary volume V, however small then

$$\frac{\partial \rho}{\partial t} + \operatorname{div}(\rho \mathbf{V}) = 0 \qquad (6.65)$$

which is the **equation of continuity**.

In component form this can be written as

$$\frac{\partial \rho}{\partial t} + \frac{\partial}{\partial x_i}(\rho u_i) = 0$$

or in full, assuming that ρ does not depend on x_i,

$$\frac{\partial \rho}{\partial t} + u_1 \frac{\partial \rho}{\partial x_1} + u_2 \frac{\partial \rho}{\partial x_2} + u_3 \frac{\partial \rho}{\partial x_3} + \rho \left(\frac{\partial u_1}{\partial x_1} + \frac{\partial u_2}{\partial x_2} + \frac{\partial u_3}{\partial x_3} \right) = 0$$

giving

$$\frac{d\rho}{dt} + \rho \operatorname{div} \mathbf{V} = 0 \qquad (6.66)$$

as an alternative form.

If the fluid is *incompressible* then, from (6.66), the equation of continuity becomes

$$\operatorname{div} \mathbf{V} = 0 \qquad (6.67)$$

and the equation of motion takes the form

$$\rho \frac{d\mathbf{V}}{dt} = \rho \mathbf{F} - \operatorname{grad} p + \mu \nabla^2 \mathbf{V} \qquad (6.68)$$

oblems

Show that the strain energy given by equation (6.57) can be expressed in terms of the invariants $J_1 = \sigma_1 + \sigma_2 + \sigma_3$ and $J_2 = \sigma_1 \sigma_2 + \sigma_2 \sigma_3 + \sigma_3 \sigma_1$ in the form $U = ½E\left\{J_1^2 - 2J_2(1+\nu)\right\}$.

2. Find the strains which arise from the stresses $\begin{bmatrix} \sigma_1 & 0 & 0 \\ 0 & \sigma_2 & 0 \\ 0 & 0 & \sigma_3 \end{bmatrix}$ and deduce the strains arising from an **isotropic** stress tensor $\begin{bmatrix} \sigma & 0 & 0 \\ 0 & \sigma & 0 \\ 0 & 0 & \sigma \end{bmatrix}$ where $\sigma = J_1/3$ (J_1 as defined in Problem 1). Find the strain energy in the latter case and show it is equal to $J_1^2/18K$ where $K = E/3(1 - 2\nu)$.

3. Find the strains resulting from the stresses $\begin{bmatrix} \sigma_1 - \sigma & 0 & 0 \\ 0 & \sigma_2 - \sigma & 0 \\ 0 & 0 & \sigma_3 - \sigma \end{bmatrix}$ where σ is defined as in Problem 2 and find the strain energy. Show that this can be written in the forn $(J_1^2 - 3J_2)/6G$ and $\left\{(\sigma_1 - \sigma_2)^2 + (\sigma_2 - \sigma_3)^2 + (\sigma_3 - \sigma_1)^2\right\}/12G$ where $G = E/2(1 + \nu)$.

4. Consider the effect of an isotropic stress tensor on a cube of material. What happens to the cube?

5. The stress tensor used in Problem 3 is called the **deviator stress tensor.** What is the sum of th principal strains arising? What effect then does the deviator stress tensor have?

6. Show that U can be expressed in the form $U = (J_1^2/18K) + (J_1^2 - 3J_2)/6G$ and deduce th the strain energy per unit volume is the sum of (a) the energy associated with change of volume and not shape (isotropic stress tensor effect) and (b) the energy associated with change of shape and not volume (deviator stress tensor effect).

7. Assuming a fluid flow where the density is constant under no body forces, show that

$$\frac{\partial u_i}{\partial t} + u_j \frac{\partial u_i}{\partial x_j} = -\frac{1}{\rho}\frac{\partial p}{\partial x_i} + \frac{1}{\rho}\frac{\partial}{\partial x_j}\left(\mu \frac{\partial u_i}{\partial x_j}\right)$$

8. Given an incompressible viscous fluid for which $p = p(\rho)$ show that the Navier-Stokes equations of motion can be written in the form

$$\frac{\partial u_i}{\partial t} + \frac{\partial}{\partial x_i}(\tfrac{1}{2} u_j u_j) - \epsilon_{jki} u_j \xi_k = F_i - \frac{\partial}{\partial x_i}\left(\int \frac{dp}{\rho}\right) + \mu \nabla^2 u_i$$

where $\xi \equiv \xi_k$ is the vorticity vector defined by

$$\xi_k = \epsilon_{ijk} \frac{\partial u_j}{\partial x_i}$$

hapter Seven

'he Finite Element Method

1 INTRODUCTION

In many engineering problems the equivalent *mathematical model* can be defined in rms of a system of differential equations together with given boundary conditions. If e model is *continuous* and therefore possesses an infinite number of degrees of freedom, is usually very difficult, if not impossible, to find an analytical solution. Thus we are d to look for an approximate solution and the continuous mathematical model is placed by an approximate *discrete* or numerical model which has a finite number of grees of freedom and a simpler solution.

Today's engineering problems, however, such as the construction of supersonic rcraft, oil-rigs in the North Sea, space satellites, large turbines for power stations, etc., ake it imperative that we approximate very closely the real behaviour of such structures. is necessary to formulate discrete models with up to several thousand degrees of eedom and new techniques of analysis have been developed to capitalise on the very eat power of modern computers. One such technique is the **finite element method** hich has evolved over the last twenty to thirty years. It was devised originally for use structural problems where previously, for simplicity and economy of effort, one-mensional line element models had been used. The finite element method of analysis troduced elements of more than one dimension and used these to build up a shape from number of such elements connected together in some prescribed manner. Furthermore, ge numbers of these simple elements could be taken to model geometrically complex uctures which could then be analysed, not by sophisticated analytical methods, t by fairly simple numerical methods. The price that had to be paid was the amount numerical computation to be carried out since very large sets of simultaneous algebraic uations have to be solved (often as many as 400 simultaneous equations).

The elements in common use are polygonal in shape, such as the triangle, rectangle d tetrahedron; more recently, elements with curved sides have been devised for use in ecific applications.

Although the finite element method was devised originally for use in structural oblems it is now used in fluid mechanics, heat transfer and general continuum mechanics ce the method is based on variational concepts and in effect minimises some nctional such as the total potential energy.

In this chapter we aim to give a feeling for the power of the method. We cannot pe to show the full theory on which the method is based; it is not necessary to know s in order to use the method. Armed with an existing computer program, it is possible to use the method purely as a "black box" whereby all the elements are assembled, their characteristics stored and the equations solved. Only by actually using the black box a few times will you realise the power and accuracy of the method. There are many book written on the subject (see the Bibliography) which give a detailed analysis of the metho and a justification of the use of different elements but by their very nature they tend to deter the student from actually using the method.

Your first step should be to use an existing program a few times. Students to whom we have spoken have said that it was only when they realised how good the methc was that they felt motivated to study the theory in more detail.

7.2 GLOBAL VIEW OF THE FINITE ELEMENT TECHNIQUE

In the finite element technique the body (structure or continuum) is imagined to broken up into a number of *elements* of finite dimensions. An example is given in Figure 7.1 where the problem is to find the stresses near a hole in a square flat plate (a quarter of the flat plate being shown).

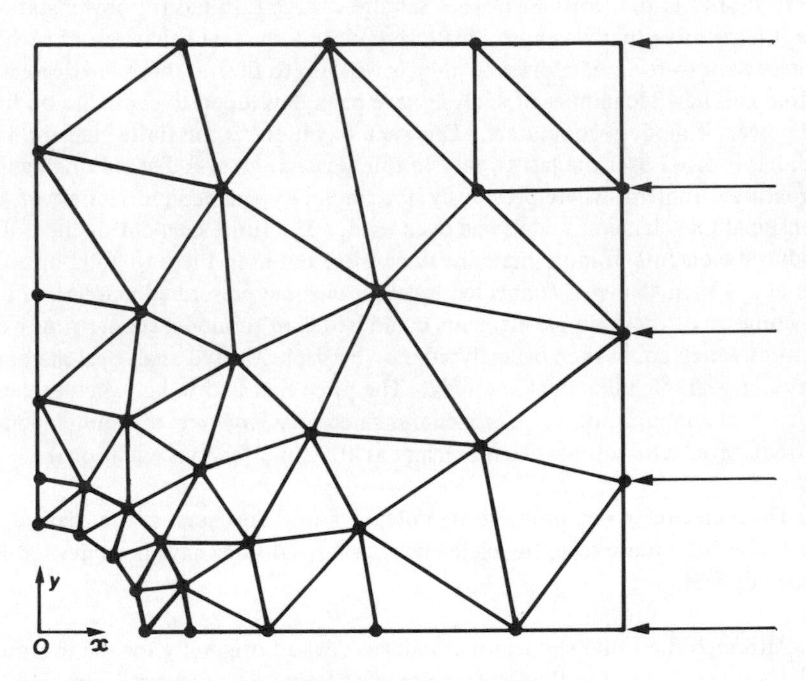

Figure 7.1

A further example is given in Figure 7.2 where it is required to find the stresses in cantilever beam subjected to an end load.

Figure 7.3 shows the breakdown carried out for a diesel engine crank case analys (Wakely (1970)).

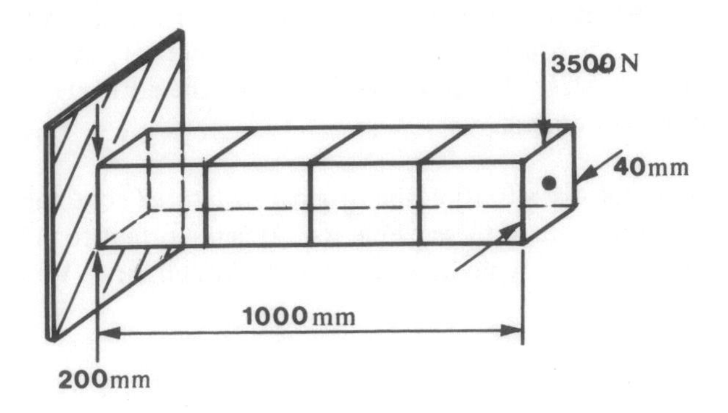

Figure 7.2

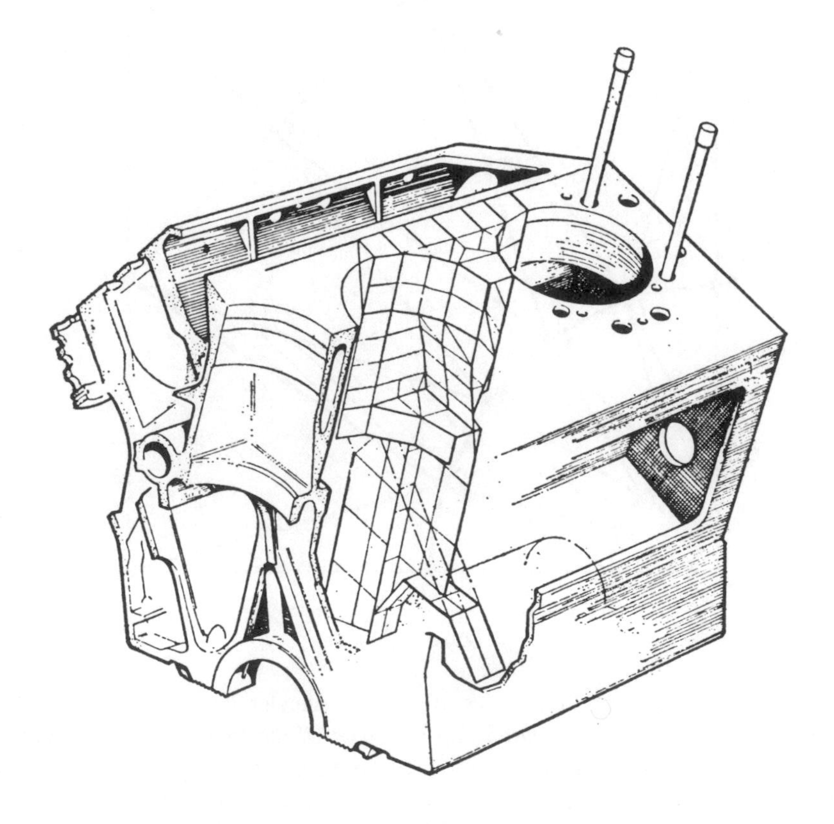

Figure 7.3

A variety of element shapes may be used and with care, different shapes may be
ed in the same solution region. The number and type of elements to be used in a given

problem are matters of engineering judgement but in general, regions in which the parameters (stress, velocity, etc.) are varying rapidly require many smaller elements to model accurately the true situation. This is seen in Figure 7.1 where the elements neare the **hole** are smaller in size. The process of choosing the elements is called **discretization of the continuum.**

Having chosen the discretization the original structure (continuum) is assumed to be replaced by an assemblage of these elements. It is important to note that these elements are only connected by **nodes** and not along boundaries. We therefore no longe have a continuum.

Consider again the problem given in Figure 7.1; for clarity we divide the region int a smaller number of elements as shown in Figure 7.4.

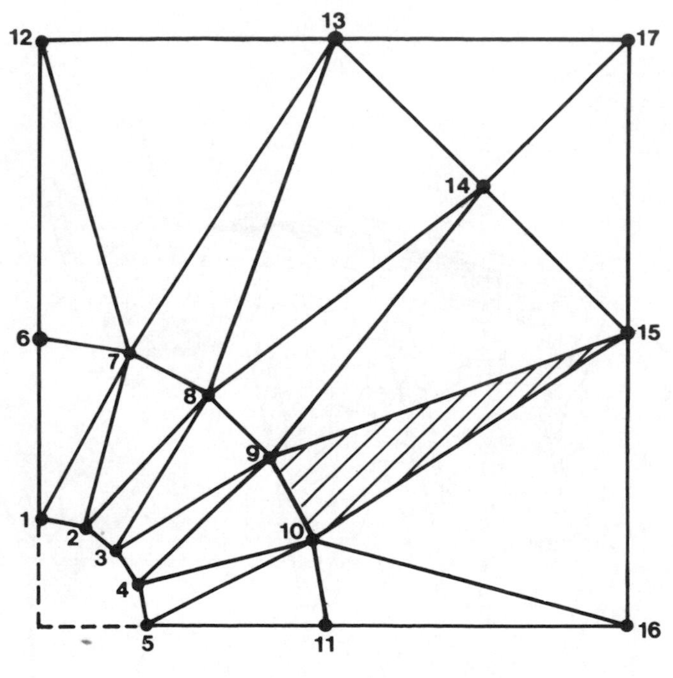

Figure 7.4

The plate is divided into 20 triangles, but these triangles are connected to each other *only at the nodes shown and not along the edges of the triangles.* The triangle wit nodes 9, 10, 15 (shaded), is only connected to the adjacent triangles at these nodes an for example, the displacement inside this element will depend only upon the displacements at the nodes 9, 10 and 15. In Figure 7.2, the elements are rectangular blocks and these blocks are connected only at the vertices of the blocks. Thus the continuum has been replaced by finite elements. What determines the choice of nodes and shape of elements will be considered later.

We have said that the elements are connected at the nodes only and that the ontinuum is discretized into a number of finite element shapes. This assembly of lements could not possibly behave like the original continuum unless we also ensure nat some other field variable, such as displacement or temperature, is approximated in ich a way that continuity is maintained along the element boundaries as well as equality ccurring at the nodes. This is accomplished by choosing a suitable approximation for ne selected field variable in the form of a polynomial which is defined using the nodal alues. The polynomial defined will be different for each element but will be selected give continuity across element boundaries. Thus the **discrete model** of the original ontinuum consists of a collection of elemental shapes, connected at the nodes, in each f which we approximate the variation of some continuous quantity, such as displacement.

The next step therefore is to **select interpolation polynomials** which will represent ne variation of the field variable over the element. This field variable can be a scalar or ector (or a higher order tensor). The degree of the polynomial chosen will depend upon ne number of nodes assigned to the element, the nature and number of unknowns at ach node and certain continuity requirements imposed at the nodes and along element oundaries.

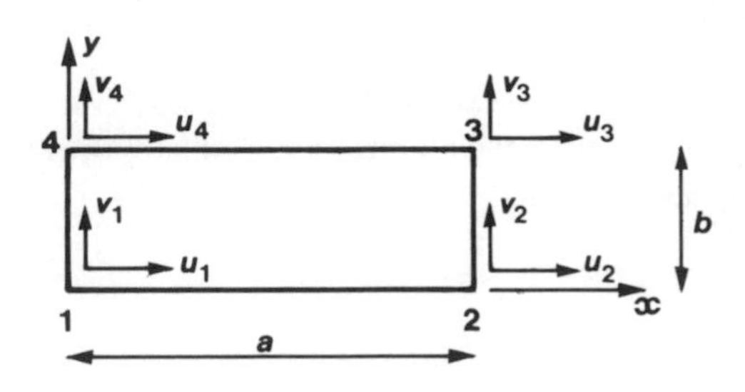

Figure 7.5

For example, the displacement of the rectangular element shown in Figure 7.5 night be given by

$$u = \alpha_1 + \alpha_2 x + \alpha_3 y + \alpha_4 xy$$
$$v = \alpha_5 + \alpha_6 x + \alpha_7 y + \alpha_8 xy$$

nd the displacement at each point inside the rectangle will be determined by these olynomials. The α_i will be fixed by the values of u and v at the nodes.

Having selected the interpolating polynomials, the next step is to **find the element roperties.** These are obtained from direct physical reasoning; for example, in stress/ train problems, we require merely the calculation of the stiffness properties (stiffness natrix) of the individual elements based on the interpolating polynomials chosen. The lement properties can be found alternatively using a variational approach, which is most onvenient for continuum problems; this approach will not be considered here.

Once the element properties are determined, we **assemble the element properties** to obtain the equations of the whole system. In other words, the matrix equations expressing the behaviour of the elements are combined to form the matrix equations expressing the behaviour of the whole assemblage of elements.

Finally, appropriate boundary conditions (external forces, fixing of points of the structure etc.) are applied and we then **solve the system equations.**

To summarise, the steps taken in a finite element analysis are

(1) Discretize the continuum

(2) Select interpolation polynomials

(3) Find the element properties

(4) Assemble the elements

(5) Solve the system equations.

In practice, using a "black box" approach, we only have to undertake step 1; the other steps are taken care of by a computer program package. We then have to interpret the solution and pick out the information required.

In the next section a simple case study is presented. We show you the input and output for such a problem, then we interpret and compare the results obtained with those obtained by other methods.

7.3 CASE STUDY

Consider a steel cantilever beam subjected to an end load of 35000 N (Figure 7.6).

(a) Analytical solution

The solution to the problem is well known analytically. The deflection y of the beam at a distance x along the beam is given by

$$EIy = \frac{Wlx^2}{2} - \frac{Wx^3}{6} \tag{7.1}$$

where E, the modulus of elasticity, is taken as 210000 N/mm^2, I is the moment of inertia of the cross-section about the neutral axis, namely $bd^3/12 = (40)(200)^3/12$ mm^4, b being the breadth and d the depth of the beam, W is the load on the beam which is taken as 35000 N and l is the overall length. The maximum deflection occurs at the end at which the load is applied and is given by

$$y_{\max} = Wl^3/3EI = 2.0833\dot{3} \text{ mm}$$

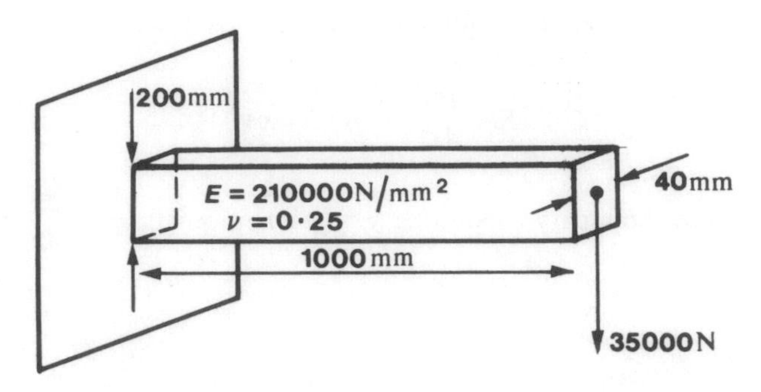

Figure 7.6

In the above analysis it was assumed that the applied load produced a distortion in pure bending only. However, shear stresses act across the cross-section and these produce an additional deflection of the centroid of the end section equal to $Wa^2(1+\nu)l/E$, where $2a$ is the depth of the beam. Hence we must add an additional deflection of 0.07813 mm due to shear, so that the total deflection of the centroid of the end section is $(2.0833\dot{3} + 0.07813)$ mm which gives

$$y_{max} = 2.16146 \text{ mm}$$

The bending moment M is given by

$$M = -EI\frac{d^2y}{dx^2} = -W[l-x]$$

and at $x = 0$, $M = -Wl = 35 \times 10^6$ Nmm.

The maximum stress occurs when $x = 0$ and is given by $\sigma_{max} = 6M/bd^2$ as illustrated in Figure 7.7. Thus $\sigma_{max} = 131.25$ N/mm².

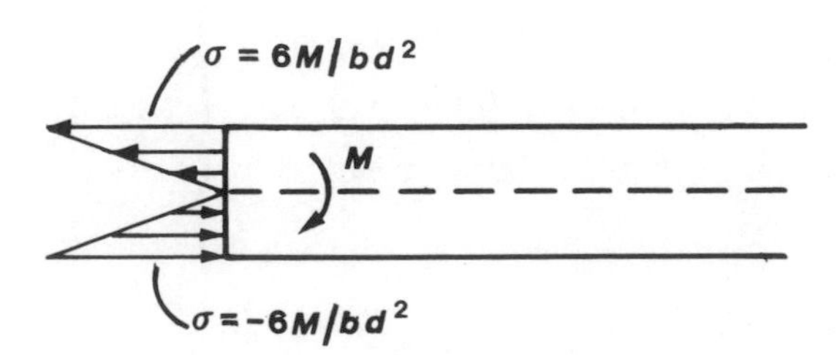

Figure 7.7

(b) Finite Difference solution

A finite difference approximation is made to solve the bending moment equation

$$EI \frac{d^2y}{dx^2} = W(l-x) \tag{7.2}$$

subject to the boundary conditions $y = 0$ when $x = 0$ and $\frac{dy}{dx} = 0$ when $x = 0$

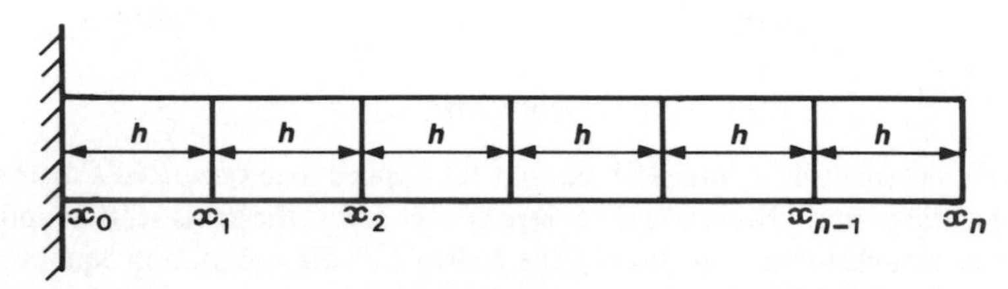

Figure 7.8

The beam is divided into n equal parts each of length h as illustrated in Figure 7.8; the deflection at the ends of the portions are $y_0, y_1, y_2, \ldots\ldots, y_n$. The term $\frac{d^2y}{dx^2}$ is approximated at x_i by the expression

$$(y_{i+1} - 2y_i + y_{i-1})/h^2$$

giving the approximate bending moment equation

$$y_{i+1} - 2y_i + y_{i-1} = \frac{Wh^2}{EI}(l - x_i)$$

Remembering that $y_0 = 0$ and adding a fictitious deflection $y_{-1} = y_1$ to give $\frac{dy}{dx} = 0$ at $x = 0$ we obtain the simultaneous equations

$$\left.\begin{aligned}
y_0 &= 0 \\
2y_1 &= G(l - x_0) \\
y_2 - 2y_1 &= G(l - x_1) \\
y_3 - 2y_2 + y_1 &= G(l - x_2) \\
&\vdots \\
y_n - 2y_{n-1} + y_{n-2} &= G(l - x_{n-1})
\end{aligned}\right\} \tag{7.3}$$

where $G = Wh^2/EI$.

The solution of these equations for various values of n is given in Table 7.1 and compared with the exact value given by equation (7.1) correct to 5 decimal places. Selected values of x are shown in the table.

Table 7.1

x	$y(n = 4)$	$y(n = 8)$	$y(n = 16)$	$y(n = 32)$	$y(n = 64)$	y (exact) pure bending
0	0	0	0	0	0	0
250	0.19531	0.18311	0.18005	0.17929	0.17910	0.17904
500	0.68359	0.65918	0.65308	0.65155	0.65117	0.65104
750	1.36719	1.33057	1.32141	1.31912	1.31855	1.31836
1000	2.14844	2.09961	2.08740	2.08435	2.08359	2.08333

It is seen that we need to take n equal to 64 before we get close to the exact values. (There are, of course, more accurate approximations using other finite difference formulae.)

Finite Element method

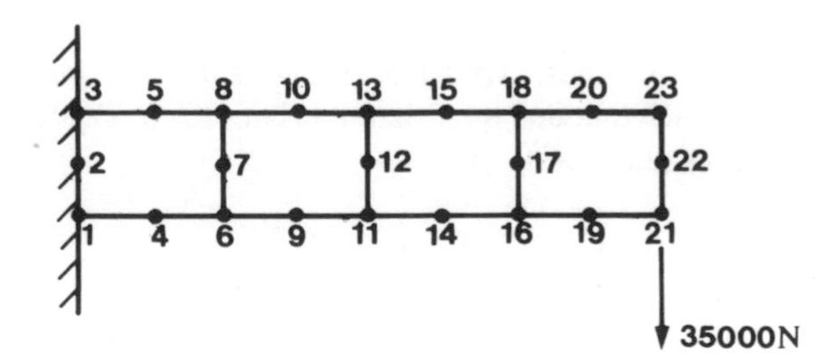

Figure 7.9

Figure 7.9 shows the beam approximated by four finite elements, the elements being rectangular blocks each with eight nodes. The nodes are numbered as shown. Notice that the finite elements are connected one to the next by three nodes. For example, nodes 6, 7 and 8 connect the first two finite elements of the beam. We used a "black box" package (called "LUFE") which is available at Loughborough University of Technology. The input data was as follows:

```
        'COORDS'
LINE
0001    NODE            COORDS                   REP   ADDN    ADD
0002    I=1,21,5) I     (250*(I–1)/5),0          2     1       0,100
0003    I=4,19,5) I     (125+(250*(I–4)/5)),0    1     1       0,200
        'QUAD8SM'
LINE
0001    NODES               MATERIAL        SECTION   REP   ADDN
0002    1,4,6,7,8,5, 3, 2   210000, 0.25    40        3     5
        'RESTRAINTS'
LINE
0001    NODES           SPRINGS
0002    1,2,3           –1,–1
        'LOAD'
LINE
0001    NODE            VALUES
0002    21              0,–5833.333333
0003    22              0,–23333.333333
0004    23              0,–5833.333333
```

The first part of the input gives the coordinates of the nodes: line 0002 says for nodes 1,6,11,16,21, the x-coordinates are 0,250,500,750,1000. The formula is repeated twice with I increased by 1 each time to give the nodes 2,7,12,17,22 and 3,8,13,18,23. ADDXY then gives the amount to be added on for each repea to get the appropriate x and y coordinates. Line 0003 determined the coordinates of nodes 4,9,14,19 and is repeated with I increased by 1 to give th same x coordinates for nodes 5,10,15,20 with corresponding y coordinate.

The second part of the input, 'QUAD8SM', gives the ordering of the nodes anticlockwise for each 8-noded quadrilateral element - notice it is repeated 3 time each time 5 being added on to give

1, 4, 6, 7, 8, 5, 3, 2
6, 9, 11, 12, 13, 10, 8, 7
11, 14, 16, 17, 18, 15, 13, 12
16, 19, 21, 22, 23, 20, 18, 17

The material properties $E = 210\,000$ N/mm² and $\nu = 0.25$ are also specified in this line together with the width of the beam of 40 mm. The third part, 'RESTRAINTS', says that nodes 1, 2, 3 are fixed in position (the term 'SPRINGS' is peculiar to this program and ensures fixing of these nodes) and the fourth part gives the load on the beam. The load of 35000 N has been split into 3 parts in the ratio 1 : 4 : 1 to represent more accurately the distribution of the load over the end of the beam at nodes 21, 22 and 23, acting vertically downwards. We now input to the computer the position of all nodes for each finite element, the type of element, the restraints and the load applied.

The next part of the program specifies the type of problem and results required and is simply

```
MASTER
PROBLEM TYPE IS 'STRESS-2D'                (2-dimension stress problem)
USE 'COORDS','QUAD8SM','RESTRAINTS'
USE 'LOAD' AS CASE 1
CALC K                                     (stiffness matrix)
ASSEMBLE AND REDUCE                        (Instruction to assemble system
SOLVE FOR CASE 1                           equation and reduce to suitable
PRINT REACTIONS FOR CASE 1                 matrix form)
PRINT DISPLACEMENTS FOR CASE 1
PRINT STRESSES FOR ELEMENTS 'QUAD8SM' FOR CASE 1
```

The computer now produces the requested output giving first of all the reactions at the nodes 1, 2, 3, then the displacements at each node, the stresses for each element at the nodes in anticlockwise order, and the average nodal stresses. A section of the output is reproduced below, this gives the calculated displacements and stresses.

ISPLACEMENTS FOR LOAD CASE LOAD

ODE	DISPLACEMENTS
1	−0.00000000, −0.00000000
2	−0.00000000, 0.00000000
3	0.00000000, −0.00000000
4	−0.06990728, −0.05354839
5	0.06990728, −0.05354839
6	−0.13314474, −0.19031225
7	−0.00000000, −0.18235880
8	0.13314474, −0.19031225
9	−0.18774919, −0.39694815
10	0.18774919, −0.39694815
11	−0.23175743, −0.66521698
12	−0.00000000, −0.66187452
13	0.23175743, −0.66521698
14	−0.26569801, −0.98322129
15	0.26569801, −0.98322129
16	−0.29009949, −1.33768938
17	−0.00000000, −1.33557517
18	0.29009949, −1.33768938
19	−0.30481085, −1.71631230
20	0.30481085, −1.71631230
21	−0.30968108, −2.10717371
22	−0.00000000, −2.10725425
23	0.30968108, −2.10717371

NODAL AVERAGE STRESSES

NODE,	SXX,	SYY,	TXY
1	−131.250	−32.813	−4.375
2	−0.000	−0.000	−4.375
3	131.250	32.813	−4.375
4	−114.844	−12.009	−4.375
5	114.844	12.009	−4.375
6	−98.438	8.795	−4.375
7	−0.000	−0.000	−4.375
8	98.438	−8.795	−4.375
9	−82.031	3.214	−4.375
10	82.031	−3.214	−4.375
11	−65.625	−2.368	−4.375
12	−0.000	0.000	−4.375
13	65.625	2.368	−4.375
14	−49.219	−0.846	−4.375
15	49.219	0.846	−4.375
16	−32.813	0.677	−4.375
17	−0.000	−0.000	−4.375
18	32.813	−0.677	−4.375
19	−16.406	0.169	−4.375
20	16.406	−0.169	−4.375
21	−0.000	−0.338	−4.375
22	−0.000	−0.000	−4.375
23	−0.000	0.338	−4.375

As regards the displacement at node 22, the deflection is 2.10725 mm corresponding to the theoretical value of 2.16146 mm and is in error by 2.6%. However, the maximum stress at nodes 1 and 3, namely 131.25 N/mm², is exactly the same as that given by the analytical theory, with zero stress at node 2. Remember that the beam has been split into 4 elements only and yet even this coarse division gives excellent results. We could have used more elements and obtained more accurate results for the deflection. Notice also that we have no real idea of what is happening inside the program and yet we can immediately use its results. This is a danger of course, and therefore in the following sections we shall examine the process in sufficient detail to permit a good understanding of the background theory.

7.4 DEFINING ELEMENTS AND THEIR PROPERTIES

Having decided on the finite element mesh (discretization) the next step is to select interpolation polynomials which represent the variation of the field variable within an individual element. In structural mechanics problems the field variable will be displacement.

First of all consider one of the simplest finite elements: **triangular element for a thin plate.** To determine the stiffness matrix for plane stress problems requires 6 steps.

Step 1

Assume a functional form for the displacement field within an element. Figure 7.10 shows a plane elastic continuum, subjected to plane stress, divided into triangular elements.

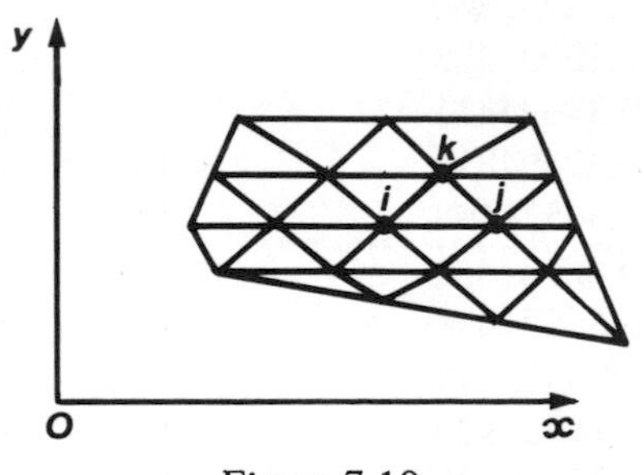

Figure 7.10

Consider a typical element with nodes i, j, k and with nodal coordinates (x_i, y_i), (x_j, y_j), (x_k, y_k). The displacement at any point (x, y) within the element has two components, u and v in the x and y directions respectively. We choose a displacement function which will uniquely define the displacement at any point in the element, given the displacement at the nodes, i.e. the 6 values u_i, v_i, u_j, v_j, u_k, v_k. A suitable displacement function is given by

$$\left.\begin{aligned} u &= \alpha_1 + \alpha_2 x + \alpha_3 y \\ v &= \alpha_4 + \alpha_5 x + \alpha_6 y \end{aligned}\right\} \qquad (7.4)$$

here the α_i are constants which have to be selected to give the appropriate nodal isplacements. It is seen that the displacements vary linearly within the element and long element boundaries and, consequently, continuity of the displacement field is btained throughout the structure modelled by the mesh of adjacent elements. (This ollows since the displacement along a boundary between two adjacent elements is niquely determined by the displacement of the two nodes at the end of such a oundary.) To find the α_i use the equations

$$\begin{aligned} u_i &= \alpha_1 + \alpha_2 x_i + \alpha_3 y_i \qquad & v_i &= \alpha_4 + \alpha_5 x_i + \alpha_6 y_i \\ u_j &= \alpha_1 + \alpha_2 x_j + \alpha_3 y_j \qquad & v_j &= \alpha_4 + \alpha_5 x_j + \alpha_6 y_j \\ u_k &= \alpha_1 + \alpha_2 x_k + \alpha_3 y_k \qquad & v_k &= \alpha_4 + \alpha_5 x_k + \alpha_6 y_k \end{aligned}$$

hich can be written

$$\begin{bmatrix} u_i \\ v_i \\ u_j \\ v_j \\ u_k \\ v_k \end{bmatrix} = \begin{bmatrix} 1 & x_i & y_i & 0 & 0 & 0 \\ 0 & 0 & 0 & 1 & x_i & y_i \\ 1 & x_j & y_j & 0 & 0 & 0 \\ 0 & 0 & 0 & 1 & x_j & y_j \\ 1 & x_k & y_k & 0 & 0 & 0 \\ 0 & 0 & 0 & 1 & x_k & y_k \end{bmatrix} \begin{bmatrix} \alpha_1 \\ \alpha_2 \\ \alpha_3 \\ \alpha_4 \\ \alpha_5 \\ \alpha_6 \end{bmatrix}$$

or $\qquad \boldsymbol{\delta} = \mathbf{A}\boldsymbol{\alpha}$

hich can be solved to find the α_i in the form

$$\boldsymbol{\alpha} = \mathbf{A}^{-1}\boldsymbol{\delta} \qquad (7.5)$$

riting equation (7.4) as

$$\begin{bmatrix} u \\ v \end{bmatrix} = \begin{bmatrix} 1 & x & y & 0 & 0 & 0 \\ 0 & 0 & 0 & 1 & x & y \end{bmatrix} \begin{bmatrix} \alpha_1 & \alpha_2 & \alpha_3 & \alpha_4 & \alpha_5 & \alpha_6 \end{bmatrix}^{\mathrm{T}} \qquad (7.6)$$

r $\mathbf{f} = \mathbf{N}\boldsymbol{\alpha}$, using equation (7.5) we obtain

$$\mathbf{f} = \mathbf{N}\mathbf{A}^{-1}\boldsymbol{\delta} \qquad (7.7)$$

tep 2

Next we find the element strains corresponding to the assumed displacements given by (7.3) (or (7.7)). For plane stress, the only strains that exist are those in the x–y plane $(\epsilon_x, \epsilon_y, \gamma_{xy})$ given for small displacements, by

$$\epsilon = \begin{bmatrix} \epsilon_x \\ \epsilon_y \\ \gamma_{xy} \end{bmatrix} = \begin{bmatrix} \dfrac{\partial u}{\partial x} \\ \dfrac{\partial v}{\partial y} \\ \dfrac{\partial u}{\partial y} + \dfrac{\partial v}{\partial x} \end{bmatrix} \tag{7.8}$$

Using equations (7.3) we obtain $\dfrac{\partial u}{\partial x} = \alpha_2$, $\dfrac{\partial v}{\partial y} = \alpha_6$, $\dfrac{\partial u}{\partial y} + \dfrac{\partial v}{\partial x} = \alpha_3 + \alpha_5$. Hence

$$\boldsymbol{\epsilon} = \begin{bmatrix} 0 & 1 & 0 & 0 & 0 & 0 \\ 0 & 0 & 0 & 0 & 0 & 1 \\ 0 & 0 & 1 & 0 & 1 & 0 \end{bmatrix} \begin{bmatrix} \alpha_1 & \alpha_2 & \alpha_3 & \alpha_4 & \alpha_5 & \alpha_6 \end{bmatrix}^{\mathrm{T}} \tag{7.9}$$

which we shall write as $\boldsymbol{\epsilon} = \mathbf{B}\boldsymbol{\alpha}$ and using equation (7.5) we obtain

$$\boldsymbol{\epsilon} = \mathbf{BA}^{-1}\boldsymbol{\delta} \tag{7.10}$$

The strains at any point within the element are now given in terms of the displacement at the nodes; notice from equation (7.9) that the strains are constant throughout the element, a consequence of course of choosing linear displacements u and v.

Step 3

We can now write down the constitutive equations relating the strains and stresses for plane stress, which are known to be given by

$$\begin{bmatrix} \sigma_x \\ \sigma_y \\ \tau_{xy} \end{bmatrix} = \frac{E}{(1-\nu^2)} \begin{bmatrix} 1 & \nu & 0 \\ \nu & 1 & 0 \\ 0 & 0 & \dfrac{1-\nu}{2} \end{bmatrix} \begin{bmatrix} \epsilon_x \\ \epsilon_y \\ \gamma_{xy} \end{bmatrix} \tag{7.11}$$

where E is Young's modulus and ν is Poisson's ratio for the material which is assumed to be isotropic. Equation (7.11) is simply

$$\sigma = \mathbf{D}\epsilon \tag{7.12}$$

One of the advantages of the finite element method is apparent – we can insert ifferent form for the matrix **D** when the stress-strain characteristics of the ement are not linear.

om equation (7.9) on substituting for ϵ,

$$\sigma = \mathbf{DB}\boldsymbol{\alpha} \tag{7.13}$$

hich can be written

$$\sigma = \mathbf{DBA}^{-1}\delta \tag{7.14}$$

ote that equation (7.14) gives the stresses at any point within the element in terms of e deflections at the nodes. Furthermore, from equation (7.11) we note that, since the rains are constant, the stresses are also constant: a further consequence of choosing a ear displacement field.

ep 4

The stiffness matrix can now be derived by the principle of virtual work. Assume rces act at the nodes of the element of the form U_i, V_i, U_j, V_j, U_k, V_k as shown Figure 7.11.

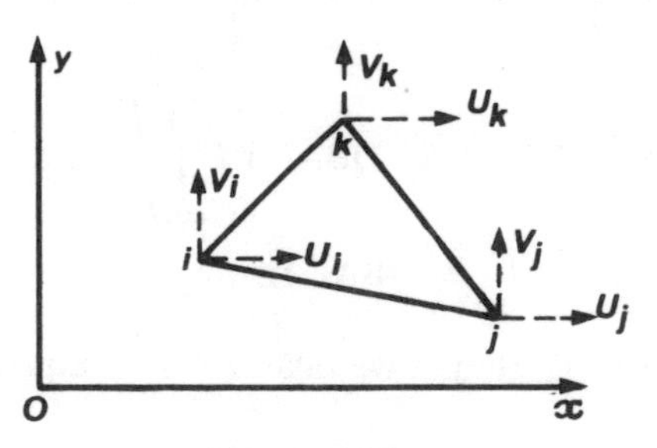

Figure 7.11

This set of nodal forces must be statically equivalent to the internal stresses of the ement. We represent the nodal forces in the form

$$\mathbf{P} = [U_1 \quad V_1 \quad U_2 \quad V_2 \quad U_3 \quad V_3]^{\mathrm{T}} \tag{7.15}$$

he internal work done by the stress field is equated to the external work done by the odal forces **P** due to a virtual displacement $\bar{\boldsymbol{\delta}}$ of the nodes $= [\bar{u}_i, \bar{v}_i, \bar{u}_j, \bar{v}_j, \bar{u}_k, \bar{v}_k]^{\mathrm{T}}$

For the element, the external work done by the nodal forces, W_E, as a result of the rtual displacement is given by

$$\begin{aligned} W_E &= \bar{u}_i U_i + \bar{v}_i V_i + \bar{u}_j U_j + \bar{v}_j V_j + \bar{u}_k U_k + \bar{v}_k V_k \\ &= [\bar{u}_i, \bar{v}_i, \bar{u}_j, \bar{v}_j, \bar{u}_k, \bar{v}_k][U_i, V_i, U_j, V_j, U_k, V_k]^{\mathrm{T}} \end{aligned}$$

i.e.

$$W_E = \bar{\boldsymbol{\delta}}^T \mathbf{P} \tag{7.16}$$

Let the virtual strains produced in the element by the virtual displacement be $\bar{\boldsymbol{\epsilon}}$. Then the internal work done, W_i = (internal stress) x (internal strain) x (volume), i.e.

$$W_i = \int_V (\bar{\epsilon}_x \sigma_x + \bar{\epsilon}_y \sigma_y + \bar{\gamma}_{xy} \tau_{xy}) \, dV \tag{7.17}$$

where we integrate throughout the volume of the element. (Notice the use of $\boldsymbol{\sigma}$ since the set of nodal forces $\mathbf{P}$ is statically equivalent to the internal stress field $\boldsymbol{\sigma}$.) Thus

$$W_i = \int_V \bar{\boldsymbol{\epsilon}}^T \boldsymbol{\sigma} \, dV = T \iint \bar{\boldsymbol{\epsilon}}^T \boldsymbol{\sigma} \, dx \, dy \tag{7.18}$$

where T is the thickness of the element. Now $\boldsymbol{\sigma} = \mathbf{DBA}^{-1}\boldsymbol{\delta}$ and $\bar{\boldsymbol{\epsilon}} = \mathbf{BA}^{-1}\bar{\boldsymbol{\delta}}$, so that

$$W_i = T \iint \bar{\boldsymbol{\delta}}^T [\mathbf{A}^{-1}]^T \mathbf{B}^T \mathbf{DBA}^{-1} \boldsymbol{\delta} \, dx \, dy$$

$$= \bar{\boldsymbol{\delta}}^T [\mathbf{A}^{-1}]^T . \mathbf{B}^T \mathbf{DBA}^{-1} \boldsymbol{\delta} \, T \iint dx \, dy$$

That is

$$W_i = \bar{\boldsymbol{\delta}}^T [\mathbf{A}^{-1}]^T \mathbf{B}^T \mathbf{DBA}^{-1} \boldsymbol{\delta} \Delta \tag{7.19}$$

where Δ is the area of the triangle and we take $T = 1$.† Equating W_E and W_i produces

$$\mathbf{P} = [\mathbf{A}^{-1}]^T \mathbf{B}^T \mathbf{DBA}^{-1} \Delta \boldsymbol{\delta} \tag{7.20}$$

which is of the form

$$\mathbf{P} = \mathbf{k} \boldsymbol{\delta} \tag{7.21}$$

or Force = $\mathbf{k}$ (displacement). Hence $\mathbf{k}$ is the **stiffness matrix of the element** given by

$$\mathbf{k} = [\mathbf{A}^{-1}]^T \mathbf{B}^T \mathbf{DBA}^{-1} \Delta \tag{7.22}$$

We can calculate $\mathbf{k}$ for any individual triangular element and then the nodal forces can be found, assuming that the corresponding nodal displacements are known, by using equation (7.21).

† In a full analysis T is carried through the calculations only to disappear when the stress values are determined.

Step 5

Assuming that the equations describing the characteristics of each element have been found, the next step is the **assembly of elements** in which all the element equations are combined to form a complete set governing the behaviour of the whole assembly of elements of the idealised structure.

For each element there is an equation of the form (7.21), namely $\mathbf{P} = \mathbf{k}\boldsymbol{\delta}$, giving the relationship between the nodal forces and nodal displacements. All these are assembled to give an equation for the whole structure of the form

$$\mathbf{R} = \mathbf{Kq} \tag{7.23}$$

which relates the forces $\mathbf{R}$ and the displacements $\mathbf{q}$ at the nodes by means of the **overall stiffness matrix**, $\mathbf{K}$. We give a simple illustration of the assembly procedure in Section 7.6.

Step 6

Equation (7.23) can now be solved as $\mathbf{q} = \mathbf{K}^{-1}\mathbf{R}$ and given a set of nodal forces $\mathbf{R}$, the corresponding displacement $\mathbf{q}$ can be found.

Note

Different computer packages will use different numerical procedures for solving such large sets of simultaneous equations; there is no space here to enter into a discussion on them. However, it should be appreciated that such numerical procedures are not exact and therefore give rise to some error. It is tempting to think that the more elements used the better the accuracy obtained, but this may not be so due to the errors inherent in the solution procedures.

In the preceding steps we have assumed that the nodal forces $\mathbf{R}$ are known and moreover that these nodal forces are point loads. In a general problem, the whole structure will perhaps be subjected not only to point loads but to distributed loads and to constraints of various kinds. The nodal forces will have to be made equivalent to this desired loading.

Examples of Step 4

Example 1

Consider the element shown in Figure 7.12 (coordinates are in mm).

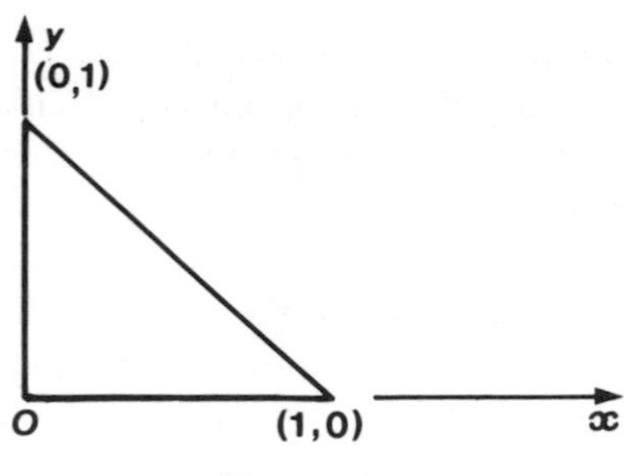

Figure 7.12

For this element

$$\mathbf{A} = \begin{bmatrix} 1 & 0 & 0 & 0 & 0 & 0 \\ 0 & 0 & 0 & 1 & 0 & 0 \\ 1 & 1 & 0 & 0 & 0 & 0 \\ 0 & 0 & 0 & 1 & 1 & 0 \\ 1 & 0 & 1 & 0 & 0 & 0 \\ 0 & 0 & 0 & 1 & 0 & 1 \end{bmatrix}$$

The inverse matrix is

$$\mathbf{A}^{-1} = \begin{bmatrix} 1 & 0 & 0 & 0 & 0 & 0 \\ -1 & 0 & 1 & 0 & 0 & 0 \\ -1 & 0 & 0 & 0 & 1 & 0 \\ 0 & 1 & 0 & 0 & 0 & 0 \\ 0 & -1 & 0 & 1 & 0 & 0 \\ 0 & -1 & 0 & 0 & 0 & 1 \end{bmatrix}$$

The matrix **B** is known and we only have to find **D** (take $\nu = 0.50$, $E = 210000$ N/mm²). Then

$$\mathbf{D} = \frac{E}{(1-\nu^2)} \begin{bmatrix} 1 & \nu & 0 \\ \nu & 1 & 0 \\ 0 & 0 & \dfrac{1-\nu}{2} \end{bmatrix} = 280000 \begin{bmatrix} 1 & \frac{1}{2} & 0 \\ \frac{1}{2} & 1 & 0 \\ 0 & 0 & \frac{1}{4} \end{bmatrix}$$

$$\text{Hence } \mathbf{DBA}^{-1} = 280000 \begin{bmatrix} -1 & -\frac{1}{2} & 1 & 0 & 0 & \frac{1}{2} \\ -\frac{1}{2} & -1 & \frac{1}{2} & 0 & 0 & 1 \\ -\frac{1}{4} & -\frac{1}{4} & 0 & \frac{1}{4} & \frac{1}{4} & 0 \end{bmatrix}$$

Δ = the area of the triangle = 0.5 mm². Hence

$$\mathbf{k} = [\mathbf{A}^{-1}]^{\mathrm{T}}\, \mathbf{B}^{\mathrm{T}}\, \mathbf{D}\mathbf{B}\mathbf{A}^{-1}\Delta \qquad (7.22)$$

$$= 140000 \begin{bmatrix} 1 & -1 & -1 & 0 & 0 & 0 \\ 0 & 0 & 0 & 1 & -1 & -1 \\ 0 & 1 & 0 & 0 & 0 & 0 \\ 0 & 0 & 0 & 0 & 1 & 0 \\ 0 & 0 & 1 & 0 & 0 & 0 \\ 0 & 0 & 0 & 0 & 0 & 1 \end{bmatrix} \begin{bmatrix} 0 & 0 & 0 \\ 1 & 0 & 0 \\ 0 & 0 & 1 \\ 0 & 0 & 0 \\ 0 & 0 & 1 \\ 0 & 1 & 0 \end{bmatrix} \begin{bmatrix} -1 & -\tfrac{1}{2} & 1 & 0 & 0 & \tfrac{1}{2} \\ -\tfrac{1}{2} & -1 & \tfrac{1}{2} & 0 & 0 & 1 \\ -\tfrac{1}{4} & -\tfrac{1}{4} & 0 & \tfrac{1}{4} & \tfrac{1}{4} & 0 \end{bmatrix}$$

$$= 140000 \begin{bmatrix} 5/4 & 3/4 & -1 & -1/4 & -1/4 & -1/2 \\ 3/4 & 5/4 & -1/2 & -1/4 & -1/4 & -1 \\ -1 & -1/2 & 1 & 0 & 0 & 1/2 \\ -1/4 & -1/4 & 0 & 1/4 & 1/4 & 0 \\ -1/4 & -1/4 & 0 & 1/4 & 1/4 & 0 \\ -1/2 & -1 & 1/2 & 0 & 0 & 1 \end{bmatrix}$$

(You should check this.)

Therefore, given displacements at the nodes, we can easily find the forces at the nodes using the equation $\mathbf{P} = \mathbf{k}\boldsymbol{\delta}$ or vice-versa. We must emphasise again that all the above calculations would take place within the "black box" program and all we need to input are the coefficients of the nodes and the values of E and ν.

Example 2 Rectangular Element

To reinforce understanding we now repeat the analysis with another simple element used in two-dimensional plane stress problems, the **rectangular four noded element**. The structure is divided into rectangular elements (a typical element is shown in Figure 7.13).

The element with nodes i, j, k, l is assumed to have nodal coordinates (x_i, y_i), (x_j, y_j), (x_k, y_k), (x_l, y_l). Now a displacement function is chosen which will provide the displacements within the element, given the displacements at the nodes: $(u_i, v_i, u_j, v_j, u_k, v_k, u_l, v_l)$.

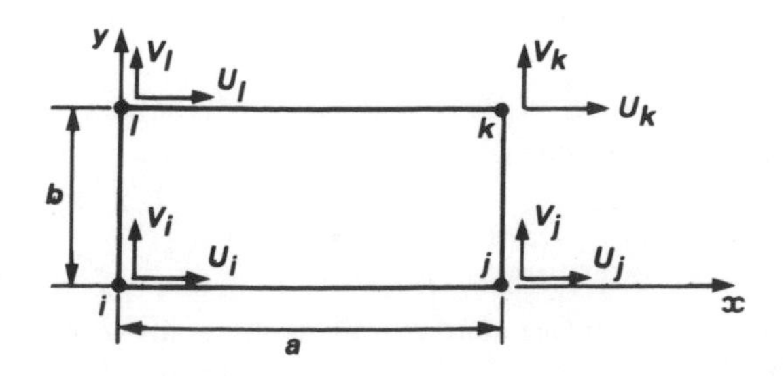

Figure 7.13

For convenience we take the element as having sides parallel to the x and y axes of length a and b as shown in Figure 7.13 and we choose the displacement function given by the polynomials

$$\begin{aligned} u &= \alpha_1 + \alpha_2 x + \alpha_3 y + \alpha_4 xy \\ v &= \alpha_5 + \alpha_6 x + \alpha_7 y + \alpha_8 xy \end{aligned} \tag{7.24}$$

The coefficients α_i in the equation (7.24) can be obtained by substituting in turn the nodal coordinates and solving the resulting simultaneous equations. To simplify the arithmetic we shall take the point i as (0, 0) so that the points j, k, l become $(a, 0)$, (a, b) and $(0, b)$ respectively. (We note that any change of coordinates will be dealt with by the computer program and so the selection of the above coordinates is not unrealistic.)

Substitution gives

$$\begin{bmatrix} u_i \\ v_i \\ u_j \\ v_j \\ u_k \\ v_k \\ u_l \\ v_l \end{bmatrix} = \begin{bmatrix} 1 & 0 & 0 & 0 & 0 & 0 & 0 & 0 \\ 0 & 0 & 0 & 0 & 1 & 0 & 0 & 0 \\ 1 & a & 0 & 0 & 0 & 0 & 0 & 0 \\ 0 & 0 & 0 & 0 & 1 & a & 0 & 0 \\ 1 & a & b & ab & 0 & 0 & 0 & 0 \\ 0 & 0 & 0 & 0 & 1 & a & b & ab \\ 1 & 0 & b & 0 & 0 & 0 & 0 & 0 \\ 0 & 0 & 0 & 0 & 1 & 0 & b & 0 \end{bmatrix} \begin{bmatrix} \alpha_1 \\ \alpha_2 \\ \alpha_3 \\ \alpha_4 \\ \alpha_5 \\ \alpha_6 \\ \alpha_7 \\ \alpha_8 \end{bmatrix} \tag{7.25}$$

or $\boldsymbol{\delta} = \mathbf{A}\boldsymbol{\alpha}$ which can be solved to give the values of α_i in the form

$$\boldsymbol{\alpha} = \mathbf{A}^{-1}\boldsymbol{\delta} \tag{7.26}$$

As before we form the equations

$$\begin{bmatrix} u \\ v \end{bmatrix} = \begin{bmatrix} 1 & x & y & xy & 0 & 0 & 0 & 0 \\ 0 & 0 & 0 & 0 & 1 & x & y & xy \end{bmatrix} \begin{bmatrix} \alpha_1 & \alpha_2 & \alpha_3 & \alpha_4 & \alpha_5 & \alpha_6 & \alpha_7 & \alpha_8 \end{bmatrix}^T \tag{7.27}$$

or $\mathbf{f} = \mathbf{N}\boldsymbol{\alpha}$ giving on substitution

$$\mathbf{f} = \mathbf{N}\mathbf{A}^{-1}\boldsymbol{\delta} \tag{7.28}$$

which again represents the displacement (u, v) at any point within the rectangle in terms of the nodal displacements. The second step is to find the element strains which were given by (7.8) which we write as $\epsilon = \mathbf{B}\boldsymbol{\alpha}$ where

$$\mathbf{B} = \begin{bmatrix} 0 & 1 & 0 & y & 0 & 0 & 0 & 0 \\ 0 & 0 & 0 & 0 & 0 & 0 & 1 & x \\ 0 & 0 & 1 & x & 0 & 1 & 0 & y \end{bmatrix} \tag{7.29}$$

giving as before

$$\epsilon = \mathbf{B}\mathbf{A}^{-1}\delta \tag{7.30}$$

The third step is to write down the constitutive equations connecting strains and stresses. These are $\sigma = \mathbf{D}\epsilon$ and therefore, by substitution, $\sigma = \mathbf{D}\mathbf{B}\boldsymbol{\alpha} = \mathbf{D}\mathbf{B}\mathbf{A}^{-1}\delta$. We now write down the stiffness matrix of the element

$$\mathbf{k} = T\,[\mathbf{A}^{-1}]^{\mathrm{T}} \left(\iint \mathbf{B}^{\mathrm{T}}\,\mathbf{D}\mathbf{B}\,\mathrm{d}x\,\mathrm{d}y \right) \mathbf{A}^{-1} \tag{7.31}$$

We have to perform a double integral since $\mathbf{B}$ is a function of x and y. Hence

$$\mathbf{B}^{\mathrm{T}}\mathbf{D}\mathbf{B} = \begin{bmatrix} 0 & 0 & 0 \\ 1 & 0 & 0 \\ 0 & 0 & 1 \\ y & 0 & x \\ 0 & 0 & 0 \\ 0 & 0 & 1 \\ 0 & 1 & 0 \\ 0 & x & y \end{bmatrix} \frac{E}{(1-\nu^2)} \begin{bmatrix} 1 & \nu & 0 \\ \nu & 1 & 0 \\ 0 & 0 & \frac{1-\nu}{2} \end{bmatrix} \begin{bmatrix} 0 & 1 & 0 & y & 0 & 0 & 0 & 0 \\ 0 & 0 & 0 & 0 & 0 & 0 & 1 & x \\ 0 & 0 & 1 & x & 0 & 1 & 0 & y \end{bmatrix}$$

$$= \frac{E}{(1-\nu^2)} \begin{bmatrix} 0 & 0 & 0 & 0 & 0 & 0 & 0 & 0 \\ 0 & 1 & 0 & y & 0 & 0 & \nu & \nu x \\ 0 & 0 & (\frac{1-\nu}{2}) & (\frac{1-\nu}{2})x & 0 & (\frac{1-\nu}{2}) & 0 & (\frac{1-\nu}{2})y \\ 0 & y & (\frac{1-\nu}{2})\,x & y^2 + \frac{x^2}{2}(1-\nu) & 0 & (\frac{1-\nu}{2})\,x & \nu x & (\frac{1+\nu}{2})\,xy \\ 0 & 0 & 0 & 0 & 0 & 0 & 0 & 0 \\ 0 & 0 & (\frac{1-\nu}{2}) & (\frac{1-\nu}{2})\,x & 0 & (\frac{1-\nu}{2}) & 0 & (\frac{1-\nu}{2})\,y \\ 0 & \nu & 0 & \nu x & 0 & 0 & 1 & x \\ 0 & \nu x & (\frac{1-\nu}{2})\,y & (\frac{1+\nu}{2})\,xy & 0 & (\frac{1-\nu}{2})\,y & x & x^2 + y^2(\frac{1-\nu}{2}) \end{bmatrix}$$

Therefore (7.31) involves integrals of the form

(i) $\displaystyle\int_0^b \int_0^a x \, dx \, dy = \frac{a^2 b}{2}$ (ii) $\displaystyle\int_0^b \int_0^a y \, dx \, dy = \frac{ab^2}{2}$

(iii) $\displaystyle\int_0^b \int_0^a xy \, dx \, dy = \frac{a^2 b^2}{4}$ (iv) $\displaystyle\int_0^b \int_0^a x^2 \, dx \, dy = \frac{a^3 b}{3}$

(v) $\displaystyle\int_0^b \int_0^a y^2 \, dx \, dy = \frac{a\, b^3}{3}$

It can easily be seen that $\iint \mathbf{B}^T \mathbf{D} \mathbf{B} \, dx \, dy$ is then given by

$$
\mathbf{C} = \frac{E}{(1-\nu^2)} \begin{bmatrix}
0 & 0 & 0 & 0 & 0 & 0 & 0 & 0 \\
0 & ab & 0 & \frac{ab^2}{2} & 0 & 0 & \nu ab & \frac{\nu a^2 b}{2} \\
0 & 0 & (\frac{1-\nu}{2})ab & (\frac{1-\nu}{4})a^2 b & 0 & (\frac{1-\nu}{2})ab & 0 & (\frac{1-\nu}{4})ab^2 \\
0 & \frac{ab^2}{2} & (\frac{1-\nu}{4})a^2 b & \frac{1}{3}[ab^3 + (\frac{1-\nu}{2})a^3 b] & 0 & (\frac{1-\nu}{4})a^2 b & \frac{\nu ab^2}{2} & (\frac{1+\nu}{8})a^2 b^2 \\
0 & 0 & 0 & 0 & 0 & 0 & 0 & 0 \\
0 & 0 & (\frac{1-\nu}{2})ab & (\frac{1-\nu}{4})a^2 b & 0 & (\frac{1-\nu}{2})ab & 0 & (\frac{1-\nu}{4})ab^2 \\
0 & \nu ab & 0 & \frac{\nu ab^2}{2} & 0 & 0 & ab & \frac{a^2 b}{2} \\
0 & \frac{\nu a^2 b}{2} & (\frac{1-\nu}{4})ab^2 & (\frac{1+\nu}{8})a^2 b^2 & 0 & (\frac{1-\nu}{4})ab^2 & \frac{a^2 b}{2} & \frac{1}{3}[a^3 b + (\frac{1-\nu}{2})ab^3]
\end{bmatrix} \qquad (7.32)
$$

The stiffness matrix $\mathbf{k} = T[\mathbf{A}^{-1}]^T \mathbf{C}\mathbf{A}^{-1}$ can then be found although we shall not perform the arithmetic here.

We reiterate that in the application of the finite element method it is not required to perform the above calculations but merely to input the nodal coordinates and the type of element we wish to use. Nevertheless it is necessary to be aware of the displacement functions that are being assumed inside the computer package and the consequences involved. The element strains for the 4-noded rectangular element are linear functions of x and y. Furthermore if we solve equation (7.25) to find $\boldsymbol{\alpha}$ and substitute back into equation (7.24) we find that along the element boundaries $y = 0$, $x = a$, $y = b$ and $x = 0$ respectively

$$u = u_i + \frac{1}{a}(u_j - u_i)x \qquad v = v_i + \frac{1}{a}(v_j - v_i)x \tag{7.33a}$$

$$u = u_j + \frac{1}{b}(u_k - u_j)y \qquad v = v_j + \frac{1}{b}(v_k - v_j)y \tag{7.33b}$$

$$u = u_l + \frac{1}{a}(u_k - u_l)x \qquad v = v_l + \frac{1}{a}(v_k - v_l)x \tag{7.33c}$$

$$u = u_i + \frac{1}{b}(u_l - u_i)y \qquad v = v_i + \frac{1}{b}(v_l - v_i)y \tag{7.33d}$$

It is clear from these equations that the displacements u and v along each oundary depend only on the components of displacement of the nodes on that particular oundary. As the components of displacement at the connecting nodes on the common oundary of two adjacent elements are equal , the components of displacement at all oints along that boundary must be the same for each element. We thus have continuity f displacement across the inter-element boundaries.

.5 CHOICE OF INTERPOLATING FUNCTIONS

When considering the **triangular** element for a thin plate restricted to plane stress roblems you will remember that the displacement functions used were

$$u = \alpha_1 + \alpha_2 x + \alpha_3 y$$
$$v = \alpha_4 + \alpha_5 x + \alpha_6 y$$

aving 6 unknown α_i. The 3 nodes of the element each had 2 components of isplacement so that, given nodal displacement, the values α_i could be found. The otal set of nodal displacements are commonly known as the **degrees of freedom** of the roblem and we see that the above triangular element had 6 degrees of freedom.

We next examined the rectangular element with four nodes, each node having 2 omponents of displacement so that this element had 8 degrees of freedom. We onsequently chose the displacement functions

$$u = \alpha_1 + \alpha_2 x + \alpha_3 y + \alpha_4 xy$$
$$v = \alpha_5 + \alpha_6 x + \alpha_7 y + \alpha_8 xy$$

which has 8 unknowns α_i that can be uniquely determined by the 8 known displacements t the nodes.

The displacement functions chosen are taken from the set of complete nth order olynomials

$$P_n(x, y) = \sum_{k=1}^{(n+1)(n+2)/2} \alpha_k x^i y^j, \quad i+j \leqslant n$$
$$i, j = 0, 1, 2, ...$$

For $n = 1$,

$$P_1(x, y) = \sum_{k=1}^{3} \alpha_k x^i y^j, \qquad i+j \leqslant 1$$

i.e.

$$P_1(x, y) = \alpha_1 + \alpha_2 x + \alpha_3 y$$
(linear interpolating function)

For $n = 2$,

$$P_2(x, y) = \sum_{k=1}^{6} \alpha_k x^i y^j, \qquad i+j \leqslant 2$$

$$= \alpha_1 + \alpha_2 x + \alpha_3 y + \alpha_4 x^2 + \alpha_5 xy + \alpha_6 y^2$$
(quadratic interpolating function)

for $n = 3$,

$$P_3(x, y) = \sum_{k=1}^{10} \alpha_k x^i y^j, \qquad i+j \leqslant 3$$

$$= \alpha_1 + \alpha_2 x + \alpha_3 y + \alpha_4 x^2 + \alpha_5 xy + \alpha_6 y^2 + \alpha_7 x^3$$
$$+ \alpha_8 x^2 y + \alpha_9 xy^2 + \alpha_{10} y^3$$
(cubic interpolating function)

Ignoring for the moment any consideration of continuity between elements, the order of the polynomial chosen to represent the field variable within an element is that having the same number of coefficients as the degrees of freedom. If we omit any terms from the complete polynomial, these must be *balanced*, otherwise the element will have directional properties. For example we could omit the pairs (x^2, y^2), $(x^2 y, xy^2)$, (x^3, y^3) etc. and suitable 8 term cubic polynomials would be

$$P(x, y) = \alpha_1 + \alpha_2 x + \alpha_3 y + \alpha_4 x^2 + \alpha_5 xy + \alpha_6 y^2 + \alpha_7 x^3 + \alpha_8 y^3,$$
$$P(x, y) = \alpha_1 + \alpha_2 x + \alpha_3 y + \alpha_4 x^2 + \alpha_5 xy + \alpha_6 y^2 + \alpha_7 x^2 y + \alpha_8 xy^2$$
etc.

Thus for triangular elements which have nodes at the mid-points of each side as shown in Figure 7.14(a), for plane stress problems, there are 12 degrees of freedom and suitable interpolating polynomials would be the quadratic functions

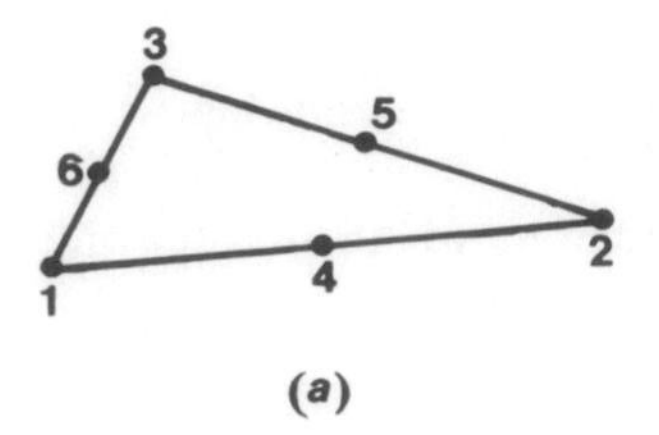

(a)

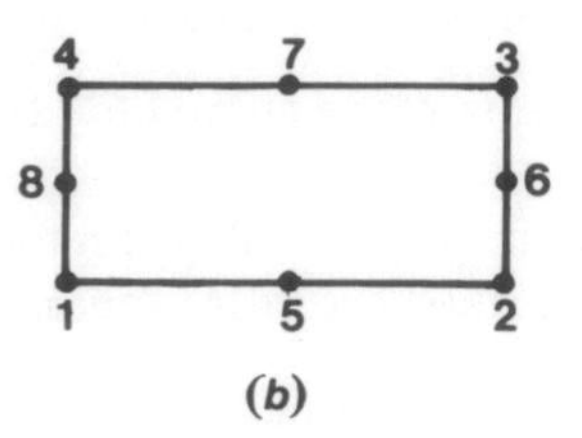

(b)

Figure 7.14

$$u = \alpha_1 + \alpha_2 x + \alpha_3 y + \alpha_4 x^2 + \alpha_5 xy + \alpha_6 y^2$$
$$v = \alpha_7 + \alpha_8 x + \alpha_9 y + \alpha_{10} x^2 + \alpha_{11} xy + \alpha_{12} y^2$$

while for the 8-noded rectangular element shown in Figure 7.14(b) we could choose the interpolating polynomials

$$u = \alpha_1 + \alpha_2 x + \alpha_3 y + \alpha_4 x^2 + \alpha_5 xy + \alpha_6 y^2 + \alpha_7 x^2 y + \alpha_8 xy^2$$
$$v = \alpha_9 + \alpha_{10} x + \alpha_{11} y + \alpha_{12} x^2 + \alpha_{13} xy + \alpha_{14} y^2 + \alpha_{15} x^2 y + \alpha_{16} xy^2$$

Of course, there is much more to the choice of interpolating polynomial than is given here. As shown (take the case with the 3-noded triangular element and the 4-noded rectangular element) it is desirable to have at least continuity of the field variable across element boundaries.

For a fuller explanation read the book by Zienkiewicz (1971) after studying this chapter and trying out several problems using the 'black box' technique. From a simplified viewpoint, the choice of element and associated interpolation polynomials must satisfy certain conditions if the finite element method is to converge on the exact solution. These are:–

(1) the displacement field within the element and on its boundaries must be continuous

(2) constant strain is produced in the element when the nodes are appropriately displaced

(3) rigid body displacements should be possible without causing internal strains and hence giving zero nodal forces

(4) compatibility across element boundaries should exist; that is, boundary displacements (and slopes) should be identical for the two elements with this common boundary.

The errors caused by the non-satisfaction of these requirements vary from problem to problem, and from mesh to mesh, but are frequently quite small and indeed many successful solutions can be obtained which use *discontinuous* displacement functions which satisfy only some but not all of the above conditions.

Alternative criteria to (2), (3) and (4) are embodied in the following:

(a) the polynomial used must be complete up to order n, where n is the order of the highest derivative appearing in the strain energy integral

(b) derivatives of the polynomial of order $(n-1)$ must be continuous across element boundaries (elements satisfying this condition are called **conforming** or **compatible elements**).

Further complications

(a) So far we have only considered displacements in the plane of the structure with two degrees of freedom at each node. However, when considering the bending of plates we shall have to introduce a displacement normal to the plate so that each node will have 3 degrees of freedom. This implies that we shall have to modify the shape functions to cope with the extra degrees of freedom of the element.

(b) Higher order elements exist, both 2-dimensional and 3-dimensional, which are of use in the finite element method; you are referred to the Bibliography for consideration of such elements. We show in Figure 7.15 some of the more commonly used types.

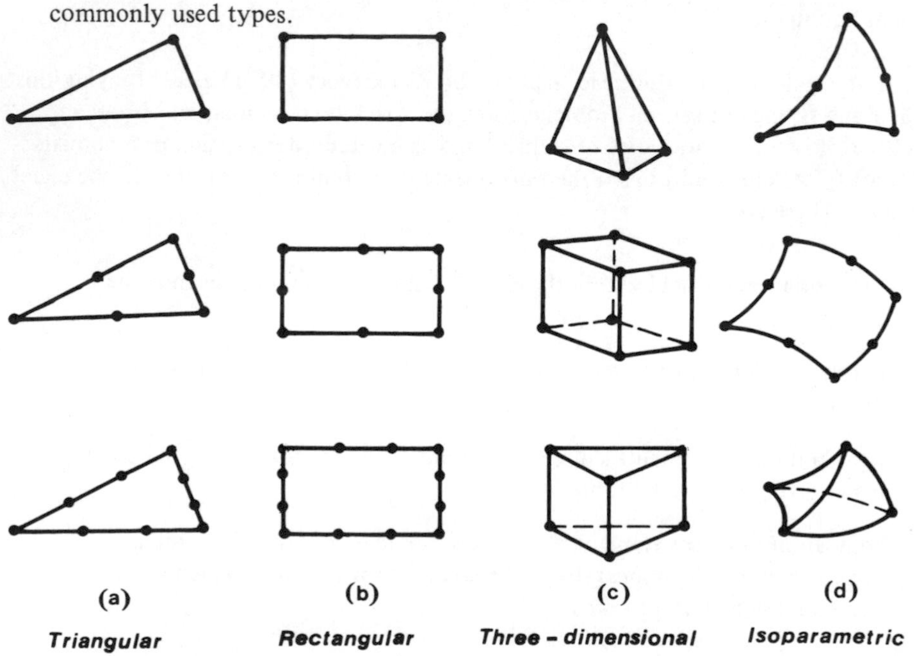

Figure 7.15

.6 FURTHER STEPS IN THE METHOD

In this section we consider the assembly of elements, nodal forces, boundary onditions, convergence and the numbering of the nodes.

ssembly of Elements

A flat plate is subdivided into two elements as shown in Figure 7.16, forces and isplacements being as indicated.

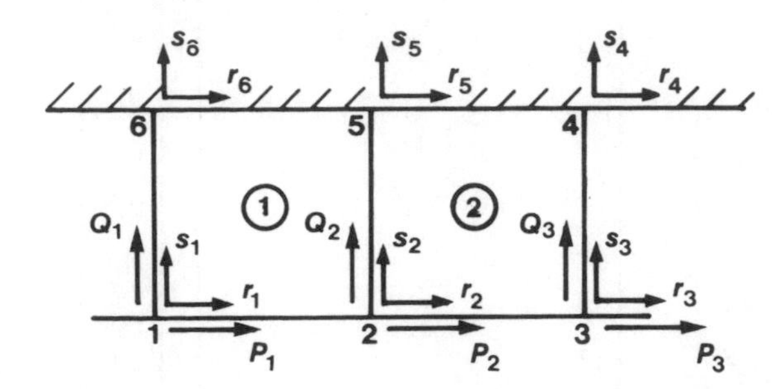

Figure 7.16

We number the nodes as depicted, the elements being 'separated' in Figure 7.17 where the nodal forces U, V and displacements u, v are indicated, the superscripts ndicating the element number.

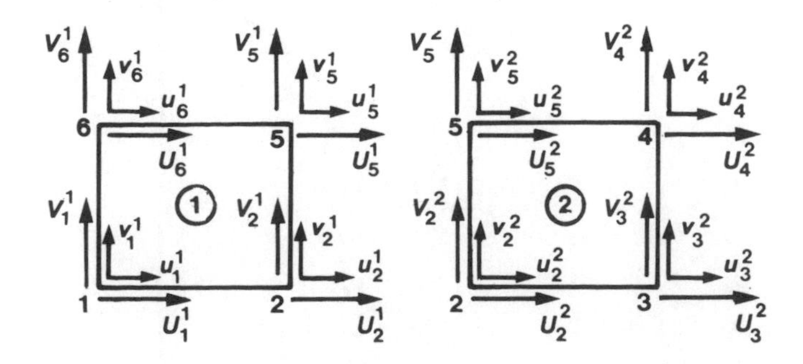

Figure 7.17

For each element there are equations of the form $\mathbf{P} = \mathbf{k}\boldsymbol{\delta}$, namely

$$
\begin{bmatrix} U_1^1 \\ V_1^1 \\ U_2^1 \\ V_2^1 \\ U_5^1 \\ V_5^1 \\ U_6^1 \\ V_6^1 \end{bmatrix} = \begin{bmatrix} k_{11}^1 & k_{12}^1 & \cdots\cdots & k_{18}^1 \\ k_{21}^1 & k_{22}^1 & \cdots\cdots & k_{28}^1 \\ \vdots & & & \\ \vdots & & & \\ k_{81}^1 & k_{82}^1 & \cdots\cdots & k_{88}^1 \end{bmatrix} \begin{bmatrix} u_1^1 \\ v_1^1 \\ u_2^1 \\ v_2^1 \\ u_5^1 \\ v_5^1 \\ u_6^1 \\ v_6^1 \end{bmatrix} \qquad (7.34)
$$

and

$$
\begin{bmatrix} U_2^2 \\ V_2^2 \\ U_3^2 \\ V_3^2 \\ U_4^2 \\ V_4^2 \\ U_5^2 \\ V_5^2 \end{bmatrix} = \begin{bmatrix} k_{11}^2 & k_{12}^2 & \cdots\cdots & k_{18}^2 \\ k_{21}^2 & k_{22}^2 & \cdots\cdots & k_{28}^2 \\ \vdots & & & \\ \vdots & & & \\ k_{81}^2 & k_{82}^2 & \cdots\cdots & k_{88}^2 \end{bmatrix} \begin{bmatrix} u_2^2 \\ v_2^2 \\ u_3^2 \\ v_3^2 \\ u_4^2 \\ v_4^2 \\ u_5^2 \\ v_5^2 \end{bmatrix} \qquad (7.35)
$$

The components of the forces at the structural nodes must be the vector sum of the forces at the element nodes and the component of displacement at the structural nodes must equal the component of displacement at the elemental nodes. Thus we find:-

$$
\left.
\begin{aligned}
P_1 &= U_1^1 & r_1 &= u_1^1 \\
Q_1 &= V_1^1 & s_1 &= v_1^1 \\
P_2 &= U_2^1 + U_2^2 & r_2 &= u_2^1 = u_2^2 \\
Q_2 &= V_2^1 + V_2^2 & s_2 &= v_2^1 = v_2^2 \\
P_3 &= U_3^2 & r_3 &= u_3^2 \\
Q_3 &= V_3^2 & s_3 &= v_3^2 \\
P_4 &= U_4^2 & r_4 &= u_4^2 \\
Q_4 &= V_4^2 & s_4 &= v_4^2 \\
P_5 &= U_5^1 + U_5^2 & r_5 &= u_5^1 = u_5^2 \\
Q_5 &= V_5^1 + V_5^2 & s_5 &= v_5^1 = v_5^2 \\
P_6 &= U_6^1 & r_6 &= u_6^1 \\
Q_6 &= V_6^1 & s_6 &= v_6^1
\end{aligned}
\right\} \qquad (7.36)
$$

'he two elements are assembled by writing down the equations corresponding to (7.23) or the whole structure in the form

$$\mathbf{R} = \mathbf{K}\,\mathbf{q}$$

vhere $\mathbf{R} = [P_1, Q_1, P_2, \ldots\ldots\ldots, Q_6]^{\mathrm{T}}$ and $\mathbf{q} = [r_1, s_1, r_2, \ldots\ldots\ldots, s_6]^{\mathrm{T}}$. From equations (7.36) we see that $\mathbf{K}$ is given by (7.37) on the next page.

We are in a position to use the equation $\mathbf{R} = \mathbf{Kq}$ to find the deflections at all the odes of the structure if the nodal forces are specified.

Nodal Forces

As mentioned previously, the finite element method uses forces at the nodes and ny force system acting on a structure has to be replaced by equivalent point loads which ct at the nodes of the chosen finite element mesh. Of course it is best to choose nodes f the mesh to coincide with the points where external point loads are applied. Iowever, in the case of distributed loads, an equivalent nodal force system is obtained by ssuming that the work done by them is equal to the work done by the distributed load esulting from a virtual displacement. Two forms of distributed loading on an element re possible: body forces where the loading is distributed throughout the volume of the lement and surface forces where the loading is distributed on one or more of the urfaces of the element. We shall work in 2-dimensions to simplify the analysis.

Suppose the components of body force on the element are (X, Y) in the direction f the (x, y) axes. We can represent the body force by the vector $\mathbf{Q} = [X, Y]^{\mathrm{T}}$. Assume that the element undergoes a virtual displacement $\mathbf{f} = [u, v]^{\mathrm{T}}$. Then the work lone by the body forces is given by

$$W_D = \iiint_{\text{volume}} (uX + vY)\,\mathrm{d}V = \iiint_{\text{volume}} \mathbf{f}^{\mathrm{T}} . \mathbf{Q}\,\mathrm{d}V$$

But from equations of the form (7.7) or (7.28) we know that $\mathbf{f} = \mathbf{N}\mathbf{A}^{-1}\boldsymbol{\delta}$, so that

$$W_D = \iiint_{\text{volume}} \boldsymbol{\delta}^{\mathrm{T}}\,[\mathbf{A}^{-1}]^{\mathrm{T}}\,\mathbf{N}^{\mathrm{T}} . \mathbf{Q}\,\mathrm{d}V$$

The work done by the nodal forces $\mathbf{L}_B$ which replace the distributed loading is given by

$$W_N = \boldsymbol{\delta}^{\mathrm{T}} . \mathbf{L}_B$$

$$
\mathbf{K} = \begin{bmatrix}
k^1_{11} & k^1_{12} & k^1_{13} & k^1_{14} & 0 & 0 & 0 & 0 & k^1_{15} & k^1_{16} & k^1_{17} & k^1_{18} \\
k^1_{21} & k^1_{22} & k^1_{23} & k^1_{24} & 0 & 0 & 0 & 0 & k^1_{25} & k^1_{26} & k^1_{27} & k^1_{28} \\
k^1_{31} & k^1_{32} & (k^1_{33}+k^2_{11}) & (k^1_{34}+k^2_{12}) & k^2_{13} & k^2_{14} & k^2_{15} & k^2_{16} & (k^1_{35}+k^2_{17}) & (k^1_{36}+k^2_{18}) & k^1_{37} & k^1_{38} \\
k^1_{41} & k^1_{42} & (k^1_{43}+k^2_{21}) & (k^1_{44}+k^2_{22}) & k^2_{23} & k^2_{24} & k^2_{25} & k^2_{26} & (k^1_{45}+k^2_{27}) & (k^1_{46}+k^2_{28}) & k^1_{47} & k^1_{48} \\
0 & 0 & k^2_{31} & k^2_{32} & k^2_{33} & k^2_{34} & k^2_{35} & k^2_{36} & k^2_{37} & k^2_{38} & 0 & 0 \\
0 & 0 & k^2_{41} & k^2_{42} & k^2_{43} & k^2_{44} & k^2_{45} & k^2_{46} & k^2_{47} & k^2_{48} & 0 & 0 \\
0 & 0 & k^2_{51} & k^2_{52} & k^2_{53} & k^2_{54} & k^2_{55} & k^2_{56} & k^2_{57} & k^2_{58} & 0 & 0 \\
0 & 0 & k^2_{61} & k^2_{62} & k^2_{63} & k^2_{64} & k^2_{65} & k^2_{66} & k^2_{67} & k^2_{68} & 0 & 0 \\
k^1_{51} & k^1_{52} & (k^1_{53}+k^2_{71}) & (k^1_{54}+k^2_{72}) & k^2_{73} & k^2_{74} & k^2_{75} & k^2_{76} & (k^1_{55}+k^2_{77}) & (k^1_{56}+k^2_{78}) & k^1_{57} & k^1_{58} \\
k^1_{61} & k^1_{62} & (k^1_{63}+k^2_{81}) & (k^1_{64}+k^2_{82}) & k^2_{83} & k^2_{84} & k^2_{85} & k^2_{86} & (k^1_{65}+k^2_{87}) & (k^1_{66}+k^2_{88}) & k^1_{67} & k^1_{68} \\
k^1_{71} & k^1_{72} & k^1_{73} & k^1_{74} & 0 & 0 & 0 & 0 & k^1_{75} & k^1_{76} & k^1_{77} & k^1_{78} \\
k^1_{81} & k^1_{82} & k^1_{83} & k^1_{84} & 0 & 0 & 0 & 0 & k^1_{85} & k^1_{86} & k^1_{87} & k^1_{88}
\end{bmatrix} \tag{7.37}
$$

nd equating W_D to W_N we find that

$$\mathbf{L}_B = [\mathbf{A}^{-1}]^{\mathrm{T}} \iiint_{\text{volume}} \mathbf{N}^{\mathrm{T}} . \mathbf{Q} \, \mathrm{d}V \qquad (7.38)$$

y a similar analysis for surface forces $\mathbf{T} = [F_x, F_y]^{\mathrm{T}}$ distributed along the edges of e element we find that

$$W_D = \iint_{\text{surface}} \boldsymbol{\delta}^{\mathrm{T}} [\mathbf{A}^{-1}]^{\mathrm{T}} \mathbf{N}^{\mathrm{T}} \mathbf{T} \, \mathrm{d}S$$

ving equivalent nodal forces

$$\mathbf{L}_s = [\mathbf{A}^{-1}]^{\mathrm{T}} \iint_{\text{surface}} \mathbf{N}^{\mathrm{T}} . \mathbf{T} \, \mathrm{d}S \qquad (7.39)$$

As an example of the procedure you should try to find $\mathbf{L}_B$ and $\mathbf{L}_s$ for the two-imensional element shown in Figure 7.18, which is acted on by constant planar istributed loadings of F_x and F_y per unit area and by a distributed gravity load from onstant density ρ. The element is of uniform thickness h. The expressions for $\mathbf{A}$ nd $\mathbf{N}$ are given in equations (7.25) and (7.27) and you should find that

$$\mathbf{L}_s = \frac{1}{2} h \quad [0, bF_x, bF_x, 0, -aF_y, -aF_y, 0, 0]^{\mathrm{T}}$$

$$\mathbf{L}_B = \frac{1}{4} abh[0, 0, 0, 0, -g\rho, -g\rho, -g\rho, -g\rho]^{\mathrm{T}}$$

hich give the equivalent forces at the nodes as depicted in Figures 7.19(a) and (b).

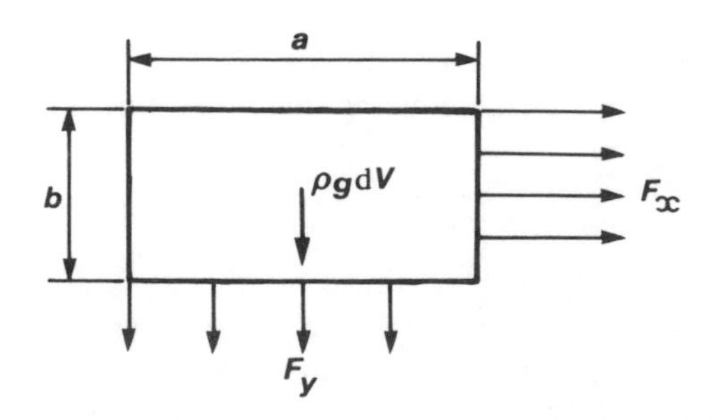

Figure 7.18

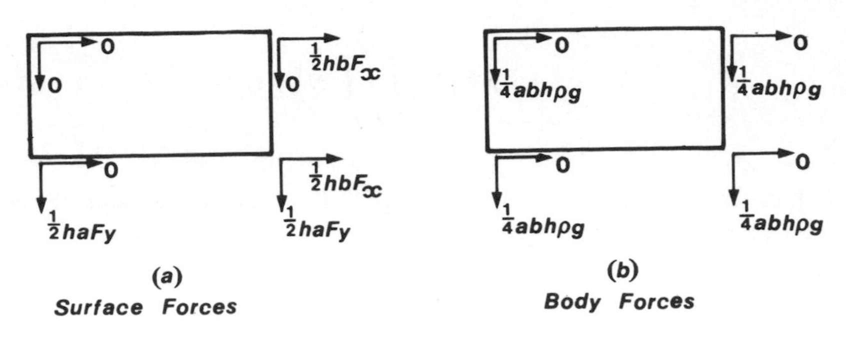

Figure 7.19

Many standard results exist in the literature for other element shapes and other loadings and you are referred to the Bibliography.

In the assembly stage, the nodal forces are the sum of the equivalent nodal forces from adjacent elements.

Boundary Conditions

We need to solve the matrix equation

$$\mathbf{R} = \mathbf{Kq}$$

to determine the nodal displacements of the structure for a given set of loads. The abov equations have been established without any consideration of the supports or constraint which the structure may have. Consequently, the displacements would normally includ rigid body motions of the structure and the deflection pattern of the structure cannot b defined. In mathematical terms this means that the set of equations is singular and cannot be solved. To obtain a solution, constraints must be applied so that rigid body motions are eliminated. Most packages will take care of the programming difficulties and we need merely to specify zero displacements (fixed points) or other specified displacements. In our case study example, you will have noticed in the program that there was a heading SPRINGS with a value -1. This is a technique used in that progra to give zero displacements; it assigns a value of 10^{20} to the load at that node so havin the effect of a very stiff spring and reducing the displacement to zero there.

Convergence

In practice, if we are to have confidence in a solution it should not change when t mesh of elements is altered. Ideally, we would plot one of the characteristics of the solutions (e.g. maximum deflection) against the number of elements used and obtain a curve like one of the curves A, B, C or D shown in Figure 7.20, which show a tendency wards an asymptote (the exact solution hopefully) as the number of elements rises.

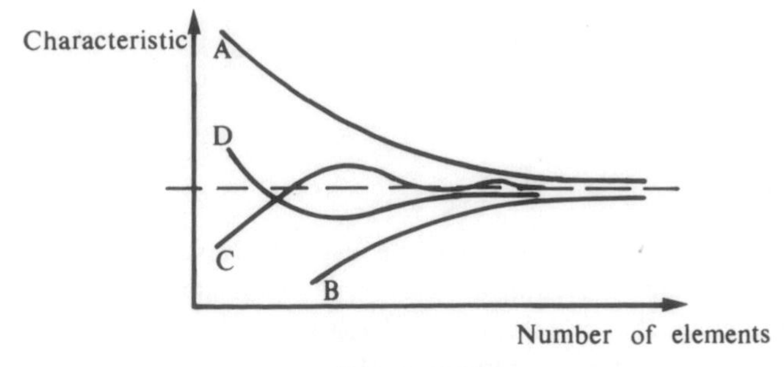

Figure 7.20

emember, however, that errors are inherent in the method arising from

) approximations made in the development of the stiffness matrices such as assumed displacement functions

) numerical inaccuracies arising from the numerical solution of large sets of equations within the program.

Finally, it is worth mentioning that it is a simple matter to calculate the total tential energy by summing the strain energy for each element. This total potential ergy should be a minimum for the exact solution and thus the summed strain energy ould show a monotonic convergence towards the minimum as the mesh is refined.

In large practical problems, it is not feasible, from an economic viewpoint, to oduce solutions from many refined meshes and so much judgement and experience is ed in establishing suitable meshes. Expert users of the finite element method have a od knowledge of element performance in various situations and are able to arrive at a itable mesh from the outset.

umbering of Elemental Nodes

Structural stiffness matrices are symmetric matrices and the typical assembled iffness matrix obtained in the finite element procedure is both symmetric and **banded** . all the non-zero terms of the matrix fall between two lines which can be drawn rallel to the main diagonal. A 6 x 6 banded matrix is shown on the next page ith a bandwidth of five (of course some of the a_{ij} could themselves be zero). It is lvantageous to obtain the overall stiffness matrix with a bandwidth as small as possible, nce a reduction in bandwidth produces a reduction in the required storage space plus a duction in the computational time taken to solve the corresponding system of multaneous equations. We could minimise the bandwidth of the assembled stiffness

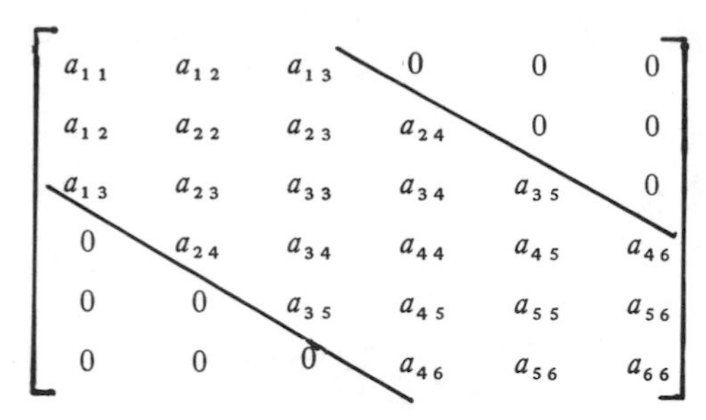

matrix by labelling the nodes in the best possible way. However, for a complicated mes
the finding of the best possible labelling of nodes can be a major computational task.
Therefore we do not endeavour to achieve the absolute minimum bandwidth but try to g
some way towards this by ordering the nodes so that those on an individual element are
not numerically too far away from each other. For example, consider two alternative
ways of labelling the mesh shown in Figure 7.21(a) and (b).

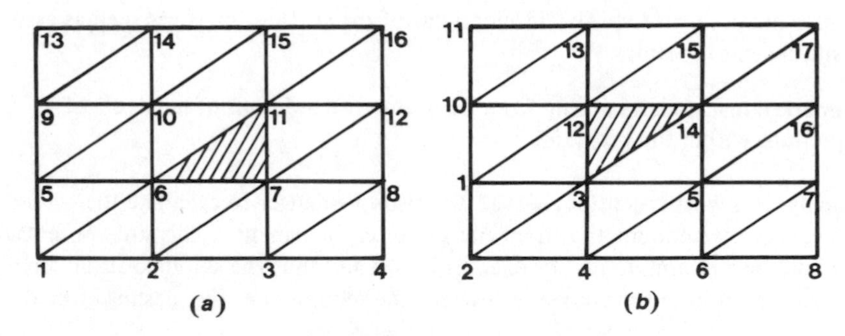

Figure 7.21

For an element in the mesh shown in (a) the node numbers differ by at most 5, whereas in (b) the node numbers can differ by 11 (consider the shaded elements in eac case). The labelling given in (a) would give a bandwidth in the overall stiffness matrix less than half that given in (b) with a consequent saving in storage and computation. Tr alternative labellings and see whether you can improve on (a).

7.7 FURTHER CASE STUDIES – SAMPLES OF INPUT AND OUTPUT

We now consider examples of two-dimensional stress/strain problems which have been solved by finite element analysis and give the input and output so that you can get a feel for the method and what to look for when analysing results.

Example 1

We return to the simple case study of the cantilever beam subjected to an end load and now take 8 elements as shown in Figure 7.22.

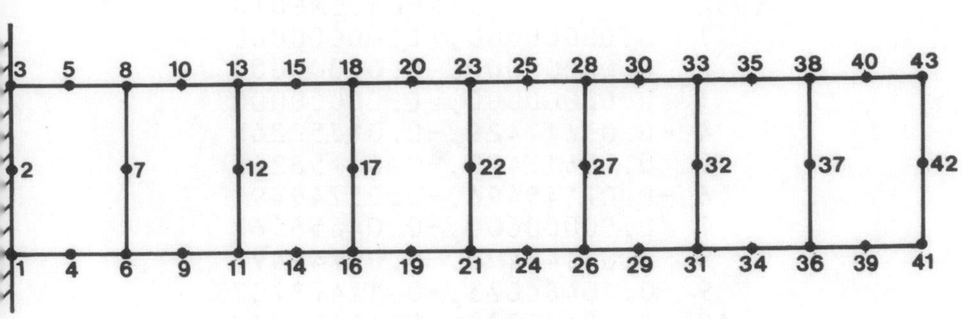

Figure 7.22

The actual load is made equivalent to loads of −5833.333333, −23333.333333, -5833.333333 at nodes 41, 42, 43 respectively (in the ratio 1 : 4 : 1). A section of the input and output is given below, and on following pages respectively.

nput 'COORDS'

LINE					
0001	NODE	COORDS	REP	ADDN	ADDXY
0002	I=1,41,5) I	(125*(I−1)/5),0	2	1	0,100
0003	I=4,39,5) I	(62.5+(125/(I−4)/5)),0	1	1	0,200

'QUAD8SM'

LINE					
0001	NODES	MATERIAL	SECTION	REP	ADDN
0002	1,4,6,7,8,5,3,2	210000,0.25	40	7	5

'RESTRAINTS'

LINE		
0001	NODES	SPRINGS
0002	1,2,3	−1,−1

'LOAD'

LINE		
0001	NODE	VALUES
0002	41	0,−5833.333333
0003	42	0,−23333.333333
0004	43	0,−5833.333333

The deflection at node 42, the centroid of the end section is −2.12132 mm as gainst −2.10725 mm for the four-element analysis and −2.16146 mm theoretical lue. The error this time is of order 1.89%. The maximum shear stress given at nodes and 2 are again exact, namely 131.25 N/mm².

Output

```
NODE              DISPLACEMENTS
   1   0.00000000, 0.00000000
   2   0.00000000, 0.00000000
   3   0.00000000, 0.00000000
   4  -0.03617429,-0.01758265
   5   0.03617429,-0.01758265
   6  -0.07145496,-0.05749497
   7   0.00000000,-0.04856568
   8   0.07145496,-0.05749497
   9  -0.10486075,-0.11478732
  10   0.10486075,-0.11478732
  11  -0.13541041,-0.19206575
  12   0.00000000,-0.18676728
  13   0.13541041,-0.19206575
  14  -0.16336686,-0.28872177
  15   0.16336686,-0.28872177
  16  -0.18899303,-0.40212236
  17   0.00000000,-0.39708926
  18   0.18899303,-0.40212236
  19  -0.21221847,-0.53049583
  20   0.21221847,-0.53049583
  21  -0.23297273,-0.67261296
  22   0.00000000,-0.66874698
  23   0.23297273,-0.67261296
  24  -0.25127468,-0.82701372
  25   0.25127468,-0.82701372
  26  -0.26714321,-0.99209277
  27   0.00000000,-0.98915228
  28   0.26714321,-0.99209277
  29  -0.28057325,-1.16630655
  30   0.28057325,-1.16630655
  31  -0.29155974,-1.34815049
  32   0.00000000,-1.34620027
  33   0.29155974,-1.34815049
  34  -0.30010404,-1.53610348
  35   0.30010404,-1.53610348
  36  -0.30620753,-1.72863391
  37   0.00000000,-1.72765651
  38   0.30620752,-1.72863391
  39  -0.30986980,-1.92421451
  40   0.30986980,-1.92421451
  41  -0.31109047,-2.12132103
  42   0.00000000,-2.12132144
  43   0.31109047,-2.12132103
```

NODAL AVERAGE STRESSES

NODE,	SXX,	SYY,	TXY
1	-131.250	-32.813	-4.375
2	0.000	0.000	-4.375
3	131.250	32.813	-4.375
4	-123.047	-12.010	-4.375
5	123.047	12.010	-4.375
6	-114.844	8.792	-4.375
7	0.000	0.000	-4.375
8	114.844	-8.792	-4.375
9	-106.641	3.218	-4.375
10	106.641	-3.218	-4.375
11	-98.438	-2.356	-4.375
12	0.000	0.000	-4.375
13	98.438	2.356	-4.375
14	-90.234	-0.862	-4.375
15	90.234	0.862	-4.375
16	-82.031	0.631	-4.375
17	0.000	0.000	-4.375
18	82.031	-0.631	-4.375
19	-73.828	0.231	-4.375
20	73.828	-0.231	-4.375
21	-65.625	-0.169	-4.375
22	0.000	0.000	-4.375
23	65.625	0.169	-4.375
24	-57.422	-0.062	-4.375
25	57.422	0.062	-4.375
26	-49.219	0.045	-4.375
27	0.000	0.000	-4.375
28	49.219	-0.045	-4.375
29	-41.016	0.017	-4.375
30	41.016	-0.017	-4.375
31	-32.813	-0.012	-4.375
32	0.000	0.000	-4.375
33	32.813	0.012	-4.375
34	-24.609	-0.004	-4.375
35	24.609	0.004	-4.375
36	-16.406	0.030	-4.375
37	0.000	0.000	-4.375
38	16.406	-0.030	-4.375
39	-8.203	0.000	-4.375
40	-8.203	0.000	-4.375
41	0.000	-0.002	-4.375
42	0.000	0.000	-4.375
43	0.000	0.002	-4.375

Example 2 Stress Concentration near a hole in a flat plate

Consider the stresses in a square flat plate with a small circular hole at its centre when subjected to a uniformly distributed load of 600 N/mm. The dimensions of the plate are shown in Figure 7.23(a).

We assume that the thickness of the plate is sufficiently small for the plane stress approximation to be applied. Since the problem possesses symmetry we need only consider the quadrant of the plate shaded which for clarity is drawn in Figure 7.23(b).

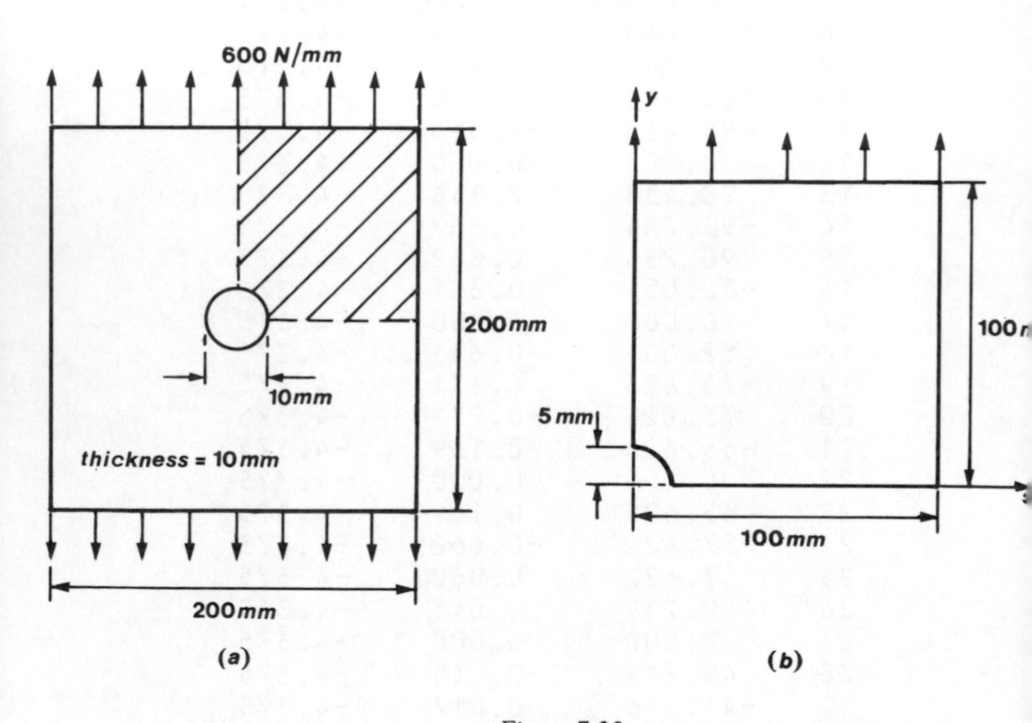

Figure 7.23

The magnitudes of the stresses near the hole are large and diminish rapidly with distance from the hole to the values they would attain in its absence. We have to devise a suitable mesh which takes account of this and employs smaller and smaller elements the nearer we are to the hole. Many possibilities exist but we chose one that, while probably not being ideal, is fairly easy to program. It was decided to use triangular elements and the nodal points were generated as follows :

(a) Taking the inner circle we generate 6 nodes given by the formulae $x = 5 \sin (I - 1)(\pi/10)$, $y = 5 \cos (I - 1)(\pi/10)$, where $I = 1, 2, \ldots\ldots, 6$.

(b) We now generate 7 nodal points on a circle of radius 5.9 given by the formulae $x = 5.9 \sin (I - 7)(\pi/12)$, $y = 5.9 \cos (I - 7)(\pi/12)$, where $I = 7, 8, \ldots\ldots, 13$.

) Six nodes on a circle of radius 6.962 (i.e. 1.18 x 5.9) are now generated from the formulae $x = 6.962 \sin (I - 14)(\pi/10)$, $y = 6.962 \cos (I - 14)(\pi/10)$, where $I = 14, 15, \ldots\ldots, 19$.

) The process is repeated, increasing the radius by 18% each time, and generating 6 or 7 nodes as appropriate, up to $x = 83.361 \sin (I - 111)(\pi/12)$, $y = 83.361 \cos (I - 111)(\pi/12)$, where $I = 111, 112, \ldots\ldots, 117$.

) We specify the other nodes individually by

I	118	119	120	121	122	123	124	125	126	127
x	0	25	50	75	100	100	100	75	100	100
y	100	100	100	75	50	25	0	100	75	100

'his generation produces the mesh shown in Figure 7.24 (not to scale) where for clarity nly a selection of nodes are shown.

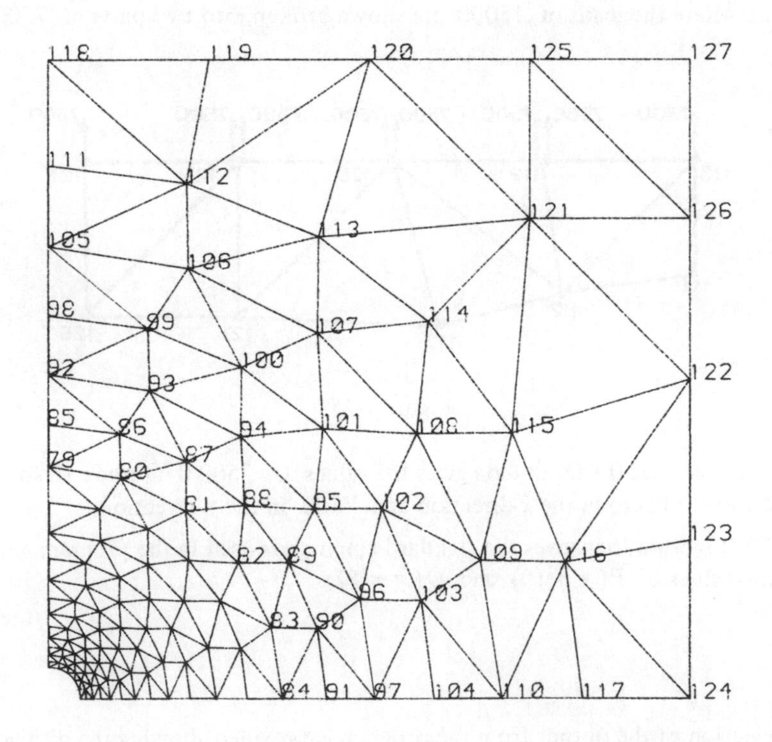

Figure 7.24

Input

The first section 'COORDS' reads in the coordinates of the nodes. 'TRI' gives the chosen type of element followed by the rows which define the elements, give the material properties (Young's Modulus and Poisson's Ratio) and the thickness of the plate. For example, the first line takes $I = 1, 14, 27, ...,$ 105; defines the element nodes as (1, 2, 8), (14, 15, 21), ..., (105, 106, 112);
Young's Modulus = 200000 N/mm², Poisson's Ratio = 0.25; the thickness = 10 mm; repeats the line 4 times adding 1, 1, 1 each time to give element nodes (2, 3, 9), (15, 16, 22), ..., (106, 107, 113) and so on.

'BOUNDARIES' gives the nodes which are restrained from movement, the −1 under 'SPRINGS' indicating that a very stiff spring is to be used and the program substitutes the value 10^{20}. Thus the nodes on the x-axis are restrained from movement in the y-direction while those on the y-axis are restrained from movement in the x-direction.

'UD LOAD' specifies that a uniformly distributed load is to be applied. We can apply loads only at the nodes of the mesh and so the UD load must be approximated by loads applied at the nodes 118, 119, 120, 125 and 127 on the upper boundary. We do this by applying the loads 7500, 15000, 15000, 7500 respectively as shown in Figure 7.25 where the loads of 15000 are shown broken into two parts of 7500 N.

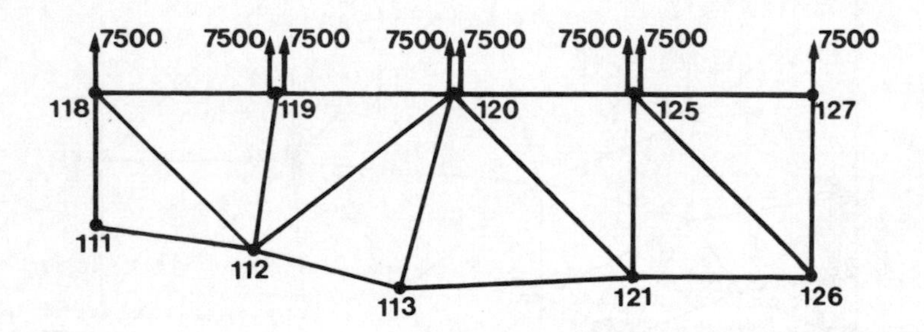

Figure 7.25

Note for example that the input data gives the values 0, 7500.0 at node 18 so specifying a load of zero in the x-direction and 7500 in the y-direction.

The MASTER section then gives the standard commands used in the program and also specifies the values of $P(=\pi/10)$ and $Q(=\pi/12)$.

Output

A selection of the output from the program is provided showing the displacements of each node, the stresses for the individual elements and the nodal average stresses.

```
'COORDS'

                NODE                    COORDS
I=1,6,1)         I     5*SIN((I-1)*P),5*COS((I-1)*P)
I=7,13,1)        I     5.9*SIN((I-7)*Q),5.9*COS((I-7)*Q)
I=14,19,1)       I     6.962*SIN((I-14)*P),6.962*COS((I-14)*P)
I=20,26,1)       I     8.215*SIN((I-20)*Q),8.215*COS((I-20*Q)
I=27,32,1)       I     9.694*SIN((I-27)*P),9.694*COS((I-27)*P)
I=33,39,1)       I     11.439*SIN((I-33)*Q),11.439*COS((I-33)*Q)
I=40,45,1)       I     13.498*SIN((I-40)*P),13.498*COS((I-40)*P)
I=46,52,1)       I     15.927*SIN((I-46)*Q),15.927*COS((I-46)*Q)
I=53,58,1)       I     18.794*SIN((I-53)*P),18.794*COS((I-53)*P)
I=59,65,1)       I     22.177*SIN((I-59)*Q),22.177*COS((I-59)*Q)
I=66,71,1)       I     26.169*SIN((I-66)*P),26.169*COS((I-66)*P)
I=72,78,1)       I     30.880*SIN((I-72)*Q),30.880*COS((I-72)*Q)
I=79,84,1)       I     36.438*SIN((I-79)*P),36.438*COS((I-79)*P)
I=85,91,1)       I     42.997*SIN((I-85)*Q),42.997*COS((I-85)*Q)
I=92,97,1)       I     50.736*SIN((I-92)*P),50.736*COS((I-92)*P)
I=98,104,1)      I     59.869*SIN((I-98)*Q),59.869*COS((I-98)*Q)
I=105,110,1)     I     70.645*SIN((I-105)*P),70.645*COS((I-105)*P)
I=111,117,1)     I     83.361*SIN((I-111)*Q),83.361*COS((I-111)*Q)
               118      0,100
               119     25,100
               120     50,100
               125     75,100
               127    100,100
               121     75,75
               126    100,75
               122    100,50
               123    100,25
               124    100,0
```

```
      'TRI'

               NODES                    MATERIAL        SECTION  REP   ADDN
I=1,105,13)    I,I+1,I+7                200000,0.25       10      4   1,1,1
I=7,98,13)     I,I+1,I+7                   =               =      5   1,1,1
I=7,111,13)    I,I-6,I+1                   =               =      5   1,1,1
I=14,105,13)   I,I-6,I+1                   =               =      4   1,1,1
               111,112,118                 =               =      1   1,7,0
               112,113,120                 =               =      1   7,-1,0
               113,114,121                 =               =      1   0,7,-1
               114,115,121                 =               =      1   1,7,0
               115,116,122                 =               =      1   1,7,0
               116,117,124                 =               =      1   0,7,-1
               120,121,125                 =               =      1   1,1,1
               121,126,125                 =               =      1   4,0,2

      'BOUNDARIES'

               NODES                                                     SPRINGS
6,13,19,26,32,39,45,52,58,65,71,78,84,91,97,104,110,117,124               0,-1
1,7,14,20,27,33,40,46,53,59,66,72,79,85,92,98,105,111,118                  -1,0

      'UD LOAD'

            NODE      VALUES
        118          0,7500.0
        119          0,15000.0
        120          0,15000.0
        125          0,15000.0
        127          0,7500.0
```

```
*MASTER
  PROBLEM TYPE IS 'STRESS-2D'
  P=0.314159
  Q=3.141592654/12.0
  USE 'COORDS','TRI','BOUNDARIES'
  USE 'UD LOAD' AS CASE 1
  CALC K
  DRAW MESH WITH SHRINKAGE
  ASSEMBLE AND REDUCE
  SOLVE FOR CASE 1
  PRINT REACTIONS FOR CASE 1
  PRINT DISPLACEMENTS FOR CASE 1
  PRINT STRESSES FOR ELEMENTS 'TRI' FOR CASE 1
```

```
'DISPLACEMENTS FOR LOAD CASE UD LOAD'
   NODE            DISPLACEMENTS
      1  0.00000000, 0.00430768
      2 -0.00039681, 0.00406531
      3 -0.00077986, 0.00343646
      4 -0.00108276, 0.00247749
      5 -0.00127447, 0.00128500
      6 -0.00136445, 0.00000000
      7  0.00000000, 0.00432103
      8 -0.00017448, 0.00413003
      9 -0.00041932, 0.00355276
     10 -0.00110582, 0.00182947
     11 -0.00137420, 0.00091086
     12 -0.00142985, 0.00000000
     13  0.00000000, 0.00439732
     14 -0.00013552, 0.00406727
     15 -0.00075034, 0.00273898
```

```
STRESSES FOR ELEMENTS TRI   LOAD CASE UD LOAD
ELEMENT       ---STRESSES---
  1  2  8 SXX= -43.462 SYY=   2.292 TXY=   6.795
  2  3  9 SXX= -15.649 SYY=  18.758 TXY=  -4.291
  3  4 10 SXX=  13.750 SYY=  58.069 TXY= -25.902
  4  5 11 SXX=  24.032 SYY= 114.504 TXY= -39.087
  5  6 12 SXX=  18.756 SYY= 158.952 TXY= -27.707
 14 15 21 SXX=  -4.908 SYY=  25.911 TXY=  -7.492
 15 16 22 SXX=  -8.018 SYY=  49.762 TXY= -16.428
 16 17 23 SXX=  -7.552 SYY=  73.307 TXY= -13.116
 17 18 24 SXX=   1.719 SYY=  87.653 TXY=  -5.978
 18 19 25 SXX=  14.268 SYY=  94.238 TXY=  -3.503
 27 28 34 SXX=   2.147 SYY=  41.574 TXY=  -7.015
 28 29 35 SXX=  -4.282 SYY=  57.819 TXY= -12.158
 29 30 36 SXX=  -7.658 SYY=  70.629 TXY=  -6.744
 30 31 37 SXX=  -2.153 SYY=  74.436 TXY=   0.609
```

```
NODAL AVERAGE STRESSES
 NODE,       SXX,        SYY,       TXY
    1     -33.523      -0.319    -1.527
    2     -23.952      10.042    -3.062
    3       0.640      33.612   -16.430
    4      15.741      76.644   -30.716
    5      15.572     123.818   -28.530
    6      17.532     141.153   -12.528
    7     -22.065       3.150    -9.550
    8     -21.101       8.306    -7.819
    9      -9.988      27.239   -14.822
   10       2.884      56.957   -20.436
   11      10.649      91.625   -19.786
   12      14.289     118.016    -9.711
   13      14.982     123.043     2.085
   14      -9.683      13.559    -7.744
   15     -10.047      26.354   -14.054
```

Discussion of Results

The absolute maximum stress is to be expected at the position of node 6 and we see that the average nodal stresses at node 6 are $\sigma_{xx} = 17.532$, $\sigma_{yy} = 141.153$, $\tau_{xy} = -12.528$. Analytical solutions for a plate of infinite width predict a maximum stress of $\sigma_{yy} = 3$ times the uniform tensile stress applied, which is 3 x 60 N/mm = 180 N/mm.

If we examine the stresses output for the element with centroid nearest to the hole and having node 6 as a node element, namely (5, 6, 12), we find $\sigma_{xx} = 18.756$, $\sigma_{yy} = 158.932$, $\tau_{xy} = -27.207$, which gives a closer approximation to the analytical value. The predicted stresses are in error but they could be improved by taking a finer mesh near the hole and in fact the total of 128 nodes and 203 elements is rather small for such a complex problem. In Figure 7.26 are plotted the values of $\sigma_{yy}/60$ against distance from the centre line and compared with the analytical solution.

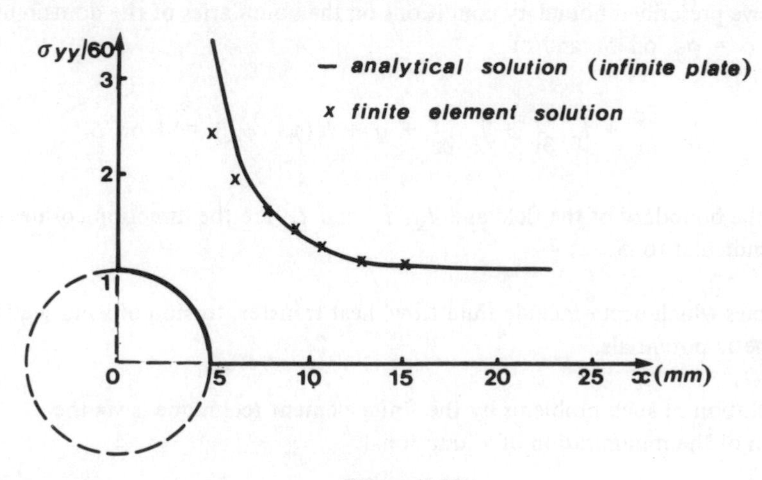

Figure 7.26

The results are excellent for $x > 1.5$ x radius of the hole and it is only nearer the hole that a much finer mesh is required.

7.8 EXTENSIONS OF THE METHOD

We have considered so far the finite element method applied to 2-dimensional stress problems where the 2-dimensional components subjected to in-plane loadings were analysed. Once you have a feel for how the method works in such situations and in what form the results are obtained then it is possible to see the extension of the method to other problems. Examples of these problems are (i) flat plate subjected to bending only, (ii) flat plate subjected to both transverse and in-plane forces, (iii) the calculation of natural frequencies, (iv) field problems. We examine briefly only the last-named.

Field Problems

Numerous physical phenomena are governed by what are known as *field* equations. The most commonly occurring equations are Laplace's equation

$$\frac{\partial^2 \phi}{\partial x^2} + \frac{\partial^2 \phi}{\partial y^2} + \frac{\partial^2 \phi}{\partial z^2} = 0$$

and Poisson's equation

$$\frac{\partial^2 \phi}{\partial x^2} + \frac{\partial^2 \phi}{\partial y^2} + \frac{\partial^2 \phi}{\partial z^2} + Q = 0$$

These will have prescribed boundary conditions on the boundaries of the domain usually of the form $\phi = \phi_B$ on S and/or

$$l_x \frac{\partial \phi}{\partial x} + l_y \frac{\partial \phi}{\partial y} + l_z \frac{\partial \phi}{\partial z} + q + h(\phi - \phi_\infty) = 0 \text{ on } S$$

where S is the boundary of the field and l_x, l_y and l_z are the direction cosines of a vector perpendicular to S.

Examples which occur include fluid flow, heat transfer, torsion of solid sections, electro-magnetic potentials.

The solution of such problems by the finite element technique is via the consideration of the minimisation of a functional

$$\chi = \iiint \frac{1}{2}\left[\left(\frac{\partial \phi}{\partial x}\right)^2 + \left(\frac{\partial \phi}{\partial y}\right)^2 + \left(\frac{\partial \phi}{\partial z}\right)^2 - 2Q\phi\right] \mathrm{d}V$$

$$+ \iint \left[q\phi + \frac{1}{2} h(\phi - \phi_\infty)^2\right] \mathrm{d}S \tag{7.40}$$

which can be shown by the calculus of variations to be equivalent to Poisson's equation and its boundary conditions.

Equation (7.40) must be minimised with respect to a set of nodal values of ϕ after hoosing suitable element characteristics (i.e. shape functions). It can be shown that the ninimisation of (7.40) produces an equation of the form

$$\frac{\partial x}{\partial(\phi)} = \sum\left([\mathbf{k}^e]\{\phi\} + \{\mathbf{f}^e\}\right) = 0 \qquad (7.41)$$

where $[\mathbf{k}^e]$ is of a similar form to our stiffness matrix for each individual element and ^{e} is equivalent to the force vector we had previously, involving integrals over the lement regions.

Equation (7.41) can be written after 'assembly' by the summation sign in the form

$$[\mathbf{K}]\{\mathbf{\Phi}\} = \{\mathbf{F}\} \qquad (7.42)$$

where $[\mathbf{K}] = \Sigma\,[\mathbf{k}^e]$ and $\{\mathbf{F}\} = -\Sigma\{\mathbf{f}^e\}$.

Equation (7.42) is of exactly the same form as equation (7.23) and can be solved ɔ give the value of ϕ which minimises the functional χ and hence solve the field quations.

Chapter Eight

Design of Experiments

8.1 CASE STUDY

At the U.S. National Bureau of Standards an experiment was conducted on the evaluation of the tensile strength of steel.[†] Three factors were taken into account: carbon content (A), temperature of tempering the steel (B), the method of cooling (C). Each factor was considered at two levels; the levels of temperature were B_1 (400°F) and B_2 (600°F) but the particular two levels of carbon content (A_1 and A_2) and those of cooling method (C_1 and C_2) are not disclosed. The data quoted in Table 8.1 are the results of a strength test and were recorded in pounds/sq. in. ÷ 1000. (The actual units do not matter for the purposes of this discussion.)

Table 8.1

	A_1		A_2	
	B_1	B_2	B_1	B_2
C_1	169	145	167	135
C_2	173	143	165	134

At this stage it is important to ask questions about the experiment itself before embarkin on the analysis of the results. Was there a better way of carrying out the experiment? Did it achieve what it set out to test? In other words, was the experiment *designed* properly?

The first stage in any experimental procedure should be to state carefully the aim of the investigation and then to consider the proposed experiment. What is the **response variable** (dependent variable) to be studied, can it be measured and if so how accurately on the instruments available to the experimenter? Might there be more than one respon variable? What are the independent variables (**factors**) which may influence the respons variable? Should each factor be kept at a constant value or given different levels, either specified fixed values or chosen at random? Can the response variable be measured quantitatively or only qualitatively (e.g. 'yes' or 'no') and are the factors qualitative (e.g. method of cooling) or quantitative (e.g. temperature)?

The next stage is the design of the experiment. Far too often an experiment is

† Zelen & Connor, "Multifactor experiments", Industrial Quality Control, 15, 1959.

erformed and the results presented for analysis with little thought to *how* the experiment hould have been performed. The number of observations to be taken should be etermined by considering how large are the differences that the experiment is supposed o measure, how much variation or experimental error is present and what magnitude of tatistical risk is expected. When the independent variables are selected and their levels pecified, other variables which are not so controlled in this manner cannot be taken into ccount. In order to attempt an 'averaging out' of their possible effects, **randomization** f the order of experimentation should be sought. It will allow the experimenter to ssume that the errors in measurement are statistically independent. In some cases, a ompletely random order of the observations might not be possible. Perhaps a mperature factor would allow only randomization of the observations at a particular mperature, before proceeding to the next level.

The final stage is to carry out the experiment, and to analyse the results.

As the experiment in our example stands we shall be able only to analyse the odel

$$x_{ijk} = \mu + \alpha_i + \beta_j + \gamma_k + \epsilon_{ijk} \tag{8.1}$$

here x_{ijk} is the response variable, μ the mean of the population from which all data re assumed to come, α_i is the **effect** of the carbon content, β_j is the effect of mperature, γ_k the effect of cooling method, and ϵ_{ijk} is the random error in the xperiment. The indices i, j and k take values 1 or 2.

The model (8.1) does not allow investigation of the **interaction** between the factors, hat is, the change in the response between the levels of one factor may not be the same or both levels of the other factors. In order to allow for such a study, it would have een necessary to repeat the experiment completely at least once (i.e. to **replicate** the xperiment). The discrepancies between observations for the same combination of actors provide a measure of the unexplained error. This feature will be dealt with in ection 8.5.

Table 8.2

Source of Variation	Sum of squares (3 dp)	Degrees of Freedom	Component of Variance (3 dp)	*F* Ratio	Critical Values of *F*
Carbon Content	105.125	1	105.125	18.7	5% $F = 7.71$
Temperature	1711.125	1	1711.125	304.2	
					1% $F = 21.2$
Cooling Method	0.125	1	0.125	0.022	
Error	22.5	4	5.625		
Total	1838.875	7			

For the moment we assume that the order of the eight observations was completely randomized, that the levels of the factors was fixed (so that α_i are fixed constants with $\alpha_1 + \alpha_2 = 0$ and similarly for β_j and γ_k) and ϵ_{ijk} are independent normally distributed random variables following $N(0, \sigma^2)$.

An analysis of variance is performed on the data and the results shown in Table 8.2 The details of the construction of the table will be dealt with in subsequent sections. We conclude that the effect of temperature is highly significant and that the effect of carbon content is significant at the 5% level. The method of cooling seems to have little effect, if any, on the tensile strength. However, we must emphasise that there are so few degrees of freedom that the accuracy of our testing procedure has to be viewed with caution. There might be some interaction between carbon content and temperature of tempering on the tensile strength, but this experimental design is of no help.

The average of the eight observations is 153.875. If we subtract the sum of the four readings at A_1 from the sum of the four readings at A_2 and divide the result by 4 we obtain the **average effect** of factor A; this is -7.25. In a similar way we can find the average effect of B to be -29.25 and that of C to be -0.25. These average effects bear out the conclusions from analysis of variance. In Figure 8.1 the graphs of the combination of A and B are shown. For example, the tensile strength at A_1 and B_1 has an average value of ½ (169 + 173) = 171.

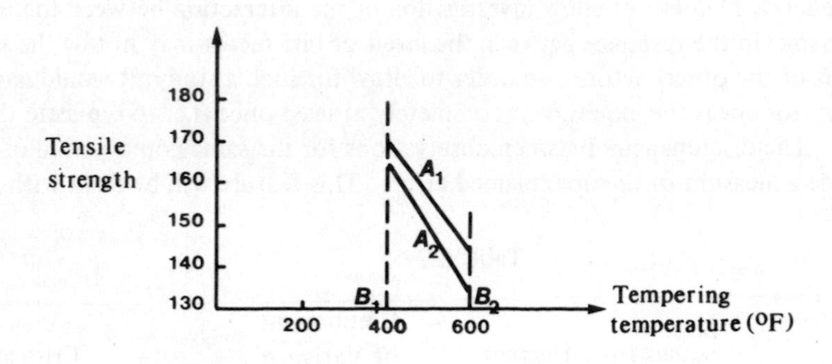

Figure 8.1

The tensile strength is higher for a lower temperature, and is higher for A_1; the lines being non-parallel implies the presence of some interaction between A and B.

Summary

The objective of a carefully designed experiment is to obtain more from that experiment than could be obtained otherwise for the same effort. There are three phases:

A. **The Experiment** The problem is stated, a response variable is selected, the factors and their levels are chosen and a decision is made as to whether the levels are at fixed values or selected at random.

B. **The Design** The mathematical model is chosen, the number of observations is decided as is the order of experimentation and technique of randomization.

C. **The Analysis** The experimental data is collected, the relevant significance tests are performed and the results of the analysis are interpreted.

In subsequent sections, several experimental designs are presented together with a selection of methods of back-up analysis.

8.2 SINGLE FACTOR EXPERIMENTS: COMPLETELY RANDOMIZED DESIGN

Consider the following example.†

The 24-hour water absorption, as a percentage of dry weight, of concrete samples taken from five different types of precast concrete slabs is shown in Table 8.3. Each type of slab is made with a particular aggregate. Is there a significant difference in the water absorption of the slabs of different types?

Table 8.3

Type	*A*	*B*	*C*	*D*	*E*
Absorption	6.8	5.2	4.5	6.8	6.6
	5.9	4.8	5.0	7.3	5.9
%	5.9	5.2	4.7	6.9	4.8
	5.6	5.3	4.6	6.4	6.0
Σ	24.2	20.5	18.8	27.4	23.3

A first glance at the table will show that there is a difference between column totals which may be due to genuine differences in absorption between the types of slab. However, the values in any column are not consistent, suggesting other, uncontrolled variation. The question to be answered is whether the differences between types of slab are significant in comparison with the uncontrolled variation. More sophisticated experiments could introduce other factors which will account for some of the residual variation, but the aim is to produce a design which caters for the important factors; there is almost certainly residual variation in any experimental design.

† Wright 'Statistical Methods in Concrete Research'. Magazine of Concrete Research. 5, 15 (1954).

In this experiment we assume that the 20 tests were carried out in a random order, perhaps by using a table of random numbers. Such a design is said to be **completely randomized.**

Certain assumptions are made about the residual variation. First, the random variables which give rise to the residual variation have zero expected value. Second, they are mutually independent. Third, they all have the same standard deviation.† Fourth, they are all normally distributed.

Of these assumptions, the fourth is the least likely to be valid; however, much of the analysis of variance can be developed without it and a reasonable departure from normality can be tolerated with little appreciable effect. The third assumption is crucial to the testing procedure, yet it is possible to have situations where different measurement techniques can produce variations of different sizes. Sometimes it is possible by logarithmic or other transformations to produce a set of variables more nearly in agreement with the assumptions.

Mathematical model

The model used is

$$x_{ij} = \mu + T_j + \epsilon_{ij} \tag{8.2}$$

where x_{ij} is the value of the ith observation of the jth treatment (type of slab), T_j is the **treatment effect,** μ is a common effect for the experiment as a whole and ϵ_{ij} is the random error which occurs in the ith observation of the jth treatment. Note that since T_j are fixed effects (i.e. the 5 classes of slab are the only classes available for consideration),

$$\sum_{j=1}^{5} T_j = 0$$

We may regard each column of observations in Table 8.3 as a sample from a population with mean μ_j; from the sample we can form a mean $\bar{x}_j$ which is an estimate of μ_j.

Consider the identity

$$x_{ij} - \mu \equiv (\mu_j - \mu) + (x_{ij} - \mu_j) \tag{8.3}$$

The first term on the right-hand side represents the treatment effect and the second term represents the residual variation.

† This is called **homoscedasticity** by some authors.

Let $\bar{x}$ be the mean of all the observations and T their grand total. If there are n observations in each column and k columns then $T = nk\bar{x}$.

If we substitute sample statistics for the population parameters in equation (8.3) we obtain

$$(x_{ij} - \bar{x}) \equiv (\bar{x}_j - \bar{x}) + (x_{ij} - \bar{x}_j)$$

Now square both sides of this identity and sum over i and j to obtain

$$\sum_{j=1}^{k} \sum_{i=1}^{n} (x_{ij} - \bar{x})^2 = \sum_{j=1}^{k} \sum_{i=1}^{n} (\bar{x}_j - \bar{x})^2 + \sum_{j=1}^{k} \sum_{i=1}^{n} (x_{ij} - \bar{x}_j)^2$$

$$+ 2 \sum_{j=1}^{k} \sum_{i=1}^{n} (\bar{x}_j - \bar{x})(x_{ij} - \bar{x}_j)$$

The last term on the right-hand side is

$$2 \sum_{j=1}^{k} (\bar{x}_j - \bar{x}) \left\{ \sum_{i=1}^{n} (x_{ij} - \bar{x}_j) \right\}$$

The term in $\{\ \}$ is zero, since it is the sum of the deviations of the observations in column about their mean. Hence

$$\sum_{j=1}^{k} \sum_{i=1}^{n} (x_{ij} - \bar{x})^2 = n \sum_{j=1}^{k} (\bar{x}_j - \bar{x})^2 + \sum_{j=1}^{k} \sum_{i=1}^{n} (x_{ij} - \bar{x}_j)^2 \qquad (8.4)$$

TOTAL SUM OF SQUARES	=	BETWEEN TREATMENTS SUM OF SQUARES	+	WITHIN TREATMENTS SUM OF SQUARES
(T.S.S.)		(B.S.S.)		(W.S.S.)

If attention is confined to the jth treatment then $\sum_{i=1}^{n} (x_{ij} - \bar{x}_j)^2 / (n-1)$ gives an unbiased estimate of σ_0^2, the variance within the jth treatment. But the variances within treatments were assumed to be the same for each treatment. Therefore, we may pool these estimates to obtain W.S.S./$(k(n-1))$ as an unbiased estimator of σ_0^2. A similar exercise shows that B.S.S./$(k-1)$ is an unbiased estimator of

$$\sigma_0^2 + \frac{n}{(k-1)} \sum_{j=1}^{k} T_j^2 \; †$$

To employ analysis of variance we set up the null hypothesis H_0: that $T_j = 0$ for all j. Under this hypothesis $\frac{\text{B.S.S.}}{(k-1)}$ is an unbiased estimator of σ_0^2. The alternate hypothesis H_1 is that at least one T_j is non-zero. It can be demonstrated that these two unbiased estimators of σ_0^2 are independent and each follows a χ^2 distribution with the appropriate number of degrees of freedom. Then as a direct consequence, their ratio, namely $\frac{\text{B.S.S.}}{(k-1)} \Big/ \frac{\text{W.S.S.}}{k(n-1)}$, follows an F distribution with $(k-1)$, $k(n-1)$ degrees of freedom. We may then set up an analysis of variance table as shown in Table 8.4. Note that for this theoretical exercise the column headed 'MEAN SQUARE' replaces that headed 'COMPONENT OF VARIANCE'.

Table 8.4

SOURCE OF VARIATION	SUM OF SQUARES	DEGREES OF FREEDOM	MEAN SQUARE	EXPECTED MEAN SQUARE
Between Treatments T_j	$n \sum_{j=1}^{k} (\bar{x}_j - \bar{x})^2$	$k-1$	$\frac{\text{B.S.S.}}{k-1}$	$\sigma_0^2 + \frac{n}{(k-1)} \sum_{j=1}^{k} T_j^2$
Within Treatments ϵ_{ij}	$\sum_{j=1}^{k} \sum_{i=1}^{n} (x_{ij} - \bar{x}_j)^2$	$k(n-1)$	$\frac{\text{W.S.S.}}{k(n-1)}$	σ_0^2
Total	$\sum_{j=1}^{k} \sum_{i=1}^{n} (x_{ij} - \bar{x})^2$	$kn-1$		

To test the hypothesis H_0: $T_j = 0$, form $F = \frac{\text{B.S.S.}}{k-1} \div \frac{\text{W.S.S.}}{k(n-1)}$ and reject H_0 with a risk α if $F >$ critical value of $F^{k-1}_{k(n-1)}$ at risk α.

Computational short-cut

Before we apply this strategy to the example of this section it is worth pointing out some computational short-cuts.

† See Hicks (1964), page 158.

It is straightforward to show that

$$\text{T.S.S.} = \sum_{j=1}^{k} \sum_{i=1}^{n} x_{ij}^2 - T^2/N \qquad (8.5)$$

here $N = kn$. Further

$$\text{B.S.S.} = \frac{1}{n} \sum_{j=1}^{k} T_j^2 - T^2/N \qquad (8.6)$$

he term T^2/N is called the **Correction Factor**. Note that W.S.S. is obtained by ıbtraction of B.S.S. from T.S.S.

nalysis

Returning to the example, we extend Table 8.3. to Table 8.5.

Table 8.5

Type	A	B	C	D	E	
Absorption	6.8	5.2	4.5	6.8	6.6	
	5.9	4.8	5.0	7.3	5.9	
%	5.9	5.2	4.7	6.9	4.8	
x_{ij}	5.6	5.3	4.6	6.4	6.0	$N = 20$
T_j	24.2	20.5	18.8	27.4	23.3	$T = 114.2$
T_j^2	585.64	420.25	353.44	750.76	542.89	$\sum_{i=1}^{k} T_j^2 = 2652.98$
$\sum_{i=1}^{n} x_{ij}^2$	147.22	105.21	88.50	188.1	137.41	$\sum_{j=1}^{k} \sum_{i=1}^{n} x_{ij}^2 = 666.44$

Correction Factor $= (114.2)^2 \div 20 = 652.082$

$$\text{T.S.S.} = 666.44 - 652.082 = 14.358$$

$$\text{B.S.S.} = \frac{1}{4} \times 2652.98 - 652.082 = 11.163$$

$$\text{W.S.S.} = 14.358 - 11.163 = 3.195$$

The analysis of variance table, Table 8.6, is constructed below.

Table 8.6

Source of Variation	Sum of Squares	Degrees of Freedom	Component of Variance	F	Critical Values of F
Between Treatments	11.163	4	2.79		5% $F^4_{15} = 3.05$
				13.1	
Within Treatments	3.195	15	0.213		1% $F^4_{15} = 4.90$
Total	14.358	19			

From the table it is clear that there is sufficient evidence at the 1% level to reject H_0. We conclude with much less than a 1% risk that there is a significant difference between types of slab as regards absorption.

Estimation of treatment effects

It is clear that an estimate of μ is provided by

$$\hat{\mu} = \frac{1}{k} \sum_{j=1}^{k} \bar{x}_j = T/N \tag{8.7}$$

so that for the slab example $\hat{\mu} = \dfrac{114.2}{20} = 5.71$

Furthermore, we may estimate each treatment effect by

$$\hat{T}_j = \bar{x}_j - \hat{\mu} = \frac{1}{n} T_j - \hat{\mu} \tag{8.8}$$

Hence

$$\hat{T}_1 = 6.05 - 5.71 = 0.34$$

$$\hat{T}_2 = 5.125 - 5.71 = -0.585$$

$$\hat{T}_3 = 4.7 - 5.71 = -1.01$$

$$\hat{T}_4 = 6.85 - 5.71 = 1.14$$

$$\hat{T}_5 = 5.825 - 5.71 = 0.115$$

Note that $\sum_{j=1}^{5} \hat{T}_j = 0$ as was suggested earlier.

Which treatment means differ?

Having established a significant difference in the treatment means, it is pertinent to ask which of these means are different from the rest. In Figure 8.2 the observations are plotted. The horizontal line in each case represents the mean of the observations for a particular type of slab.

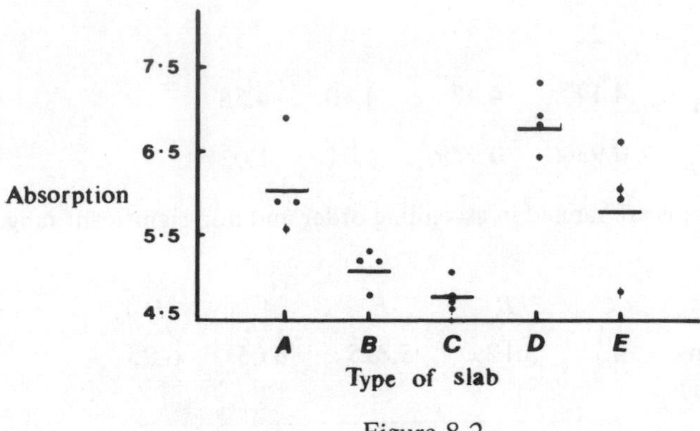

Figure 8.2

A first glance at the diagram shows that C appears somewhat below the other treatments and D seems to be somewhat above. We can put these ideas on a quantitative footing by using a technique of **contrasts analysis.** The method described here is the **Duncan Multiple Range Test.** It is used when, having rejected the null hypothesis, we seek to group the means into subsets of not significantly different means. The number of observations in each treatment sample ***must be the same.***

The idea is to rank the treatment means in ascending order. Means are first tested in adjacent pairs. To see whether gaps between adjacent values are significantly large, the procedure is as follows. First, obtain the error mean square from the analysis of variance table. This value is the one used as the denominator in the F ratio and in the present example is 0.213. This value is divided by the number of observations for each treatment, i.e. n, and the square root taken. The result is $\hat{\sigma}_{\bar{x}_j}$ (compare $\hat{\sigma}/\sqrt{n}$ for the standard error of a mean) and in this case is $\sqrt{\dfrac{0.213}{4}} = 0.224$ (3 dp). **Duncan's table** (Appendix **B**) provides values of significant ranges at the 5% and 1% level and the appropriate value from the table, r_p is multiplied by $\hat{\sigma}_{\bar{x}_j}$ to give the least significant range R_p. Here p denotes the number of adjacent means in the subsets under consideration, in this case 2. Suppose we look for significance at the 1% level. The appropriate number of degrees of freedom is that associated with the error mean square, 15. The value of r_p is 4.17. This gives a value of $R_p = 4.17 \times 0.224 = 0.934$ (3 sf). We then underline adjacent means which are not significantly different (i.e. whose difference does not exceed R_p).

Now means are tested in groups of three. The value tested against the new R_p is the range, that is (highest value – lowest value). Ranges which are not significant are underlined.

Finally the means are tested as groups of four and as a group of 5. To see the effect of this we note that at the 1% level the following values are used with 15 degrees of freedom.

p	2	3	4	5
r_p	4.17	4.37	4.50	4.58
R_p	0.934	0.979	1.01	1.03

The treatment means are ranked in ascending order and non-significant ranges underlined.

Type	C	B	E	A	D
Mean	4.7	5.125	5.825	6.05	6.85
$p = 2$	———	——— ———	——— ———	——— ———	———
$p = 3$		———	———	———	
$p = 4$					
$p = 5$					

From these results we can see that the homogeneity breaks down completely after $p = 3$. The conclusion to be drawn is that type D and type C are significantly different from the rest. Although it is possible to group C and B, and A and D, the strain on the latter from D is too great to group E, A and D. The strain on the group C, B, E comes from the gap between B and E.

This contrasts analysis was carried out after the experimentation. A decision to make specific comparisons which is taken in advance of the experiment requires a different method,one such being that of orthogonal contrasts. It will not be pursued here; reference can be made to Hicks (1964).

Extension to unequal sample sizes

The analysis of variance method gives the smallest value of the β-risk for a given sample size, provided that the number in each sample is the same. If there are missing bservations, the procedure can be amended as follows. Let n_j represent the number observations for the jth treatment. Then

$$\text{T.S.S.} = \sum_{j=1}^{k} \sum_{i=1}^{n_j} x_{ij}^2 - T^2/N$$

$$\text{B.S.S.} = \sum_{j=1}^{k} \frac{T_j^2}{n_j} - \frac{T^2}{N}$$

$$\text{W.S.S.} = \text{T.S.S.} - \text{B.S.S.}$$

ne respective components of variance from B.S.S. and W.S.S. are based on $(k - 1)$ and $\left(\sum_{j=1}^{k} n_j - k\right)$ degrees of freedom respectively.

A contrasts analysis method which can cope with unequal sample sizes has been veloped by Scheffé (1959).

sts for assumptions of normality and equal variances

The assumption of normality can be checked by plotting the data points for a mple on normal probability paper to see whether the values lie nearly on a straight ne.

The assumption of equal variance can be checked, if suspected, before analysis of riance is attempted. One such test is described here: it is not robust to departures om normality. Further, it depends upon equal sample sizes. The test is due to ochran.

The model we have used is

$$x_{ij} = \mu + T_j + \epsilon_{ij} \tag{8.2}$$

e test whether $\text{var}(\epsilon_{ij}) = \sigma_0^2$ varies from treatment to treatment.

The components of W.S.S., viz. $S_j = \sum_{i=1}^{n} (x_{ij} - \bar{x}_j)^2$ can be used to test the null pothesis of equal variances. If ϵ_{ij} are normally distributed then S_j are distributed as with $(n - 1)$ degrees of freedom. It has been suggested that substantial evidence of parture from homoscedasticity should arise before the standard procedure is abandoned d a 1% level of significance at most used (i.e. even lower % might be used).

The test calculates the quantity $[(\text{maximum } S_j)/ \sum_{j=1}^{k} S_j]$ and rejects the null hypothesis at the α level of significance if this ratio exceeds a value $C_{(1-\alpha,n,k)}$ which is tabulated in Appendix B. The number of degrees of freedom is $\upsilon = (n - 1)$.

For the example of this section, $\upsilon = 3$, $k = 5$ and $C_{(0.99,4,5)}$ is found from the table to be 0.696. The value of S_j for type A is

$$\frac{1}{4}\left(4 \times \sum_{i=1}^{n} x_{i1}^2 - T_1^2\right) = 0.81$$

Similarly, the values of S_j for the other types are obtained as 0.1475, 0.14, 0.41 and 1.6875 respectively. Then the ratio

$$\frac{\text{maximum } S_j}{\sum_{j=1}^{k} S_j} = \frac{1.6875}{3.195} = 0.526 \text{ (3 dp)}$$

Since this does not exceed the critical value obtained from the table, we do not reject the hypothesis of equal variances.

Random Effects

So far we have assumed that the treatment effects T_j were **fixed** and represented specific types of slab. Suppose, in fact, that the slab types were **randomly** selected from several which were available. Then we must assume that the T_j are independent random variables normally distributed with zero mean and variance σ_T^2. The variance σ_T^2 is the variance among the true treatment means μ_j.

Note that the T_j average to zero when all possible cases are taken into account, but for the random selection of the experiment this will not necessarily be so.

Although the mechanics of the analysis of variance test are the same, the theoretical mean square between treatments is now $\sigma_0^2 + n\sigma_T^2$. Under the null hypothesis, $\sigma_T^2 = 0$. If this is rejected we assume a difference between all the treatments from which the selection of k were randomly taken. Obviously, there are no fixed biases to estimate in this model. Note that (from Table 8.6) an estimate of the expected value of $\sigma_0^2 = 0.213$ and the expected value of $(\sigma_0^2 + 4\sigma_T^2) = 2.79$. Hence (the expected value of σ_0^2 + 4 times the expected value of $\sigma_T^2) = 2.79$, from which follows that the expected value of $\sigma_T^2 = 0.64$.

Four specimens of each of four different metals were immersed in a highly corrosive solution, and their corrosion rates were measured with the following results:

Metal	A	B	C	D
	27	25	27	26
Rates	30	20	31	26
	25	22	30	25
	26	21	32	23

(a) Test the null hypothesis that the metals have the same corrosion rate.

(b) Estimate the difference in corrosion rate between A and B.

(c) Use Duncan's test on the mean corrosion rate for each metal. What conclusions can you draw?

Random samples of 3 makes of tyres required the following braking distances in feet while going at 60 miles per hour:

Make A	Make B	Make C
250.1	248.8	254.5
249.7	243.6	261.8
251.0	253.9	255.5
248.9	251.4	253.6
241.8	247.3	260.2
242.0	249.3	256.2
248.2	253.6	264.7
246.3	248.7	262.3

(a) With a 1% level of significance, test whether the differences among sample means are attributable to chance.

(b) Estimate the parameters in the model you are using.

To study the dust deposited on a precipitator a completely randomized experiment was carried out. The results are :

Total flow (m^3/h)	Dust loading (grains per cubic metre in flue gas)		
5	1.5	1.3	1.9
10	2.3	2.1	2.8
15	2.7	2.2	2.8
20	3.4	4.1	4.7

(a) Test whether the flow through the precipitator has an effect on the dust loading.

(b) Estimate the effects corresponding to the different rates of flow.

4. It is suspected that five machines in a factory are turning out products of non-uniform weight. Using a 1% level of significance, test the hypothesis that the machines do not differ. Are any machines behaving significantly different from the rest? Use the data below, the results are measured in grams in excess of the supposed standard.

Machine *A*	Machine *B*	Machine *C*	Machine *D*	Machine *E*
0.22			0.25	
0.25			0.25	0.20
0.30	0.21		0.23	0.18
0.27	0.16		0.21	0.20
0.28	0.22		0.27	0.15
0.24	0.25	0.29	0.24	0.20
0.31	0.21	0.21	0.28	0.19
0.32	0.18	0.25	0.29	0.16

5. A completely randomized experiment was carried out to test four production lines which deposited a coating on a sheet metal surface with a view to determining whether any particular line was depositing an abnormal amount of coating. The results are given below in mg. With a 1% risk of finding spurious differences, what do you conclude?

Line 1	Line 2	Line 3	Line 4
4.45	8.03	5.08	7.03
3.02	6.15	3.23	5.46
3.88	5.89	4.51	6.36
2.69	6.25	5.00	6.23
4.42	6.34	4.27	4.53

6. A completely randomized experiment was conducted to compare the reflective properties of kinds of paint. The 20 available test specimens were allotted to the paints at random. Th response (in suitable units) was coded into the data shown below.

1	2	3	4	5
78	72	80	85	66
73	69	77	89	71
70	68	75	78	67
65	79	72	74	76

(a) State the model you use,

(b) With a 5% risk, state whether the paints differ in their response.

3 RANDOMIZED BLOCKS DESIGN

There are occasions when an environmental factor may make a major contribution
the uncontrolled variation in an experiment. For example, humidity might affect the
perimental results. If, therefore an experiment had to extend over several days,
servations taken on the same day might show a closer agreement than a selection of
ose taken on different days. It would therefore make sense to provide for an equal
mber of tests on every treatment each day. Such a procedure is known as **blocking**
d in this instance each day is a block. It allows the experimenter to improve the
ecision of comparison since the comparison is within each block instead of being made
om one block to another. As a second example, suppose that four treatments for
eparing a wooden surface prior to painting were to be compared for their effectiveness
the durability of the painted surface. It is known that the moisture content of the
ood will affect the experimental results. The woods to be used can be classified by
spection as either low moisture content or high moisture content. If 80 frames were
be tested it would be wise to arrange to have 40 with high moisture content and 40
ith low moisture content. Each of the four treatments could be assigned to 10
ames in each set.

Other examples of blocking can be found in the problems at the end of the section.
ne commonly quoted example is that of testing four brands of tyre for tread wear.
ppose that four cars are available. Table 8.7 shows three possible arrangements
r testing. The brands of tyre are denoted A, B, C, D and the four cars are denoted
II, III, IV. The order within each column (car) is front nearside, front offside, rear
arside, rear offside.

Table 8.7

I	II	III	IV		I	II	III	IV		I	II	III	IV
A	B	C	D		C	B	C	D		A	C	A	D
A	B	C	D		B	B	B	D		B	D	B	A
A	B	C	D		D	C	A	A		D	B	C	C
A	B	C	D		A	D	A	C		C	A	D	B

The arrangement on the left is clearly useless since we cannot distinguish the
fects of make of tyre and car used: the effects are said to be **confounded**. The centre
rangement is an example of a completely randomized design; it has disadvantages,
wever, since for example, brand D is never used on car III and variations within the
sults for that brand may indicate differences between cars I, II and IV. In short,
ndom error' may not include experimental error alone, but also variation within cars.
e arrangement on the right is a randomized blocks design. Each brand of tyre is used
ce on each car. The position it occupies on a car is selected at random. Each car is
w a block. Note that no regard has been given to the influence of position on the car.

ample

Four production processes manufacture a specific cylindrical component with diameter 2 mm. It is suspected that the average diameters of the components from the processes are not the same. Since it is important to maintain consistency of the produc it is decided to test the suspicion. The raw metal used in the manufacture of the components varies from batch to batch and it is therefore advisable to allow for this variation. However, it is no use restricting the experiment to one batch only, since we should not know whether the variations would apply for other batches. We suppose that there are 5 batches of metal available. Then each batch becomes a block and fro each block four amounts are taken and randomly assigned to the four processes.

The results of the experiment are shown in Table 8.8; units are mm.

Table 8.8

Batch \ Process	*A*	*B*	*C*	*D*
I	2.060	2.134	1.905	2.057
II	1.957	2.048	2.021	2.037
III	2.071	2.056	2.078	2.015
IV	2.055	2.073	2.082	2.016
V	2.051	2.098	1.956	2.061

The mathematical model employed is

$$x_{ij} = \mu + \alpha_i + \beta_j + \epsilon_{ij} \tag{8.9}$$

where x_{ij} is the value of the observation of the jth treatment in the ith block, μ is a common effect, α_i is the block effect and β_j is the treatment effect; ϵ_{ij} is the rando error.

As in the completely randomized design the total sum of squares is computed as

$$\text{T.S.S.} = \sum_{j=1}^{k} \sum_{i=1}^{n} x_{ij}^2 - \frac{T^2}{N}$$

where N is the total number of observations and T is the grand total of observed valu The between treatments sum of squares is computed as

$$\text{B.T.S.S.} = \frac{1}{n} \sum_{j=1}^{k} T_j^2 - \frac{T^2}{N}$$

Now we may compute the between blocks sum of squares in a similar way i.e.

$$\text{B.B.S.S.} = \frac{1}{k} \sum_{i=1}^{n} T_i^2 - \frac{T^2}{N}$$

The error sum of squares (E.S.S.) is the difference between the total sum of squares ıd the sum of the blocks and treatments sum of squares. Table 8.8 is first extended ter modification (subtraction of 2.0 from each observation and multiplication of the sult by 1000) to Table 8.9.

Table 8.9

Batch \ Process	A	B	C	D	T_i	T_i^2	
I	60	134	–95	57	156	24336	
II	–43	48	21	37	63	3969	
III	71	56	78	15	220	48400	
IV	55	73	82	16	226	51076	
V	51	98	–44	61	166	27556	
T_j	194	409	42	186	831	155337	Σ
T_j^2	37636	167281	1764	34596	241277		
$\sum_{i=1}^{n} x_{ij}^2$	16116	38329	24210	8820	87475		
					Σ		

e now compute the relevant sums of squares

$$\text{T.S.S.} = 87475 - (831)^2/20 = 52946.95$$

$$\text{B.T.S.S.} = \frac{1}{5} \times 241277 - (831)^2/20 = 13727.35$$

$$\text{B.B.S.S.} = \frac{1}{4} \times 155337 - (831)^2/20 = 4306.2$$

$$\text{E.S.S.} = 52946.95 - 13727.35 - 4306.2 = 34913.4$$

We have not justified these computations. The proof that the total sum of squares nd the total degrees of freedom) can be partitioned as above and that the resulting ıalysis of variance procedure is valid will be assumed by analogy with Section 8.2.

Table 8.10 is the appropriate analysis of variance table.

Table 8.10

Source of Variation	Sum of Squares	Degrees of Freedom	Component of Variance	F (1 dp)	Critical values of F
Between Treatments	13727.35	3	4575.8	1.6	5% $F^{3}_{12} = 3.49$
Between Blocks	4306.2	4	1076.5	< 1	5% $F^{4}_{12} = 3.26$
Error	34913.4	12	2909.5		
Total	52946.95	19			

In neither instance is the calculated F value in excess of the appropriate critical value. The null hypothesis that there is no difference between the processes is not rejected. It is also possible to test the hypothesis of no variation between batches and this too is not rejected. However, the purpose of the design was to test, as a single factor experiment, possible process difference; the blocking was merely a restriction on complete randomization, placed because of the nature of the problem.

Missing Values

There are occasions when an observation is invalidated for one reason or another and cannot be included in the analysis. For a randomized blocks design this is a serious matter. The most usual procedure is to replace the missing value by that value which makes the total sum of squares a minimum.

This value can be found by replacing the missing value by x, say , and differentiating to find the minimum.

Balanced Incomplete Blocks

The design that we have considered in this section has been **complete** in the sense that each treatment has been applied to each block. When for reasons of lack of time or the sheer impossibility of achieving this completeness (for example, five brands of tyre to be tested on four-wheeled cars), a modification is necessary; a design known as **balanced incomplete blocks** is employed. Since each treatment cannot be applied to each block, the design is balanced by ensuring that each pair of treatments occurs the same number of times, l, in the experiment. Consider the following example.

It is required to find the effect of four different coatings A, B, C, D on the conductivity of fluorescent tubes. Each observation takes several hours to complete ad it is found that only three observations can be carried out in a day. The balanced complete blocks design used is shown with the results of the experiment in coded form Table 8.11. This is one of several possible patterns.

Table 8.11

Blocks \ Treatments	A	B	C	D	T_i	T_i^2	
1	8	21		3	32	1024	
2	4	15	33		52	2704	
3		32	14	4	50	2500	
4	12		24	1	37	1369	
T_j	24	68	71	8	171	7597	Σ
$\sum_{i=1}^{n} x_{ij}^2$	224	1690	1861	26			

Notice that each pair of treatments occurs twice and only twice; for example, the air BC occurs on days 2 and 3. In each block, the order of the treatments applied is lected randomly.

The analysis proceeds as follows.

First, the total sum of squares is calculated as usual.

$$\text{T.S.S.} = \sum_{j=1}^{k} \sum_{i=1}^{n} x_{ij}^2 - T^2/N = 3801 - (171)^2/12 = 1364.25$$

ext, the between blocks sum of squares is calculated without regard to the 'missing' lues; t is the number of treatments in each block.

$$\text{B.B.S.S.} = \frac{1}{t} \sum_{i=1}^{n} T_i^2 - T^2/N = \frac{1}{3} \times 7597 - (171)^2/12 = 95.58 \qquad \text{(2 dp)}$$

ow, the between treatments sum of squares is determined, making allowance for the complete blocks.

The first step is to calculate $Z_j = t\,T_j - \sum_{i=1}^{n} \alpha_{ij} T_i$ where $\alpha_{ij} = 1$ if treatment j

occurs on block i and $\alpha_{ij} = 0$ otherwise.

In this problem, $Z_1 = 3 \times 24 - (32 + 52 + 37) = -49$. Similarly,

$$Z_2 = 3 \times 68 - (32 + 52 + 50) = 70$$
$$Z_3 = 3 \times 71 - (52 + 50 + 37) = 74$$
$$Z_4 = 3 \times 8 - (32 + 50 + 37) = -95$$

(Observe that $Z_1 + Z_2 + Z_3 + Z_4 = 0$)

The second step is to compute

$$\text{B.T.S.S.} = \frac{Z_1^2 + Z_2^2 + Z_3^2 + Z_4^2}{t.l.k} = \frac{21802}{3.2.4} = 908.42 \qquad \text{(2 dp)}$$

Finally, the error sum of squares is computed, in the usual way, by subtraction.

$$\text{E.S.S.} = 1364.25 - 95.58 - 908.42 = 360.25 \qquad \text{(2 dp)}$$

The analysis of variance table is shown in Table 8.12 below.

Table 8.12

Source of Variation	Sum of Squares	Degrees of Freedom	Component of Variance	F	Critical Values of F
Between Treatments	908.42	3	302.8	4.2	5% $F_5^3 = 5.41$
Between Blocks	95.58	3			1% $F_5^3 = 12.1$
Error	360.25	5	72.1		
Total	1364.25	11			

The conclusion to the experiment is that differences between treatments are n[ot] significant at the 5% level.

It is possible in cases where the number of blocks is equal to the number of treatments to adjust the between blocks sum of squares in order to test for significant differences between blocks; in this instance the component sum of squares will not ad[d] up to the total.

If a design is unbalanced, a different approach has to be employed; see Hicks (1964).

The following table gives the productivity (measured by the number of acceptable items per worker per hour) of three shifts in a given industrial plant, over a period of four weeks.

Shift	I	II	III	IV
A	20	18	27	22
B	14	15	22	15
C	22	23	30	21

Are there significant differences in productivity between shifts?

The following data gives the times taken for 5 operators to perform a specific task on three machines.

Machine	I	II	III	IV	V
A	3.9	5.8	4.2	5.9	5.1
B	4.8	6.5	6.0	5.8	5.0
C	5.6	6.1	5.9	6.5	5.8

(a) Test the results for a significant difference between machines.

(b) Analyse row means by Duncan's test.

An experiment was conducted to determine the effect of 4 treatments applied to the coils of TV tube filaments on the current flow through these coils. Because of the amount of time needed to perform this experiment, only 3 sample values could be taken each day. The data are given, in coded form, in the table below.

Day	Treatment			
(Blocks)	*A*	*B*	*C*	*D*
1	13		31	18
2		43	25	14
3	15	24	42	
4	11	34		22

Is there a significant effect among treatments?

Four specimens of rubber are sent to the laboratory for a test of flexural strength. There are 4 curing times. However, each specimen is sufficient for only 3 samples. Hence a balanced incomplete block is proposed. Specimens are considered as blocks and curing times as treatments. Investigate the effect of curing time on flexural strength, using the coded data given.

Specimen	Curing Times 1	2	3	4
A	14	11	16	
B	16	12		22
C	19		24	30
D		21	27	29

8.4 LATIN SQUARE DESIGN

In the case of randomized blocks design, the aim is to separate out an extraneous source of variation from the error sum of squares. If a second source of variation is suspected it is possible to design out this as well. At the outset it is important to emphasise that the number of levels of both restrictions must be equal to the number of levels of the treatment. The **Latin Square** design is such that each treatment appears once (and only once) in each row and each column. Lest it be thought that the design precludes any randomization, there are several Latin Squares for each size $(n \times n)$; hence the particular square used can be chosen at random from those available. An example will serve as a vehicle for the discussion.

Example

To test the most efficient of four fuels for use with a particular machine it was decided to use each fuel with each of 4 machines available. It was thought possible that the order in which the fuels were introduced into the machines might influence the results so that each fuel was the first to be used on one occasion, second on one occasion etc. The results of the test in coded form are shown in Table 8.13. Notice the order of use and the brand of fuel form the rows and columns.

Table 8.13

Order \ Machine	I	II	III	IV
1	16(A)	13(D)	19(B)	17(C)
2	18(B)	18(C)	18(A)	17(D)
3	19(C)	13(A)	16(D)	19(B)
4	14(D)	16(B)	12(C)	17(A)

The model used is

$$x_{ijk} = \mu + \alpha_i + \beta_j + \gamma_k + \epsilon_{ijk} \tag{8.10}$$

where γ_k represents the effect of the kth position (or order).

Note that the totals for treatments A, B, C and D are 64, 72, 66 and 60 espectively.

The following totals are quoted:

$$\Sigma T_i^2 = 17214, \quad \Sigma T_j^2 = 17236, \quad \Sigma\Sigma x_{ij}^2 = 4368, \quad \Sigma\Sigma x_{ij} = 262$$

Then, the total sum of squares $= 4368 - (262)^2/16 = 77.75$

the between rows sum of squares $= \frac{1}{4} \times 17214 - (262)^2/16 = 13.25$

and the between columns sum of squares $= \frac{1}{4} \times 17236 - (262)^2/16 = 18.75$

The between treatments sum of squares $= \frac{1}{4} \times [(64)^2 + (72)^2 + (66)^2 + (60)^2] - (262)^2/16 = 18.75$

The error sum of squares $= 77.75 - 13.25 - 18.75 - 18.75 = 27.00$

The analysis of variance table is shown in Table 8.14.

Table 8.14

Source of Variation	Sum of Squares	Degrees of Freedom	Component of Variance	F	Critical Values of F
Treatments	18.75	3	6.25	1.4	$5\% F_6^3 = 4.76$
Machines	18.75	3	6.25		
Positions	13.25	3	4.42		
Error	27.00	6	4.50		
Total	77.75	15			

It is clear that at the 5% level there is no significant difference between brands of uel.

xtensions

If a third environmental factor is suspected an arrangement known as a Graeco-Latin quare can be used, but this tends to leave only a few degrees of freedom associated with he error variance. Further if a Latin Square cannot be completed then in a similar way o the case of Section 8.3 a design has been developed to allow for this incompleteness. t is called a **Youden square**.

emarks

Although the Latin Square allows the partitioning off of a second source of variation from the error sum of squares, thereby improving the sensitivity of the test, there is an effect of reducing the number of degrees of freedom associated with the error sum of squares and this tends to reduce the sensitivity of the test. What the net effect is cannot be decided. Since a design must be chosen before a test is carried out, we must have a clear idea of the problem under consideration and of the variables which may be confounded *before* the experiment proceeds.

Problems

1. The amount of material lost, in hundredths of a milligram, during the grinding of crystal wafers on five production lines are measured. A Latin Square experimental design was used to test the effect on loss of wafer composition. The letters represent different wafer compositions. Analyse the results.

Position in sequence \ Line	1	2	3	4	5	Totals
I	16*A*	40*B*	50*C*	20*D*	15*E*	141
II	30*B*	25*C*	62*D*	67*E*	30*A*	214
III	50*C*	50*D*	83*E*	85*A*	45*B*	313
IV	80*D*	80*E*	95*A*	98*B*	70*C*	423
V	90*E*	92*A*	98*B*	100*C*	88*D*	468
Totals	266	287	388	370	248	1559

2. The data below shows the result of an experiment to test the performance of 4 oils in 4 cars. Each car was run for 4 months with a different oil in each month. The figures are related to oil consumption and are in coded form. Is the type of oil important? Letters represent oils.

Car \ Month	I	II	III	IV
1	218*B*	236*D*	268*A*	235*C*
2	227*D*	241*B*	229*C*	251*A*
3	274*A*	273*C*	226*B*	234*D*
4	230*C*	270*A*	225*D*	195*B*

3. The following are measurements of the breaking strength (in grams) of six threads *A*, *B*, *C*, *D*, *E* and *F* obtained by 6 different laboratory technicians on 6 different days. Analyse the experiment and use Duncan's test with a risk of 1% to study the means of the breaking strengths of the 6 threads.

Days \ Technicians	1	2	3	4	5	6	Totals
1	33*F*	34*D*	30*A*	28*D*	29*C*	33*E*	187
2	26*B*	29*F*	28*D*	32*A*	30*E*	27*C*	172
3	24*C*	32*E*	27*F*	22*B*	30*D*	24*A*	159
4	16*D*	22*C*	24*B*	18*E*	13*A*	15*F*	108
5	20*E*	15*A*	18*C*	11*F*	19*B*	10*D*	93
6	24*A*	20*D*	22*E*	17*C*	12*F*	18*B*	113
Totals	143	152	149	128	133	127	832

Threads	*A*	*B*	*C*	*D*	*E*	*F*
Totals for Threads	138	143	137	132	155	127

4. In order to study the effect of temperature on the rate of flow of oil through different types of nozzles, a Latin Square experiment was carried out by 5 operators chosen at random at 5 different temperatures on 5 different nozzles *A*, *B*, *C*, *D* and *E*. The coded results are shown below. Analyse the results with a 1% risk. What conclusion do you draw after applying Duncan's test?

Operators \ Temperature	I	II	III	IV	V
1	30.2*A*	24.3*B*	19.6*C*	21.5*D*	17.3*E*
2	21.4*B*	27.1*C*	23.4*D*	24.5*E*	31.0*A*
3	20.7*C*	26.5*D*	25.2*E*	29.1*A*	20.6*B*
4	20.7*D*	24.7*E*	32.3*A*	25.2*B*	22.2*C*
5	20.6*E*	35.8*A*	23.9*B*	23.6*C*	21.5*D*

8.5 AN INTRODUCTION TO FACTORIAL DESIGNS

The designs that have been studied in detail have been single factor designs; they provide data from which we obtain information about the variability due to one treatment. When other, unwanted, sources of variability are taken into account, we can use the technique of blocking to separate out these unwanted effects. In a completely randomized design there is one treatment variable; in a randomized blocks design there is also one blocking variable; in a Latin Square design there is one treatment variable and two blocking variables.

If we want to investigate simultaneously the effect of more than one treatment variable, and their possible interactions, without sacrificing any blocking variables we employ **factorial designs**. In such a design, each combination of levels of the treatment variables is called a **treatment**; the treatment variables are called **factors**.

As an introduction consider a two-factor design. Let the two factors be A and B there are a levels of A and b levels of B. To study the effects of interaction between A and B we must replicate the experiment by taking n observations for each combination of levels of A and B.

Example

Three types of inexpensive noise meters were suspected of responding to mechanical vibrations when used under practical conditions to monitor industrial noise. On each of four sites, where noise was accompanied by different amounts of mechanical vibration, each meter was compared with a standard meter known to respond to sound noise only; two such comparisons were made for each test meter on each site. Differences between the test meter readings and those of the standard meter were recorded; the coded results are shown in Table 8.15. An element of randomization could be introduced by organising the 6 observations at each site into random order. Site 1 had no mechanical vibration, Site 2 had slight vibration, Site 3 moderate vibration and Site 4 had high vibration. Against each pair of results is their total.

Table 8.15

Site \ Meter	I		II		III	
1	−1, 0	−1	−1, 2	1	−6, −4	−10
2	1, −2	−1	1, 0	1	−5, −2	−7
3	−2, 0	−2	0, −1	−1	2, 4	6
4	1, −1	0	2, −1	1	4, 9	13

It is required to analyse the results with a 1% risk of finding spurious differences.

Preliminary analysis

First we partition the total sum of squares into a 'between treatments' component nd a 'within treatments' component in the style of the completely randomized design. The model used is

$$x_{ijk} = \mu + t_{ij} + \epsilon_{ijk} \tag{8.11}$$

where

x_{ijk} is the kth observation of treatment A_iB_j $(k = 1, 2, \ldots, n)$

t_{ij} is the effect of treatment A_iB_j $(i = 1, 2, \ldots, a;\ j = 1, 2, \ldots, b)$

μ is the common mean of the x_{ijk}

nd

ϵ_{ijk} is the random error associated with x_{ijk} and $\sim N(0, \sigma_0^2)$

Let $N = abn$ and $T = \sum_{i=1}^{a} \sum_{j=1}^{b} \sum_{k=1}^{n} x_{ijk}$. Then the total sum of squares is iven by

$$\text{T.S.S.} = \Sigma\Sigma\Sigma x_{ijk}^2 - T^2/N$$

f $x_{ij} = \sum_{k=1}^{n} x_{ijk}^2$ then the between treatment sum of squares is given by

$$\text{B.T.S.S.} = \frac{1}{n} \sum_{i=1}^{a} \sum_{j=1}^{b} x_{ij}^2 - T^2/N$$

n this example, as you can verify, $\Sigma\Sigma\Sigma x_{ijk}^2 = 222$ and $\Sigma\Sigma x_{ij}^2 = 364$, $a = 4$, $b = 3$, $= 2$, $T = 0$. Then $N = 24$, T.S.S. = 222 and B.T.S.S. = $\frac{364}{2} = 182$.

With the total sum of squares are associated $N - 1 = 23$ degrees of freedom, with he between treatments sum of squares are associated $(ab - 1) = 11$ degrees of freedom. y subtraction it follows that the error sum of squares is given by E.S.S. = 222 – 182 = 40 nd is associated with 23 – 11 = 12 degrees of freedom.

etailed Analysis

We now break down the within treatments sum of squares into a 'between rows', a etween columns' and an interaction component. The between rows sum of squares is iven by

$$\text{B.R.S.S.} = \frac{1}{bn} \sum_{i=1}^{a} T_i^2 - T^2/N$$

and is associated with $(a - 1)$ degrees of freedom.

Similarly the between columns sum of squares is given by

$$\text{B.C.S.S.} = \frac{1}{an} \sum_{j=1}^{b} T_j^2 - T^2/N$$

and is associated with $(b - 1)$ degrees of freedom.

The interaction sum of squares is given by

$$\text{I.S.S.} = \text{B.T.S.S.} - \text{B.R.S.S.} - \text{B.C.S.S.}$$

and is associated with $(ab - 1) - (a - 1) - (b - 1) = (a - 1)(b - 1)$ degrees of freedom

The full model is given by the equation

$$x_{ijk} = \mu + \alpha_i + \beta_j + (\alpha\beta)_{ij} + \epsilon_{ijk} \tag{8.12}$$

where α_i is the effect of A_i, β_j is the effect of B_j and $(\alpha\beta)_{ij}$ is the effect of the interaction between A_i and B_j. Since this is a fixed effects problem $\sum_{i=1}^{a} \alpha_i = 0$ and $\sum_{j=1}^{b} \beta_j = 0$. When the main effects are both fixed, the interaction effects are also fixed i.e. $\sum_{i=1}^{a} \sum_{j=1}^{b} (\alpha\beta)_{ij} = 0$. Note that it can be shown that

$$\sum_{i=1}^{a} \sum_{j=1}^{b} \sum_{k=1}^{n} (x_{ijk} - \bar{x})^2 = \sum_{i=1}^{a} \sum_{j=1}^{b} \sum_{k=1}^{n} (\bar{x}_i - \bar{x})^2 + \sum_{i=1}^{a} \sum_{j=1}^{b} \sum_{k=1}^{n} (\bar{x}_j - \bar{x})^2$$

$$+ \sum_{i=1}^{a} \sum_{j=1}^{b} \sum_{k=1}^{n} (\bar{x}_{ij} - \bar{x}_i - \bar{x}_j + \bar{x})^2$$

$$+ \sum_{i=1}^{a} \sum_{j=1}^{b} \sum_{k=1}^{n} (x_{ijk} - \bar{x}_{ij})^2 \tag{8.13}$$

to justify the claim that

$$\text{T.S.S.} = \text{B.R.S.S.} + \text{B.C.S.S.} + \text{I.S.S.} + \text{E.S.S.}$$

(Here $\bar{x}_i = \frac{1}{a}T_i$, $\bar{x}_j = \frac{1}{b}T_j$, $\bar{x}_{ij} = \frac{1}{n}\sum_{k=1}^{n} x_{ijk}$.) Also

$$abn - 1 \equiv (a-1) + (b-1) + (a-1)(b-1) + ab(n-1)$$

Hypotheses tests

H_0 : there is no difference between rows (sites); $\alpha_i = 0$, all i
H_1 : at least one $\alpha_i \neq 0$
H_0': there is no difference between columns (meters); $\beta_j = 0$, all j
H_1': at least one $\beta_j \neq 0$
H_0'': there is no interaction; $(\alpha\beta)_{ij} = 0$, all i and j
H_1'': at least one $(\alpha\beta)_{ij} \neq 0$

Example revisited

You can verify that $\Sigma T_i^2 = 354$, $\Sigma T_j^2 = 24$

$$\text{B.R.S.S.} = \frac{1}{6} \times 354 - 0 = 59$$

$$\text{B.C.S.S.} = \frac{1}{8} \times 24 - 0 = 3$$

$$\text{I.S.S.} = 182 - 59 - 3 = 120$$

The appropriate analysis of variance table is presented as Table 8.16.

The hypothesis of no interaction is rejected and hence there is significant interaction present. The vibrations main effect is almost significant (and would be definitely significant had a 5% risk been used) but the main effect due to the meters is not significant. However, because of the large interaction we must consider the situation at each site separately. This is carried out via Duncan's test.

Contrasts Analysis

$\hat{\sigma}_x^2 = 3.3$ (from Table 8.16)

$\therefore \quad \hat{\sigma}_{\bar{x}}^2 = \frac{3.3}{2}$ (since there are two observations for each 'cell' in Table 8.15)

Hence $\hat{\sigma}_{\bar{x}} = 1.3$ (1 dp)

Table 8.16

Source of Variation	Sum of Squares	Degrees of Freedom	Component of Variance	F	Critical Values of F
Between Vibrations	59	3	19.7	5.9	$1\% F^{3}_{12} = 5.95$
Between Meters	3	2	1.5	0.45	$1\% F^{2}_{12} = 6.93$
Interaction	120	6	20.0	6.0	$1\% F^{6}_{12} = 4.82$
Sub Total	182	11			
Errors	40	12	3.3		
Total	222	23			

From the table in Appendix **B** the following values for 1% risk and 12 degrees of freedom are extracted.

	p	2	3
	r_p	4.32	4.50
Hence	R_p	5.61	5.85

$(R_p = r_p \cdot \hat{\sigma}_{\bar{x}}$, remember)

1) **Site 1; noise with no vibration**

Meters	III	I	II
Total	−10	−1	1
Means	−5	−0.5	0.5

Ranges < critical range underlined

The conclusion is that meter III gives a lower reading than meters I or II when there is no mechanical vibration.

2) **Site 2; noise with slight vibration**

Meters	III	I	II
Totals	−7	−1	1
Means	−3.5	−0.5	0.5

Where there is slight vibration, the meters do not give significantly different readings.

3) **Site 3; noise with moderate vibrations**

Meters	I	II	III
Totals	–2	–1	6
Means	–1	–0.5	3

Again, the meters do not give significantly different readings.

4) **Site 4; noise with high vibration**

Meters	I	II	III
Totals	0	1	13
Means	0	0.5	6.5

With high vibration, meter III gives a significantly higher reading than I or II.

It is also possible to test whether any of the meters reads significantly different from the standard meter on average.

The mean of all readings from meter I, $\bar{x}_I = \frac{-4}{8} = -0.5$, $\sigma^2_{\bar{x}_I} = \frac{3.3}{8}$, and hence $\sigma_{\bar{x}_I} = 0.64$.

To test the hypothesis $\mu_I = 0$ first calculate

$$t = \frac{\bar{x}_I - 0}{\sigma_{\bar{x}_I}} = \frac{-0.5}{0.64} = -0.78$$

At 1%, the critical value of t is 3.5, hence the conclusion is that meter I does not differ significantly from the standard meter on average.

Similar conclusions apply to meter II.

We already know that meter III reads low when there is low mechanical vibration and high when the mechanical vibration is high. It responds to mechanical vibration, whereas meters I and II do not.

Note: In the case where interaction is not significant, the main effects can be examined separately for significant differences and tested independently for contrasts.

Example of a three-factor analysis

An experiment was conducted to test the effects on the friction horsepower of an engine of lubricating oils, temperature and speed of the engine.† A completely randomized experiment was carried out using three oils, three temperatures and three speeds and the results in coded form are shown in **Table 8.17**; for each treatment three observations were made. The three oils used are labelled A_1, A_2, A_3; the three temperatures are B_1, B_2, B_3 and the three speeds are C_1, C_2, C_3.

The model used is

$$x_{ijkl} = \mu + t_{ijk} + \epsilon_{ijkl} \tag{8.14}$$

where x_{ijkl} is the lth observation of the treatment $A_iB_jC_k$ $(l = 1, 2, \ldots\ldots, n)$

μ is the common mean of the x_{ijkl}

t_{ijk} is the effect of treatment $A_iB_jC_k$ $(i = 1, 2, \ldots\ldots, a;$ $j = 1, 2, \ldots\ldots, b;$ $k = 1, 2, \ldots\ldots, c)$

and ϵ_{ijkl} is the random error associated with x_{ijkl} and $\sim N(0, \sigma_0^2)$

Let $N = abcn$, $T = \sum_{i=1}^{a} \sum_{j=1}^{b} \sum_{k=1}^{c} \sum_{l=1}^{n} x_{ijkl}$. Then the total sum of squares is given by

$$\text{T.S.S.} = \Sigma\Sigma\Sigma\Sigma\, x_{ijkl}^2 - T^2/N$$

Let $x_{ijk}^2 = \sum_{l=1}^{n} x_{ijkl}^2$ then the between treatment sum of squares is given by

$$\text{B.T.S.S.} = \frac{1}{n} \sum_{i=1}^{a} \sum_{j=1}^{b} \sum_{k=1}^{c} x_{ijk}^2 - T^2/N$$

† This example is based on Frazier, Klingel and Tupa, 'Friction and Consumption Characteristics of Motor Oils', Industrial and Engineering Chemistry, 1953.

Table 8.17

		Friction Horsepower (coded)			
		C_1	C_2	C_3	Σ
	B_1	10.8, 11.2, 11.7 33.7	17.5, 19.3, 18.9 55.7	26.6, 29.2, 29.9 85.7	175.1
A_1	B_2	9.2, 8.8, 9.4 27.4	14.7, 16.0, 16.2 46.9	23.6, 25.4, 24.0 73.0	147.3
	B_3	10.0, 10.3, 11.3 31.6	16.4, 18.2, 18.4 53.0	25.4, 27.4, 27.7 80.5	165.1
	B_1	10.4, 12.3, 13.4 36.1	19.0, 20.6, 22.4 63.0	26.7, 29.6, 32.2 88.5	187.6
A_2	B_2	9.1, 9.6, 10.0 28.7	15.6, 17.1, 18.2 50.9	24.1, 26.3, 27.5 77.9	157.5
	B_3	9.9, 10.6, 12.3 32.8	17.2, 19.5, 20.7 57.4	25.8, 28.4, 30.0 84.2	174.4
	B_1	10.0, 12.1, 12.3 34.4	17.2, 21.0, 21.5 59.7	26.4, 29.1, 30.6 86.1	180.2
A_3	B_2	8.5, 9.4, 10.2 28.1	14.7, 17.4, 17.3 49.4	23.2, 25.8, 26.7 75.7	153.2
	B_3	9.3, 11.0, 12.6 32.9	16.2, 18.9, 19.8 54.9	25.6, 28.4, 29.8 83.6	171.4
	Σ	285.7	490.9	735.2	1511.8 = T

In this example, $\Sigma\Sigma\Sigma\Sigma\, x_{ijkl}^2 = 32\,223.34$ and $\Sigma\Sigma\Sigma\, x_{ijk}^2 = 95\,429.96$; $a = 3$, $b = 3$, $c = 3$, $n = 3$, $T = 1511.8$, $N = 81$.

Then $\text{TSS} = 32\,223.34 - (1511.8)^2/81 = 32\,223.34 - 28\,216.53 = 4006.81$ and

$$\text{BTSS} = \frac{1}{3} \times 96\,403.96 - 28\,216.53 = 3918.12.$$

With the total sum of squares are associated $N - 1 = 80$ degrees of freedom and with the between treatments sum of squares are associated $(abc - 1) = 26$ degrees of freedom. By subtraction it follows that the error sum of squares is given by E.S.S. = 4006.81 − 3918.12 = 88.69 and is associated with 80 − 26 = 54 degrees of freedom.

Incidentally, we may now compute an analysis of variance as detailed in Table 8.18.

Table 8.18

Source of Variation	Sum of Squares	Degrees of Freedom	Component of Variance	F	Critical Values of F
Between Treatments	3918.12	26	150.7		5% $F \simeq 1.7$
				91.8	
Within Treatments	88.69	54	1.64		1% $F \simeq 2.1$
Total	4006.81	80			

Clearly, there are highly significant differences between treatments. To carry out a detailed analysis we use the model

$$x_{ijkl} = \mu + \alpha_i + \beta_j + \gamma_k + (\alpha\beta)_{ij} + (\beta\gamma)_{jk} + (\alpha\gamma)_{ik} + (\alpha\beta\gamma)_{ijk} + \epsilon_{ijkl} \qquad (8.15)$$

where α_i is the effect of A_i

$(\alpha\beta)_{ij}$ is the effect of the interaction between A_i and B_j

and $(\alpha\beta\gamma)_{ijk}$ is the effect of the interaction between A_i, B_j and C_k.

We now partition the between treatments sum of squares into the following components:

(i) **main effects** each with 2 degrees of freedom

- (A) oil effect sum of squares
- (B) temperature effect sum of squares
- (C) speeds effect sum of squares

(ii) **interaction effects**

$A \times B$ sum of squares
$B \times C$ sum of squares
$A \times C$ sum of squares

} each with 4 degrees of freedom

$A \times B \times C$ sum of squares with 8 degrees of freedom

Reason for yourselves why these degrees of freedom obtain.

Note that if the experiment had not been replicated, i.e. there was only one observation per treatment, the $A \times B \times C$ sum of squares would have had to serve as the error sum of squares and would not have been available to explain this third-order interaction. However, high-order interactions are not often significant and such **confounding** of these two sums of squares may not be too important.

To calculate the sum of squares for A first find the total score for A_1 i.e.

$$175.1 + 147.3 + 165.1 = 487.5$$

That for A_2 is

$$187.6 + 157.5 + 174.4 = 519.5$$

and that for A_3 is

$$180.2 + 153.2 + 171.4 = 504.8$$

Then the sum of squares for A, based on the three components each of which is associated with 27 observations, is given by

$$\frac{(487.5)^2 + (519.5)^2 + (504.8)^2}{27} - \frac{T^2}{N} = 19.01$$

Similarly, the sum of squares for B is given by

$$\frac{(175.1 + 187.6 + 180.2)^2 + (147.3 + 157.5 + 153.2)^2 + (165.1 + 174.4 + 171.4)^2}{27}$$

$$- \frac{T^2}{N} = 136.18$$

Likewise, the sum of squares for C is given by

$$\frac{(285.7)^2 + (490.9)^2 + (735.5)^2}{27} - \frac{T^2}{N} = 3751.11$$

To find the $A \times B$ sum of squares we work out the sums of the observations for each combination of $A_i B_j$.

For example, the observations for $A_1 B_1$ total 175.1.

It is advisable to construct a table for the effects $A_i B_j$ and this is done in Table 8.19 below.

Table 8.19

	B_1	B_2	B_3	Σ
A_1	175.1	147.3	165.1	487.5
A_2	187.6	157.5	174.4	519.5
A_3	180.2	153.2	171.4	504.8
Σ	542.9	458.0	510.9	1511.8

This can be thought of as a parallel to a two-factor experiment. First we calculate a 'total' sum of squares, noting that the cell values (e.g. 175.1) are each based on 9 observations. Then

$$\text{'total' sum of squares} = \frac{1}{9}\left[(175.1)^2 + (147.3)^2 + \ldots + (171.4)^2\right] - T^2/N$$

$$= 155.79$$

The sum of squares for A could be found from Table 8.19 by using the row totals (and is probably best done this way).

It was found earlier to be 19.01.

Similarly, the sum of squares for B could be found from column totals; it was 136.18.

The $A \times B$ interaction sum of squares is $155.79 - 19.01 - 136.18 = 0.60$. Table 8.20 shows the analagous totals for $B_j C_k$ and $A_i C_k$.

Table 8.20

	C_1	C_2	C_3	Σ
B_1	104.2	178.4	260.3	542.9
B_2	84.2	147.2	226.6	458.0
B_3	97.3	165.3	248.3	510.9
Σ	285.7	490.9	735.2	1511.8

	C_1	C_2	C_3	Σ
A_1	92.7	155.6	239.2	487.5
A_2	97.6	171.3	250.6	519.5
A_3	95.4	164.0	245.4	504.8
Σ	285.7	490.9	735.2	1511.8

From these tables it can straightforwardly be verified that the sum of squares for the interaction $B \times C = 0.14$ and that for the interaction $A \times C = 3.24$.

Finally, by subtraction, we obtain the three-way interaction sum of squares as $(3918.12 - 19.01 - 136.18 - 3751.11 - 0.60 - 0.14 - 3.24) = 7.84$.

In Table 8.21 is displayed the analysis of variance.

Table 8.21

Source of Variation		Sum of Squares	Degrees of Freedom	Component of Variance	F	Critical Values of F
Main Effects	A	19.01	2	9.50	5.8	5% $F^2_{54} = 3.17$
	B	136.18	2	68.09	41.5	
	C	3751.11	2	1875.56	1144	1% $F^2_{54} = 5.04$
$A \times B$		0.60	4	0.15		
$B \times C$		0.14	4	0.035		
$A \times C$		3.24	4	0.81		
$A \times B \times C$		7.84	8	0.98		
Error		88.69	54	1.64		
Total		4006.81	80			

Hence, none of the interactions is significant and all the main effects are significant at the 1% level. Speed (C) is crucial, temperatures (B) are highly significant and oils (A) are significant but not so markedly.

It is left as an exercise to perform a contrasts analysis and, to draw diagrams similar to Figure 8.1 and to find the average effects of A, B and C.

Consider again the model for the introductory example:

$$x_{ijk} = \mu + \alpha_i + \beta_j + \gamma_k + \epsilon_{ijk} \tag{8.1}$$

It is clear that there is no allowance for interaction; indeed, we were almost forced to confound the interactions with the error to get a respectably large number of degrees of freedom associated with the error sum of squares. It would have been more sensible to replicate the experiment as this would have allowed the relaxation of the assumption of zero interaction. The only justification for no replicates is if previous experience indicates that interactions are insignificant. The model (8.1) could be taken as the model for a Latin Square design; however, the difference is in the nature of the experiment: the Latin Square design is for a single factor with two restrictions on randomization, whereas the introductory example was a multi-factor completely randomized design. The former assumes no interaction and if the experimenter cannot be certain of the validity of this assumption be should replicate the experiment and test the assumption.

Problems

1. A manufacturer wished to determine the effects of various levels of voltage and temperature on the leakage current in semi-conductors he produces. He selected at random 18 semi-conductors from a day's production and randomly assigned them to the nine treatments determined by three voltage levels and three temperature levels. Each treatment was tested on two semi-conductors. The results are shown below, with the current measured in microamperes. Analyse the experiment.

Temperature \ Volts	10		15		20		Totals
	1.6		1.9		2.3		
25°C		3.4		3.6		4.3	11.3
	1.8		1.7		2.0		
	1.9		2.5		1.9		
50°C		3.9		4.8		3.7	12.4
	2.0		2.3		1.8		
	2.2		2.1		1.9		
75°C		4.3		4.5		3.6	12.4
	2.1		2.4		1.7		
Totals	11.6		12.9		11.6		36.1

2. A factorial experiment was performed with two replications to compare three fuels and two types of burners. The coded results are shown on the next page. Discuss the effects of fuel and burner on the response variable and determine whether the interaction is significant. Use a 1% level of significance.

3. The process of coating steel pipes with plastic comprises three phases: heating the pellets of plastic, forcing the hot plastic through a tube via a screw and wrapping the resulting sheet around the steel tube. To study the effect of temperature T, screw speed S and type of plastic P on output of the resultant coating, a 2 x 2 x 3 factorial experiment was performed with 3 replications with results on the next page. Analyse the results and interpret them in terms of the experiment.

(Prob. 2)

Fuel \ Burner	1		2		Totals
I	12.7	23.4	12.8	25.7	49.1
	10.7		12.9		
II	12.3	25.8	18.6	36.2	62.0
	13.5		17.6		
III	15.5	30.1	17.1	35.4	65.5
	14.6		18.3		
Totals	79.3		97.3		176.6

(Prob. 3)

Temperature	Speed	Plastic	Replication I	Replication II	Replication III
T_1	S_1	P_1	66	62	63
T_1	S_1	P_2	41	38	40
T_1	S_2	P_1	67	64	69
T_1	S_2	P_2	51	52	52
T_2	S_1	P_1	53	52	50
T_2	S_1	P_2	35	38	33
T_2	S_2	P_1	53	54	49
T_2	S_2	P_2	45	43	48
T_3	S_1	P_1	54	50	53
T_3	S_1	P_2	37	32	39
T_3	S_2	P_1	55	56	52
T_3	S_2	P_2	46	44	44

The performance of three detergents at different temperatures and different times of washing was studied by performing a factorial experiment, the results of which are shown below. Assume that the three-factor interaction is zero; analyse the experiment and apply Duncan's test to each factor.

Temperature	I			II			III		
Detergent \ Time	T_1	T_2	T_3	T_1	T_2	T_3	T_1	T_2	T_3
D_1	44.9	46.7	46.9	42.3	44.9	46.6	44.1	46.9	48.8
D_2	52.3	54.3	54.7	50.7	52.9	55.0	52.9	56.2	57.6
D_3	60.4	62.0	62.2	58.7	61.9	63.4	60.6	64.6	66.7

5. Four methods have been suggested for mixing concrete. Two batches were mixed, three specimens from each batch used with each method and tested for compression. What conclusions can you draw from the results below?

Batch \ Method	A	B	C	D
I	56	60	60	68
	58	64	61	70
	60	62	64	73
II	60	58	58	72
	61	63	66	77
	60	67	62	74

8.6 2^n FACTORIAL EXPERIMENTS

Experiments in which each of n factors is taken at only two levels have certain advantages. First, the economics of running experiments could render a many-factor prohibitively expensive if more than two levels were taken. Second, there are computational short-cuts for this case. Third, it is relatively easy to confound higher-order interactions. However, it must be stated that there are disadvantages: with only two levels it is impossible to determine whether the effects of a factor are linear or not. Often, 2^n factorial experiments are used as a preliminary step to sift out the likely significant factors, before using a design with more levels.

Consider first an example of a 2^2 experiment, i.e. a 2-factor, 2-level experiment. It was required to test the relative wear of two particular bearing materials Y and Z. The bearings could be dried in an oven before testing. The experimental results are recorded in Table 8.22 in coded form.

Table 8.22

	Y		Z	
Dried	76		84	
		147		158
	71		74	
Not Dried	64		74	
		141		154
	77		80	

Let the drying condition be factor A and the material used be factor B. The model for a completely randomized design is

$$x_{ijk} = \mu + \alpha_i + \beta_j + (\alpha\beta)_{ij} + \epsilon_{ijk} \tag{8.12}$$

(see page 320) where i and j both take values 1 or 2 (as, coincidentally does k in this example, since the experiment was replicated twice). It is customary to describe the two levels of each factor as 'low' and 'high'. In this instance, there is no physical attribute of either factor which warrants this classification, therefore we arbitrarily assign the low level of A, written A_0, to 'dried' and the high level, A_1 to 'not dried'. Similarly, we assign B_0 to material Y and B_1 to material Z. Another notation is popular: A_0 becomes a^0, B_1 becomes b^1, etc. Then if both factors are at their low level - $a^0 b^0$ - the treatment is symbolised as (1); $a^0 b^1 \equiv b$, $a^1 b^0 \equiv a$ and $a^1 b^1 \equiv ab$. The *normal* order of writing these treatments is

$$(1); \; a; \; b, \; ab$$

and each time a new factor is introduced, it appears at the end of the sequence (or order) and multiplies all the previous components. (Hence for a third factor C, also at 2 levels, the order is

$$(1); \; a; \; b, \; ab; \; c, \; ac, \; bc, \; abc$$

where the semi-colons signal the appearance of a new factor.)

Returning to the two-factor case, the treatment combinations can be represented as the vertices of a square; see Figure 8.3.

The **effect** of a factor can be defined as the change in response caused by a change in the level of the factor.

At the low level of B we see from Figure 8.3 that the effect of A is

$$141 - 147 = -6 \qquad \text{which is symbolically } a - (1)$$

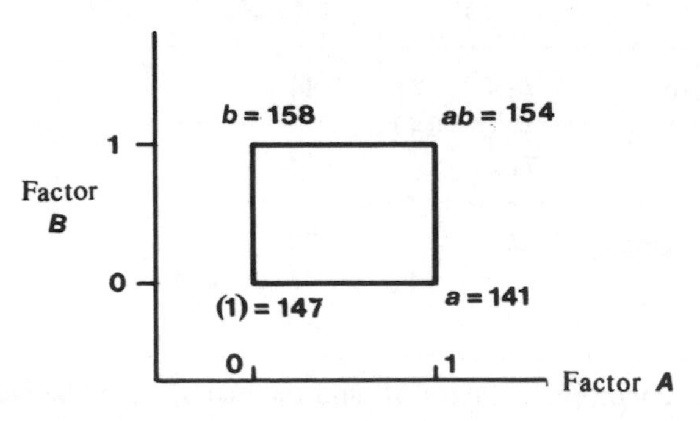

Figure 8.3

At the high level of B the effect of A is

$$154 - 158 = -4 \qquad \text{symbolically } ab - b$$

The average effect of A is given by

$$A = \tfrac{1}{4}(\mathrm{a} - (1) + ab - b) = -2.5$$

since there are 4 observations involved at each level of A. Note that

$$4A = -(1) + a - b + ab \tag{8.16}$$

This is called a **contrast**, i.e. it is a linear combination of the four treatments with the sum of its coefficients equal to zero. Similarly, the average effect of B is given by

$$B = \tfrac{1}{4}(b - (1) + ab - a) = \tfrac{1}{4}(158 - 147 + 154 - 141) = 6$$

Note that

$$4B = -(1) - a + b + ab \tag{8.17}$$

is also a contrast. Regarded as vectors: $(-1,\ 1,\ -1,\ 1)$ and $(-1,\ -1,\ 1,\ 1)$ we see that these contrasts are **orthogonal**.

Next, we determine the effect of the **interaction** between A and B. At the low level of B the effect of A is $a - (1)$ and at the high level of B the effect of A is $ab - b$ so that if these two effects are different, interaction is present. The average *difference* between these two effects is given by

$$AB = \tfrac{1}{4}[(ab - b) - (a - (1))] = 0.5$$

Note that

$$4AB = (1) - a - b + ab \tag{8.18}$$

nd that this is orthogonal to the other contrasts. (See what coefficients attach to the ertices of Figure 8.3.)

It is also possible to derive the relationship (8.18) by considering the effect of B t each level of A.

The sum of squares associated with a contrast can be shown to be given by

$$SS_{\text{contrast}} = \frac{(\text{contrast})^2}{n\Sigma c_i^2} \tag{8.19}$$

vhere c_i are the coefficients in the contrast (note that here $\Sigma c_i^2 = 4$) and n is the umber of replications (here $n = 2$).

Then from (8.16), (8.17) and (8.18) we find that

$$SS_A = \frac{[4(-2.5)]^2}{2 \times 4} = 12.5$$

$$SS_B = \frac{(4 \times 6)^2}{2 \times 4} = 72$$

$$SS_{AB} = \frac{(4 \times 0.5)^2}{8} = 0.5$$

The total sum of squares would be found in the usual way, and the error sum of quares by subtraction. Note that the total effect can be found from the relationship

$$4[\text{Total}] = (1) + a + b + ab \tag{8.20}$$

It would be instructive for you to perform the rest of the analysis of these xperimental results and also to compute the three sums of squares above by the methods f the previous section.

hree-factor case

The effect on the reliability of a switch due to lubrication, suppression of sparks and mount of dust protection was studied. A 2^3 factorial experiment was used, i.e. a ompletely randomized design with the three factors each at low and high levels.

Let A be the lubrication factor, 'no lubrication' being the low level and ubricated' the high level; B is suppression of sparks 'no' being low level, 'yes' being high level; C is dust protection with the same classification as B. The experiment had two replications and the results recorded as hours of continuous operation until failure in Table 8.23.

Table 8.23

Treatment	First results	Second results	Total
(1)	710	805	1515
a	885	987	1872
b	737	849	1586
ab	870	1070	1940
c	872	960	1832
ac	1135	1138	2273
bc	1046	1090	2136
abc	937	1082	2019
Σ	7192	7981	15173

The results can be represented as the vertices of a cube as in Figure 8.4.

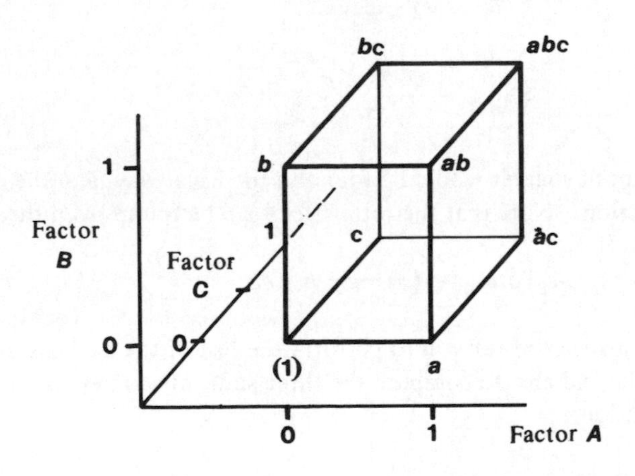

Figure 8.4

First the results are analysed as a two-way arrangement, with 8 treatments and 2 replications.

Note that $T = 15\,173$, $N = 16$.

Then the Total Sum of Squares associated with 15 degrees of freedom is given by

$$\text{TSS} = (710)^2 + (805)^2 + \ldots + (1082)^2 - T^2/N = 282\,705$$

The Between Treatments Sum of Squares associated with 7 degrees of freedom is
iven by

$$\text{BTSS} = \tfrac{1}{2}[(1515)^2 + (1872)^2 + \ldots + (2019)^2] - T^2/N = 231\,362$$

The Error Sum of Squares associated with 8 degrees of freedom is given by

$$\text{ESS} = 282\,705 - 231\,362 = 51\,343$$

The Between Treatments Sum of Squares *can* be broken down in a manner
nalogous to that used in the previous, 2^2 example. However, it can be a tedious
natter. A short-cut method due to Yates is employed and is illustrated by its application
o the present problem in Table 8.24.

Table 8.24

Treatments	Total	I	II	III	Effect	Sum of Squares
(1)	1515	3387	6913	15173		14 388 746
a	1872	3526	8260	1035	*A*	66 952
b	1586	4105	711	189	*B*	2233
ab	1940	4155	324	–561	*AB*	19 670
c	1832	357	139	1347	*C*	113 401
ac	2273	354	50	–387	*AC*	9361
bc	2136	441	–3	–89	*BC*	495
abc	2019	–117	–558	–555	*ABC*	19 251

In column I, the first half of the entries are found by adding successive pairs of
reatment totals: hence $1515 + 1872 = 3387$, $1586 + 1940 = 3526$, etc. The second
alf of the entries are obtained by subtraction in successive pairs: the first total from the
econd: hence $1872 - 1515 = 357$, etc. The entries in column II are obtained by
nalogous operations on the entries in column I and those in column III by operating on
he entries in column II.

The sum of squares entries are obtained by dividing the squares of the entries in
olumn III by $2 \times 2^3 = 16$. The first entry is the correction factor T^2/N.

In Table 8.25 is presented a table of contrasts for the three-factor case with n
eplications.

You could calculate some of these contrasts and compare with the sum of squares
s found by the Yates method. Note that in the three-factor case the sum of squares is
ound by dividing the square of the contrast by $n2^3$.

Table 8.25

Treatment \ Contrast	Total	A	B	AB	C	AC	BC	ABC
(1)	1	−1	−1	1	−1	1	1	−1
a	1	1	−1	−1	−1	−1	1	1
b	1	−1	1	−1	−1	1	−1	1
ab	1	1	1	1	−1	−1	−1	−1
c	1	−1	−1	1	1	−1	−1	1
ac	1	1	−1	−1	1	1	−1	−1
bc	1	−1	1	−1	1	−1	1	−1
abc	1	1	1	1	1	1	1	1

Problems

1. A test was conducted on the heat loss of insulating material. Three factors A, B and C wer considered each at two levels. Two replications are taken for each treatment and the coded results shown below. Via Yates' algorithm determine whether any of the factors significantly affect the response.

		A_1		A_2	
B_1	C_1	28	26	22	20
	C_2	24	25	18	19
B_2	C_1	26	29	22	23
	C_2	24	21	17	18

2. An experiment to determine the effects of three elements used in alloys on the ductility of the base metal was conducted as a 2^3 design with two replications. Analyse the results given on the next page.

3. Steel casting is heat treated by coils which may be placed on either the upper or the lower bos of the casting. A 2^5 factorial experiment with two replications was carried out to determine the effect on the hardness of casting of A heating time, B time of quenching, C time of drawing, D the position of the boss and E the position of the measurement on the boss. Each factor was studied at two levels. Analyse the coded results given on the next page.

(Prob. 2)

Mn	C	Ni	Replication 1	Replication 2
I	I	I	74	65
		II	75	82
	II	I	51	50
		II	57	64
II	I	I	78	72
		II	76	80
	II	I	53	49
		II	76	84

Prob. 3)

		A_0				A_1			
		B_0		B_1		B_0		B_1	
		C_0	C_1	C_0	C_1	C_0	C_1	C_0	C_1
D_0	E_0	3	2	0	10	3	4	7	17
		1	5	3	6	7	8	4	11
	E_1	4	11	8	6	7	19	15	5
		9	6	2	1	4	13	9	10
D_1	E_0	2	4	3	6	11	4	4	14
		8	1	7	14	5	8	11	9
	E_1	5	2	5	11	10	17	10	16
		7	4	12	15	17	23	6	11

3.7 FACTORIAL EXPERIMENTS IN A RANDOMIZED BLOCKS DESIGN

We have so far assumed that a factorial experiment was conducted in a completely randomized way. However, restrictions on time may mean that this randomization is not possible and each replication may need to be run on a separate day; in this case a randomized blocks design is used with replications acting as blocks. If space is a restricting factor then each replication may be run in one space unit, or, more severely, each replication itself may be divided into runs or blocks.

Blocking in a 2^n experiment

If a randomized blocks design is used then one or more treatment effects may become confounded with the block effects. Consider as an example a 2^3 experiment which for reasons of space has to be conducted in two separate runs. Suppose the treatments a, ab, bc, abc constitute the first run and the treatments (1), a, c, ac constitute the second. Then the block effect, i.e. the difference between the run totals is

$$b + ab + bc + abc - (1) - a - c - ac = (-1) - a + b + ab - c - ac + bc + abc$$

which, by reference to Table 8.25 can be seen to be the contrast for effect B; hence the estimate of the main effect due to factor B is confounded with block effect. All other effects are unconfounded with blocks.

Similarly, if the two runs comprised treatments a, b, c, abc and (1), ab, ac, bc respectively it can readily be seen that the interaction ABC is confounded with blocks. The decision as to which effect to confound with blocks is therefore up to the experimenter; often, he will confound a high-order interaction.

It must be stated that if the experiment comprises r replications then the problem of confounding becomes more complicated. It is slightly less complicated if the number of blocks is a power of 2, say 2^b. If the experiment is of type 2^n then $(2^b - 1)$ effects will be confounded. If the experimenter wishes to confound only b effects which do not contain any 'overlap' † he will find that $[2^b - (b + 1)]$ further effects are confounded. The extra effects are just all possible overlaps of his b chosen effects.

Consider a 2^4 experiment which is divided into four blocks so that 3 effects will be confounded. We do not wish to confound any main effects and would prefer to confound only high-order interactions. Suppose we chose $ABCD$ and ABC. The 'overlap' is $\cancel{A}\cancel{B}\cancel{C}D\ \cancel{A}\cancel{B}\cancel{C}$, i.e. D and this is undesirable. Better would be to choose ABC and ABD, for example, so that CD is also confounded.

First the treatments are distributed over two blocks in such a way that ABC is confounded with blocks. All treatments which have an odd number of letters in common with ABC are put in one block and the rest (with an even number) in the second block.

Hence Block I: a, b, c, abc, ad, bd, cd, $abcd$

Block II: (1), ab, ac, bc, d, abd, acd, bcd

† Consider AC and BD – the 'overlap' is the effect $ACBD$; consider ABC and ACD – the 'overlap' is $\cancel{A}B\cancel{C}\ \cancel{A}\cancel{C}D$ or BD.

Next the *ABD* interaction is confounded with blocks. Each of the two blocks is itself divided into two blocks:

Block 1: *a*, *b*, *cd*, *abcd*

Block 2: *c*, *abc*, *ad*, *bd*

Block 3: *ac*, *bc*, *d*, *abd*

Block 4: (1), *ab*, *acd*, *bcd*

The experiment is run with treatments randomized within blocks and the order of blocks randomized (within each replication).

In carrying out the analysis, the confounded effects are absorbed into the error due to blocks; it is possible if the experiment has been replicated to recover some information about the confounded blocks, but unless the number of replications is large the lack of sensitivity of the analysis makes it a doubtful proposition. For further details, see Hicks (1964).

Replications as blocks

Suppose that the replications are blocks. Remembering that blocking is a technique designed to remove an unwanted source of variability, it is possible to arrange the experiment so that the unwanted source (different days on which the replications were run or different locations where the experiment was performed) is the blocking variable.

The mathematical model for a two-factor randomized blocks experiment is

$$x_{ijk} = \mu + \alpha_i + \beta_j + (\alpha\beta)_{ij} + \rho_k + \epsilon_{ijk} \tag{8.21}$$

where ρ_k is the effect due to blocks (replications).

Let the following problem serve as an illustration.

An experiment is run to test the effects of two factors A and B but has to be run on separate days to allow time for replications. Each of three replications is run on a particular day to remove the effect of day-to-day variability. The coded results are shown in Table 8.26.

As usual, all effects are assumed fixed. With the usual notation, $T = 360$, $N = 18$

Total sum of squares

$$\text{TSS} = 28^2 + 13^2 + \ldots.. + 16^2 - T^2/N = 408$$

Between treatments sum of squares

$$\text{BTSS} = \frac{1}{3}(78^2 + 36^2 + \ldots.. + 51^2) - T^2/N = 354$$

Table 8.26

Level of A		1			2			Block
Level of B		1	2	3	1	2	3	Totals
Blocks	1	28	13	22	26	23	18	130
(Days)	2	25	10	18	20	22	17	112
	3	25	13	23	20	21	16	118
Treatment Totals		78	36	63	66	66	51	360

Between blocks sum of squares

$$\text{BBSS} = \frac{1}{6}(130^2 + 112^2 + 118^2) - T^2/N = 28$$

Error sum of squares

$$\text{ESS} = 26$$

The sum of squares due to A is given by

$$\text{SSA} = \frac{1}{9}\left[(78 + 36 + 63)^2 + (66 + 66 + 51)^2\right] - T^2/N = 2$$

Similarly,

$$\text{SSB} = \frac{1}{6}\left[(78 + 66)^2 + (36 + 66)^2 + (63 + 51)^2\right] - T^2/N = 156$$

Then the sum of squares due to the interaction between A and B is given by SSAB = 354 – 2 – 156 = 196. Table 8.27 presents the analysis of variance table.

Hence the effect of A is not significant, that of B is significant at the 1% level, but so is the effect of interaction and this weakens our faith in the result for B. The blocks effect is significant at the 5% level, but not at the 1% level.

We may estimate the parameters in (8.21)

$$\hat{\mu} = 360/18 = 20, \quad \hat{\sigma}_0^2 = 2.6 \quad \text{and} \quad \widehat{(\alpha\beta)}_{11} = \bar{x}_{11} - \hat{\alpha}_1 - \hat{\beta}_1 - \bar{x}$$

(But $\alpha_i = 0$ under the hypothesis for A, which was not rejected.)

$$\therefore \widehat{(\alpha\beta)}_{11} = \frac{1}{3} \times 78 - 0 - (78 + 66)/6 = 2$$

Table 8.27

Source of Variation	Sum of Squares	Degrees of Freedom	Component of Variance	F	Critical Values of F
A	2	1	2	< 1	5% F^2_{10} = 4.10
B	156	2	78	30	
$A \times B$	196	2	98	38	1% F^2_{10} = 7.56
Blocks	28	2	14	5.4	
Error	26	10	2.6		
Total	408	17			

'e have assumed that $\hat{\beta}_1 = (78 + 66)/6 - \bar{x}$.

Other interaction biases can be found in a similar way.

oblems

If a 2^4 factorial experiment is performed in two blocks with $ABCD$ confounded, what treatments appear in each block?

If a 2^4 factorial experiment has to be performed in 4 blocks and the treatments confounded are $ABCD$ and CD, what other effect is confounded?

The effect of two factors on quality of a product were to be investigated. Three levels of factor B and two levels of factor A were taken. Specimens from three different localities were used in the experiment. In order to eliminate any locality effect, the experiment was designed as a two-factor, randomized blocks design, with localities as blocks. The coded results are shown below. Analyse the results and calculate any appropriate effects.

A		1			2		
B	1	2	3	1	2	3	Totals
1	31	16	25	29	26	21	148
Blocks 2	28	13	21	23	25	20	130
3	28	16	26	23	24	19	136
Totals	87	45	72	75	75	60	414

4. Analyse the following coded results for the testing of two fixed effect factors A and B on fastness of a dye. The two levels of A are different dyestuffs and the two levels of B are methods of preparing the fabric to be dyed. The experiment was replicated five times and was run as a randomized blocks design with days as blocks. Each replication occurred on on particular day.

A	1		2		
B	1	2	1	2	Totals
1	24	31	13	24	92
2	22	29	9	20	80
Blocks 3	24	32	6	18	80
4	22	29	10	15	76
5	23	24	7	18	72
Totals	115	145	45	95	400

8.8 OTHER FEATURES

To conclude this chapter we mention some other topics in the design of experiments. Further information can be found by consulting books in the Bibliography.

Confidence Intervals

In performing a Duncan multiple range test, we compute the averages for each treatment level. These averages can be the centre of a confidence interval of the form

$$\overline{x}_i \pm t \frac{\hat{\sigma}_0}{\sqrt{n}}$$

where $\overline{x}_i$ is the average referred to, $\hat{\sigma}_0^2$ is the error variance, n is the number of items in the averaged value and t is the t-score for the level of significance chosen and for the degrees of freedom associated with the error variance.

Quantitative factors

If a factor is studied at several fixed levels we may explain its effect by fitting a linear or curvilinear model by a regression approach. (Obviously we could not fit any suitable curve to a factor if only two levels were examined.) Hicks gives several example f the regression approach applied to different designs. Such approaches regard the ariables as quantitative factors.

'ransformation of variables

The standard factorial experiment treats the effects of the factors as additive, i.e. he response is the sum of the separate effects. If the more likely model is of a ifferent kind we may be able to use a suitable transformation of the variables to obtain he standard form. For example , if

$$S = x^{\alpha} y^{\beta} e^{\gamma \sin z}$$

hen $\log S = \alpha \log x + \beta \log y + \gamma \sin z$ and the new variables would be $\log x$, $\log y$ nd $\sin z$.

'ixed, random and mixed effects

At the end of Section 8.2 we referred to the distinction between fixed and random ffects. In situations where some factors are fixed and some are random the model is aid to be **mixed**. The calculation of the theoretical components of variance is necessary o decide how the tests of hypothesis should be conducted. This decision should be made efore an experiment is carried out. Hicks gives some rules for determining these omponents from the mathematical model for the experiment.

'ested Designs

In a **crossed** design of the type we have so far studied, an observation is taken at ach possible treatment. In some circumstances this may not be practical or even not ossible. For example, samples of a metal component are sent to three laboratories for esting. At each laboratory two technicians carry out two tests each. Figure 8.5 epresents the set-up.

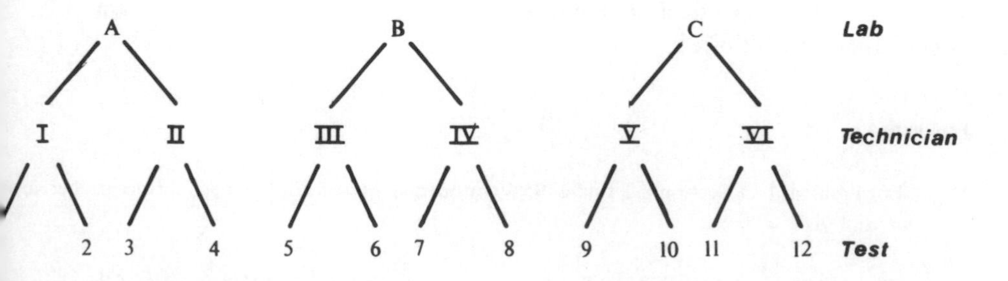

Figure 8.5

It would not have been possible to have each technician carry out his measurements at each laboratory. In this example the factor of technicians is **nested** within the factor of laboratories. Some authors use the term **'hierarchical design'**. An appropriate model is

$$x_{ijk} = \mu + \alpha_i + \beta_{j(i)} + \epsilon_{ijk} \tag{8.22}$$

where α_i is the laboratory effect, k is the measurement number and $\beta_{j(i)}$ represents the effect of the jth technician in the ith laboratory. Note that we could not detect an interaction between laboratory and technician.

Fractional Replication

With a large number of factors, the number of measurements for a full factorial experiment can be prohibitive; for example a 2^{10} experiment would necessitate 1024 runs. We can employ a technique of **fractional replication.** In these cases we try only some of the possible treatments. This will, of course, cause us to lose information about some of the factors and/or interactions. A careful design will preserve information about those effects of greatest interest.

Analysis of covariance

It may be of use to combine the analysis of variance and regression methods. Just as by blocking we use a qualitative factor to remove an unwanted source of variability (such as variations in the quality of a raw material) so we can use a quantitative factor in removal if the unwanted variable is continuous. The extra variable has been called a **concomitant variable.** Suppose the response variable depends linearly on this concomitant variable. Then an appropriate model for a single factor randomized blocks experiment is

$$x_{ij} = \mu + b_i + t_j + c(y_{ij} - \bar{y}) + \epsilon_{ij} \tag{8.23}$$

where b_i is the block effect, t_j the treatment effect, c a constant and y_{ij} the concomitant variable effect.

Problems

1. For Problem 1 of Section 8.2, find a 95% confidence interval for the mean difference between A and B.

2. Find a 95% confidence interval for the mean difference between Lines 1 and 4 of Problem 5 of that Section.

Chapter Nine

Functional Analysis

9.1 INTRODUCTION

Functional analysis, a branch of mathematics that has evolved from 'traditional' analysis, is now making itself felt in many areas of mathematics applied to engineering problems. It is certainly abstract and this, together with its highly specialised notation, makes it a daunting subject to many engineers. There is no doubt that it will become more widespread in its appearance and the engineer of the future will most probably need to be able to understand it as a language.

It is as a language that we approach functional analysis in this chapter. It is our intention to follow a selectively narrow path through the subject leading to some applications in numerical analysis. The beauty and efficiency of this abstract subject is the clarity it can bring to certain areas of mathematics, allowing the disciple to understand the real essentials of each area and to see that the *structure* of some of these areas is the same. Hence, results in one area of application can be recognised as essentially the same as those in another area. The mathematical structures we study will be concerned with collections of elements which obey specified axioms: these collections are called **spaces.** (Mention of one such collection - a vector space - was made in AEM.) After reading this chapter and working through the examples and problems, you should have the confidence to tackle further texts and be capable of understanding the phraseology of functional analysis in engineering contexts.

9.2 METRIC SPACES

Metric spaces are to functional analysis what the real line is to calculus. Essentially, a metric space is a set of elements X with a function associating a 'distance' with each pair of points in X. But how can we talk about the distance between two functions? We shall see. First the axioms for a metric space are stated.

Definition

A **metric space** is a set X, whose elements are called *points*, on which there is defined a real-valued function $d(x, y)$ on all ordered pairs (x, y) of points in X, satisfying

(i) $d(x, y) = 0$ if and only if $x = y$

(ii) $d(y, x) = d(x, y)$ (symmetry)

(iii) $d(x, y) \leqslant d(x, z) + d(z, y)$ (triangle inequality)
where z also belongs to X.

It can be deduced that

(iv) $d(x, y) \geqslant 0$ for all x and $y \in X$.

To show that a metric space generalises our concept of the real line, let X be the real line and define $d(x, y) = |x - y|$, the usual distance function for real numbers. It is straightforward to verify axioms (i) to (iv). It is customary to refer to $d(x, y)$ as the **metric**. Some authors adopt the notation (X, d) for the space.

Examples

1. **Cartesian plane, $\mathbf{R}^2$**

$X = \mathbf{R}^2$ is the set of ordered pairs of real numbers and d is the Euclidean metric defined by

$$d(x, y) = \left\{ (\xi_1 - \eta_1)^2 + (\xi_2 - \eta_2)^2 \right\}^{1/2} \tag{9.1}$$

where $x = (\xi_1, \xi_2)$ and $y = (\eta_1, \eta_2)$ are points in the plane. Again, verify the axioms for yourself. Note that a second metric could be defined by

$$d(x, y) = |\xi_1 - \eta_1| + |\xi_2 - \eta_2| \tag{9.2}$$

and this would provide a second metric space.

2. **Three-dimensional space, $\mathbf{R}^3$**

The Euclidean metric is defined on $x = (\xi_1, \xi_2, \xi_3)$ and $y = (\eta_1, \eta_2, \eta_3)$ by

$$d(x, y) = \left\{ (\xi_1 - \eta_1)^2 + (\xi_2 - \eta_2)^2 + (\xi_3 - \eta_3)^2 \right\}^{1/2} \tag{9.3}$$

3. **Sequence space, l^∞**

The set X is the set of all bounded sequences of real (or complex) numbers so that each point x is a sequence $\{\xi_1, \xi_2, \xi_3, \ldots\}$ with the property that all terms are in modulus less than or equal to a finite value (which may vary from point to point in X). Note that each point in the space is a sequence. The metric chosen is

$$d(x, y) = \sup_i |\xi_i - \eta_i| \tag{9.4}$$

where $i = \{1, 2, 3, \ldots\}$ and $\sup\limits_i$ is the least upper bound of $|\xi_i - \eta_i|$ as i takes all its possible values.

4. **Hilbert Sequence space, l^2**

We now impose the condition that X is the set of all sequences of real numbers such that $|\xi_1|^2 + |\xi_2|^2 + \ldots$ converges. This requirement is now written as

$$\sum_{i=1}^{\infty} |\xi_i|^2 < \infty \tag{9.5}$$

The metric chosen is

$$d(x, y) = \left(\sum_{i=1}^{\infty} |\xi_i - \eta_i|^2 \right)^{1/2} \tag{9.6}$$

5. **C[0, 1]**

X is the set of all real-valued functions which are continuous on the interval $[0, 1]$. Note that each point in the space is a function. The metric adopted is

$$d(x, y) = \max_{0 \leq t \leq 1} |x(t) - y(t)| \tag{9.7}$$

where both $x, y \in X$.

Perhaps the great flexibility of metric spaces becomes apparent.. If we can derive results for metric spaces in general they can be applied to such diverse areas as points in a plane, bounded sequences and continuous functions. Imagine regarding a function as a point in a space! Figure 9.1 shows the metric graphically.

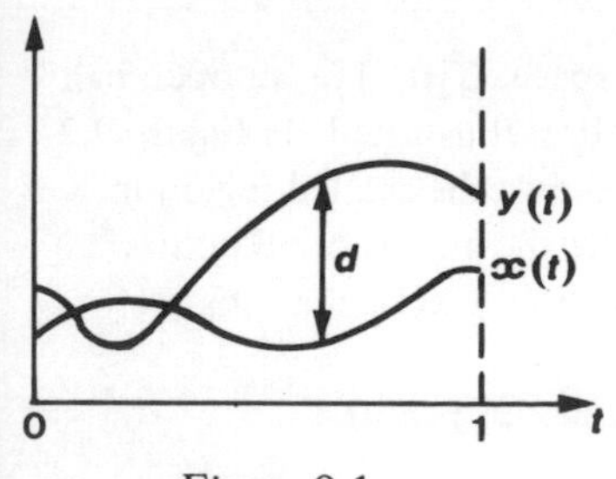

Figure 9.1

You can see that d is a measure of how closely $y(t)$ approximates $x(t)$ in the interval $[0, 1]$.

Note that $d(x, y) = 0$ only if $x(t)$ and $y(t)$ coincide and vice-versa. Clearly, if the two functions differ in at least one point, $d(x, y) > 0$. It is evident that the symmetry axiom holds. The triangle inequality is more awkward to prove: the task is therefore left to you. (The remainder of this section can be omitted without disrupting the flow of the argument towards Section 9.4.)

Open and Closed Sets

We now consider some results and concepts which are of fundamental importance in the study of functional analysis.

Assume that x_0 is a point in a set X and $\rho > 0$ is a real number. Then the set $B(x_0;\ \rho)$ which is the set of points $x \in X$ such that $d(x_0,\ x) < \rho$ is called an **open ball.**

A subset S of X is called an **open set** if every point $x \in S$ is the centre of some ball which is a subset of S, that is there is a number $\rho_x > 0$ such that $B(x;\rho_x) \subseteq S$.

It can be shown that an open ball is an open set. This follows by first denoting the ball $B(a;\ \rho)$ and then letting $x \in B(a;\rho)$. If $\rho_x = \rho - d(x,\ a)$ then for any $y \in B(x;\rho_x)$, $d(x,\ y) < \rho_x$. Then from axiom (iii) for a metric space

$$\begin{aligned} d(a,\ y) &\leqslant d(a,\ x) + d(x,\ y) \\ &< d(a,\ x) + \rho_x \\ &= \rho \end{aligned}$$

If $d(a,\ y) < \rho$ then $y \in B(a,\ \rho)$. Hence $B(x;\rho_x) \subseteq B(a;\rho)$, an open set.

Examples

(i) In Euclidean three-space, $\mathbf{R}^3$, the open ball is geometrically obvious, but remember the spherical surface of radius ρ is absent.

(ii) In the plane, $\mathbf{R}^2$, an open ball is the interior of a circular curve: $\xi^2 + \eta^2 < \rho^2$.†

(iii) On the real line, $\mathbf{R}$, an open ball is the open interval $-\ \rho < x < \rho$.†

(iv) In the space $C[0,\ 1]$, an open ball can be best illustrated via Figure 9.2, which shows the shaded region in which lie the graphs of all points (or functions) $x(t)$ such that

$$d(x,\ x_0) < 0.4$$

where $x_0(t) = t^3$.

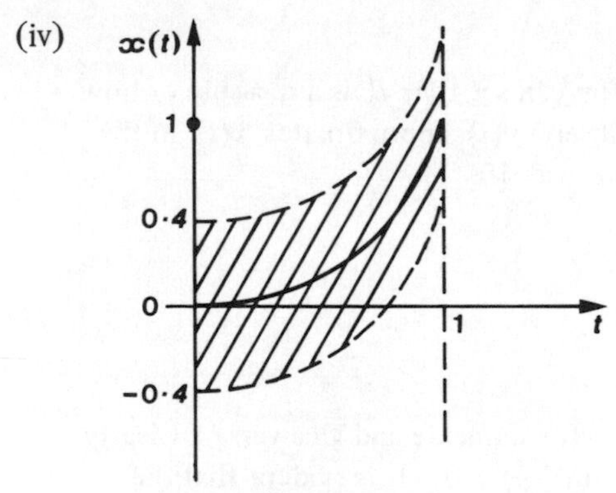

Figure 9.2

† The centre of the ball need not be the origin.

An open ball is sometimes called an **epsilon-neighbourhood of x_0** and written
$B(x_0 ; \epsilon)$.

A point x_0 is an **interior point** of a subset $S \subset X$ if S is a neighbourhood of
x_0.

A subset S of X is called a **closed set** if its complement in X, viz. X/S, is an
open set. (The set X/S consists of all those points of X which do not belong to S.)

It should be straightforward for you to apply these definitions to the spaces of
the Examples on page 350.

A **closed ball** is a set $\bar{B}(x_0 ; \rho)$ which is the set of points $x \in X$ such that
$d(x_0, x) \leqslant \rho$.

roblems

. Let (X, d) be a metric space. Show that the function

$$\rho(x, y) = \frac{d(x, y)}{1 + d(x, y)}$$

defines a metric on X.

. The diameter of subset A of a metric space is the least upper bound of the distances $d(x, y)$ where $x, y \in A$. It is denoted diam (A). If $A \cap B \neq \emptyset$ show that diam $(A \cup B) \leqslant$ diam (A) + diam (B).

. A signal consisting of 1's and zeros is transmitted. It may contain some errors. Suppose we define the distance between two signals of the same as the number of different digits, for example the distance between 01010 and 11100 is 3. Does this idea of distance define a metric space?

. In the plane $\mathbf{R}^2$, which point on the line $x + 2y = 6$ is nearest to the origin under the metric (9.1)? Repeat for metric (9.2). Repeat for the metric $d(x, y) = \max\left\{|\xi_1 - \eta_1| + |\xi_2 - \eta_2|\right\}$. What are the distances in each case? Comment.

. What are the open balls in the plane $\mathbf{R}^2$ which are centred at the origin under the three metrics of Problem 4?

.3 CONVERGENCE AND COMPLETENESS

We now extend the idea of convergence of a sequence of real numbers to
onvergence of a sequence of points in a metric space X. It will be helpful to note that
is the metric on the space $\mathbf{R}$ which allowed the definitions of convergence.

A sequence $\{x_n\}$ of points in a metric space X is **convergent** if there is a point $x \in X$ such that

$$\lim_{n\to\infty} d(x_n, x) = 0 \qquad (9.8)$$

x is said to be the **limit** of $\{x_n\}$. We also write $x_n \to x$.

Note that for any number $\epsilon > 0$, no matter how small, there is an integer N, depending on ϵ, such that all those points x_n where $n > N$ lie in the epsilon-neighbourhood $B(x;\ \epsilon)$.

Example

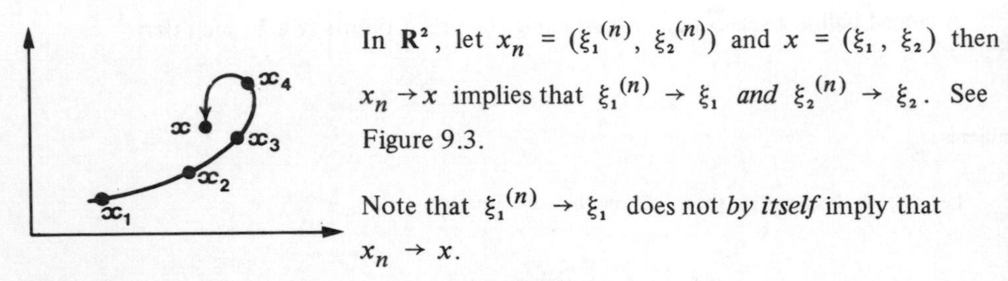

In $\mathbf{R}^2$, let $x_n = (\xi_1^{(n)}, \xi_2^{(n)})$ and $x = (\xi_1, \xi_2)$ then $x_n \to x$ implies that $\xi_1^{(n)} \to \xi_1$ *and* $\xi_2^{(n)} \to \xi_2$. See Figure 9.3.

Note that $\xi_1^{(n)} \to \xi_1$ does not *by itself* imply that $x_n \to x$.

Figure 9.3

Three important results are now considered.

(i) **If the limit x exists then it is unique.**

To see this, let $x_n \to x$ and $x_n \to x'$. Now

$$0 \leq d(x, x') \leq d(x, x_n) + d(x_n, x') \qquad \text{(triangle inequality)}$$
$$= d(x_n, x) + d(x_n, x') \qquad \text{(symmetry)}$$

As $n \to \infty$, $d(x_n, x)$ and $d(x_n, x')$ both tend to 0 and therefore $d(x, x')$ must be zero: it is bounded below by 0 and above by a value which tends to zero. But if $d(x, x') = 0$ then $x = x'$ and uniqueness is proved. A more rigorous proof is given in Kreysig (1978).

(ii) **If $x_n \to x$ then every subsequence $x_{n_k} \to x$**

In the study of iteration a more practical definition of convergence is the Cauchy criterion. That concept is now generalised. A sequence $\{x_n\}$ is said to be **fundamental** if for every value $\epsilon > 0$ an integer N can be found, dependent on ϵ such that $d(x_n, x_m) < \epsilon$ whenever n and m are both greater than N. That is, if we go far enough 'along' the sequence, the distance between any two points of the sequence is as small as we please.

(iii) **Every convergent sequence is a fundamental sequence**

This is shown as follows. We may choose N so that $d(x_n, x) < \epsilon/2$ whenever $n > N$. Then

$$\begin{aligned} d(x_n, x_m) &\leq d(x_n, x) + d(x, x_m) \quad \text{(triangle inequality)} \\ &< \epsilon/2 + \epsilon/2 \\ &= \epsilon \end{aligned}$$

and hence $\{x_n\}$ is fundamental. A fundamental sequence is also called a **Cauchy sequence.** It is important to realise that a fundamental sequence may not be convergent. A favourite counter-example is to choose X as the set of rational numbers and let $d(r_1, r_2) = |r_1 - r_2|$ if r_n is the value of $\sqrt{2}$ correct to n decimal places (which can be expressed as a rational number) then $\{r_n\}$ clearly converges to $\sqrt{2}$, but $\sqrt{2}$ is **not** a member of X.

(Now it must be admitted that the counter-example is not really the sort of engineering example we ought to be considering. However, it serves to warn of the need to specify carefully when certain conditions like convergence occur.)

However, for real numbers, i.e. $X \equiv \mathbf{R}$, every fundamental sequence is convergent. This is also the case for $\mathbf{R}^2$, $\mathbf{R}^n$ and $C[0, 1]$ to name but a few.†

A metric space in which every fundamental sequence is convergent is said to be **complete.** It is instructive to consider the proof of completeness for $C[0, 1]$. Let $\{x_m\}$ be a fundamental sequence in this space; then for any given value $\epsilon > 0$ we can find an integer N such that $d(x_n, x_m) < \epsilon$ whenever n and m are both greater than N. Hence

$$\max |x_n(t) - x_m(t)| < \epsilon \tag{9.9}$$

Therefore, for any particular t_0 satisfying $0 \leq t_0 \leq 1$, it follows that

$$|x_n(t_0) - x_m(t_0)| < \epsilon$$

This demonstrates that $\{x_1(t_0), x_2(t_0), x_3(t_0), ...\}$ is a Cauchy sequence in $\mathbf{R}$. But this space is complete and hence $\{x_n(t_0)\}$ converges; say, to $x(t_0)$.

For each $0 \leq t \leq 1$ it is possible to find a 'limiting' value $x(t)$. This defines a new function on $[0, 1]$.

First, we show that $x(t) \in C[0, 1]$. In (9.9) let $m \to \infty$, then

† Of course, a metric space is not defined without its metric $d(x, y)$; we assume, unless otherwise stated, that the metric is the "usual" one for each space.

$\max|x_n(t) - x(t)| \leqslant \epsilon$ (notice the weakening of the inequality). For all $0 \leqslant t \leqslant 1$, then, $|x_n(t) - x(t)| \leqslant \epsilon$. Hence $\{x_n(t)\}$ converges uniformly to $x(t)$ on the interval $[0, 1]$. It is a result of calculus that the function $x(t)$ is continuous. But $x_m \to x$, therefore $C[0, 1]$ is complete.

A metric space can be **completed** in a way analogous to the completion of the rational numbers by extension to the real numbers.

Problems

1. Show that any closed subset of a complete space is itself a complete space.

2. Let $a_1, a_2, a_3, \ldots\ldots$ be a fundamental sequence in a metric space (X, d) and $b_1, b_2, b_3, \ldots\ldots$ be another sequence in X such that $d(a_n, b_n) < 1/n$.

   Show that $\{b_n\}$ is also a fundamental sequence and that $\{b_n\}$ converges to $x_0 \in X$ if and only if $\{a_n\}$ converges to x_0.

3. Show that every finite metric space is complete.

4. Show that the open interval (a, b) on the real line is incomplete.

5. In the space $C[0, 1]$ suppose we consider the **subset** of those x for which $x(0) = x(1)$. Show that it is a subspace and moreover is complete.

6. If the space (X, d) is complete, show that (X, d') is complete when $d' = d/(1 + d)$.

9.4 CONTRACTION MAPPINGS AND THEIR APPLICATIONS

Let T: $X \to X$ be a mapping of a set onto itself, i.e. to each element $x \in X$ is associated a single element $y \in X$ written $y = Tx$. The mapping is said to be **continuous** if $x_n \to x$ implies $Tx_n \to Tx$. Let (X, d) be a metric space. Then T is called a **contraction mapping** if whenever $x, y \in X$

$$d(Tx, Ty) \leqslant \alpha d(x, y) \tag{9.10}$$

where α is a real number satisfying $0 \leqslant \alpha < 1$. (The same number α holds for all x and y.) In other words, the images Tx and Ty are "closer" together than the original points x and y. (Definition (9.10) is not quite saying this; how does it differ?)

As an example, consider X to be the set of all real numbers in the interval $(-1, 1)$ and let the mapping T be defined by $Tx = x^2$. Note that T maps X onto a subset of itself. The metric placed on X is the 'usual metric': $d(x, y) = |x - y|$. We now establish four results.

i) The iterative sequence $\{x_n\}$, defined by x_0, $x_1 = Tx_0$, $x_2 = Tx_1$, ..., $x_{n+1} = Tx_n$, ... can be shown to be a fundamental sequence when T is a contraction mapping. To see this, suppose that $m > n$. Then

$$\begin{aligned}
d(x_m, x_n) &= d(Tx_{m-1}, Tx_{n-1}) \\
&\leqslant \alpha d(x_{m-1}, x_{n-1}) \qquad (*) \qquad \text{(by (9.10))} \\
&= \alpha d(Tx_{m-2}, Tx_{n-2}) \\
&\leqslant \alpha^2 d(x_{m-2}, x_{n-2}) \qquad \text{(by (9.10))} \\
&\vdots \\
&\leqslant \alpha^n d(x_{m-n}, x_0) \\
&\leqslant \alpha^n [d(x_0, x_1) + d(x_1, x_{m-n})] \qquad \text{(triangle inequality)} \\
&\leqslant \alpha^n [d(x_0, x_1) + \alpha d(x_0, x_{m-n-1})] \qquad \text{(by (*))} \\
&\vdots \\
&\leqslant \alpha^n [d(x_0, x_1) + \alpha d(x_0, x_1) + \ldots\ldots \\
&\qquad + \alpha^{m-n-1} d(x_0, x_1)] \\
&\leqslant \alpha^n d(x_0, x_1)[1 + \alpha + \ldots + \alpha^{m-n-1}] \\
&= \alpha^n d(x_0, x_1) \frac{(1 - \alpha^{m-n})}{1 - \alpha} \\
&\leqslant \alpha^n d(x_0, x_1) \frac{1}{1 - \alpha} \qquad (0 \leqslant \alpha)
\end{aligned}$$

Collecting together the results, we obtain

$$d(x_m, x_n) \leqslant \alpha^n d(x_0, x_1) \frac{1}{1 - \alpha} \tag{9.11}$$

Now the right-hand side of (9.11) is effectively a constant multiplied by α^n and hence tends to zero as $n \to \infty$, since $\alpha < 1$. This means that $\{x_n\}$ is a fundamental sequence.

ii) If (X, d) is a complete metric space, then $\{x_n\}$ converges. A point $x \in X$ is called a **fixed point** of a mapping T: $X \to X$ if $Tx = x$. (9.12)
Perhaps you see a glimpse of the destination to which we are heading.

(iii) Suppose the sequence $\{x_n\}$ converges to x; we then show that x is a fixed point of T. Now

$$d(x, Tx) \leqslant d(x, x_n) + d(x_n, Tx) \qquad \text{(triangle inequality)}$$

$$\leqslant d(x, x_n) + \alpha . d(x_{n-1}, x) \qquad \text{(by (9.10))}$$

Since $x_n \to x$ (and therefore $x_{n-1} \to x$) the right-hand side can be made arbitrarily small by taking n large enough. Hence $d(x, Tx) = 0$ which implies that $Tx = x$.

(iv) Finally, we show that x is the only fixed point of T. Now if x' were also a fixed point, then

$$d(x, x') = d(Tx, Tx') \qquad \text{(fixed points)}$$

$$\leqslant \alpha d(x, x') \qquad \text{(by (9.10))}$$

Since $\alpha < 1$ then it must be the case that $d(x, x') = 0$ so that $x' = x$. Hence the fixed point is unique.

The foregoing results may be collected as follows:

Theorem 1

If (X, d) is a complete metric space and T: $X \to X$ is a contraction mapping, then T has one and only one fixed point.

Application to basic iteration

The equation $x^3 + 2x - 1 = 0$ has one real root, near 0.45. Suppose we set up the iteration sequence

$$x_{n+1} = F(x_n) = 1/(2 + x_n^{\ 2})$$

(Put $x_{n+1} = x_n = x$ and check that we have a rearrangement of the given equation.)

From the mean value theorem we know that if $F(x)$ is differentiable in an open interval (a, b) and continuous in $[a, b]$ then, for $x, y, \xi \in (a, b)$ we have

$$|F(x) - F(y)| = |x - y| . |F'(\xi)| \qquad (9.13)$$

In this particular instance $F(x)$ is differentiable everywhere with $F'(x) = -2x/(2 + x^2)^2$. Notice that $|F'(x)| < 1$ for all x.

[We show this for $x > 0$ and leave you to prove the result for $x \leqslant 0$ if you wish. Now $(2 + x^2)^2 > 2x$ if $4 + 4x^2 + x^4 > 2x$ i.e. if $3 + 3x^2 + x^4 + 1 - 2x + x^2 > 0$, i.e.

if $3 + 3x^2 + x^4 + (1 - x)^2 > 0$ which is clearly true.]

If $|F'(x)| < 1$ for all x then $|F'(\xi)| < 1$ wherever ξ is. Translating these particular results into metric space terminology, we take $X = \mathbf{R}$, $d(x, y) = |x - y|$ and define $Tx = 1/(2 + x^2)$. Equation (9.13) becomes $d(Tx, Ty) \leqslant \alpha\, d(x, y)$ where $0 \leqslant \alpha < 1$. But $(\mathbf{R}, d)$ is complete and therefore the iteration sequence $x_{n+1} = Tx_n$ converges since T is a contraction mapping.

The argument may be generalised to the familiar result that a sufficient condition for the iteration sequence $x_{n+1} = F(x_n)$ to converge to the (unique) root of the equation $x = F(x)$ is that $F(x)$ be continuously differentiable and $|F'(x)| \leqslant r < 1$. In practice, however, it is often the case that $|F'(x)| \leqslant r < 1$ only in a finite interval enclosing the sought-for root of $x = F(x)$. The following theorem for metric spaces helps.

Theorem 2

Suppose the mapping T of Theorem 1 is a contraction mapping on the closed ball $\bar{B}(x_0; \rho)$ and

$$d(x_0, Tx_0) < (1 - \alpha)\rho \tag{9.14}$$

Then $x_{n+1} = Tx_n$ converges to the only fixed point x of T which lies in $\bar{B}(x_0; \rho)$.

To see this we show that all the x_n lie in $\bar{B}(x_0; \rho)$. In the inequality (9.11) put $m = 0$ so that

$$\begin{aligned} d(x_n, x_0) &\leqslant \frac{1}{1-\alpha}\, d(x_0, x_1) \\ &= \frac{1}{1-\alpha}\, d(x_0, Tx_0) \\ &< \rho \qquad \text{by (9.14)} \end{aligned}$$

By definition, all the x_n are in $\bar{B}(x_0; \rho)$.

Now $\bar{B}(x_0; \rho)$ is complete if X is complete and hence the conditions of Theorem 1 are applied to $\bar{B}(x_0; \rho)$.

Error Bounds

So far, we have merely reformulated the basic iteration method. A spin-off is that we can estimate the error in each successive iterate. This is embodied in the following theorem, which is really a corollary to Theorem 1.

Theorem 3

Suppose the conditions of Theorem 1 hold. Then the sequence $x_{n+1} = Tx_n$ gives successive approximations to the fixed point x. The errors may be estimated by

$$d(x_n, x) \leq \frac{\alpha^n}{1-\alpha} d(x_0, x_1) \tag{9.15}$$

or by

$$d(x_n, x) \leq \frac{\alpha}{1-\alpha} d(x_n, x_{n-1}) \tag{9.16}$$

Proof

The first result is a consequence of letting $m \to \infty$ in (9.11). For the second result, let $n = 1$ in (9.15) so that

$$d(x_1, x) \leq \frac{\alpha}{1-\alpha} d(x_0, x_1)$$

We now effectively relabel the sequence. Put $x_0 = y_{n-1}$ so that $x_1 = Tx_0 = Ty_{n-1} = y_n$ then

$$d(y_n, x) \leq \frac{\alpha}{1-\alpha} d(y_{n-1}, y_n)$$

It is easy to see that this gives result (9.16). Check that you understand how to use y's instead of x's and vice-versa.

Example 1

We may use (9.15) as an error bound to estimate at the outset how many steps are necessary to achieve a specified accuracy.

For the iteration sequence $x_{n+1} = 1/(2 + x_n^2)$, $F''(x) = \dfrac{-(4 - 6x^2)}{(2 + x^2)^3}$ and therefore $|F'(x)|$ has a maximum, at $x = \sqrt{(2/3)}$, of $3\sqrt{6}/32 < 0.23 = \alpha$.

Suppose we require an accuracy of 0.00005, then (9.15) advises us that if we start with $x_0 = 0.5$, so that $x_1 = 1/(2 + 0.25) = 4/9$ and $d(x_0, x_1) = 1/18$ then

$$\left(\frac{(0.23)^n}{0.77}\right) \frac{1}{18} < 0.00005$$

from which we find that $n \geqslant 5$. Try the iteration yourself and check this lower bound on n.

Example 2

We may use (9.16) at later stages in the calculations. Obtain x_2 and x_3 and compare the estimate of the error in x_3 with the actual error.

Application to simultaneous linear equations

In order to apply Theorem 1 to the solution of simultaneous linear equations, it is necessary to provide a complete metric space together with a contraction mapping. First of all we rewrite the simultaneous equations in a form analagous to the basic iteration rearrangement. We cite a two-equation case as a working base. The system of equations

$$\left.\begin{aligned} 3\xi_1 + 4\xi_2 &= 6 \\ 2\xi_1 - \xi_2 &= 2 \end{aligned}\right\} \tag{9.17}$$

can be written

$$\begin{bmatrix} \xi_1 \\ \xi_2 \end{bmatrix} = \begin{bmatrix} -2 & -4 \\ -2 & 2 \end{bmatrix} \begin{bmatrix} \xi_1 \\ \xi_2 \end{bmatrix} + \begin{bmatrix} 6 \\ 2 \end{bmatrix} \tag{9.18}$$

i.e.

$$x = Cx + b \tag{9.19}$$

where $x = \begin{bmatrix} \xi_1 \\ \xi_2 \end{bmatrix}$ etc.

Suppose we define $y = Cx + b$ for an arbitrary x. Then x will be a solution of the equations if $y = x$. This suggests the basis of an iterative scheme.

Let X be the set of vectors with n real components, e.g. $x = (\xi_1, \xi_2, \ldots\ldots, \xi_n)$, $y = (\eta_1, \eta_2, \ldots\ldots, \eta_n)$. Really, we should write $x = (\xi_1, \xi_2, \ldots\ldots, \xi_n)^{\mathrm{T}}$, etc. since they are column vectors.

Consider a mapping on X defined by

$$y = Tx = Cx + b \tag{9.20}$$

where b is a fixed vector in X and $C = [C_{ij}]$ is a specified $n \times n$ matrix with real elements. Further, let $w = Tz$ where $w = (\omega_1, \omega_2, \ldots\ldots, \omega_n)$ and $z = (\zeta_1, \zeta_2, \ldots\ldots, \zeta_n)$. The metric chosen is defined by the Euclidean metric

$$d(x, z) = \left(\sum_{i=1}^{n} (\xi_i - \zeta_i)^2 \right)^{1/2} \tag{9.21}$$

It is well known that the space (X, d) is complete. We need a condition under which T is a contraction mapping. Equation (9.20) is now written in component form, taking $n = 2$.

$$\eta_1 = C_{11}\xi_1 + C_{12}\xi_2 + \beta_1$$

$$\eta_2 = C_{21}\xi_1 + C_{22}\xi_2 + \beta_2$$

i.e. $$\eta_i = C_{i1}\xi_1 + C_{i2}\xi_2 + \beta_i$$

similarly,

$$\omega_i = C_{i1}\zeta_1 + C_{i2}\zeta_2 + \beta_i$$

We return to the general n-dimensional case and then

$$d(y, w) = d(Tx, Tz)$$

$$= \left(\sum_{i=1}^{n} (\eta_i - \omega_i)^2 \right)^{1/2}$$

$$= \left(\sum_{i=1}^{n} \left\{ [C_{i1}\xi_1 + C_{i2}\xi_2 + \ldots + C_{in}\xi_n + \beta_i] - [C_{i1}\zeta_1 + C_{i2}\zeta_2 + \ldots + C_{in}\zeta_n + \beta_i] \right\}^2 \right)^{1/2}$$

$$= \left(\sum_{i=1}^{n} \left\{ \sum_{j=1}^{n} C_{ij} [\xi_j - \zeta_j] \right\}^2 \right)^{1/2}$$

For $n = 2$

$$d(y, w) = [(C_{11}[\xi_1 - \zeta_1] + C_{12}[\xi_2 - \zeta_2])^2 + (C_{21}[\xi_1 - \zeta_1] + C_{22}[\xi_2 - \zeta_2])^2]^{1/2}$$

It has been verified, after much effort by the authors, that

$$d(y, w) \leqslant \left(C_{11}^2 + C_{12}^2 + C_{21}^2 + C_{22}^2 \right)^{1/2} \left((\xi_1 - \zeta_1)^2 + (\xi_2 - \zeta_2)^2 \right)^{1/2}$$

$$= \left(\sum_{i=1}^{2} \sum_{j=1}^{2} C_{ij}^2 \right)^{1/2} d(x, z)$$

After even more effort the more general result below can be verified.

$$d(y,\ w) \leqslant \left(\sum_{i=1}^{n} \sum_{j=1}^{n} C_{ij}^{2} \right)^{1/2} d(x,\ z) \qquad (9.22)$$

Hence if

$$\sum_{i=1}^{n} \sum_{j=1}^{n} C_{ij}^{2} < 1 \qquad (9.23)$$

T is a contraction mapping. (Where has the square root gone?)

Theorem 4

If the system of linear equations $x = Cx + b$ where C is a fixed $n \times n$ matrix with real elements, b is a fixed vector with n components and $x = (\xi_1, \xi_2, \ldots\ldots, \xi_n)$ is such that

$$\alpha = \sum_{i=1}^{n} \sum_{j=1}^{n} C_{ij}^{2} < 1$$

then the system possesses one and only one solution x. The solution is the limit of the iterative sequence $x_0,\ x_1 = Cx_0 + b,\ x_2 = Cx_1 + b,\ \ldots\ldots$ where x_0 is an arbitrary vector with n real elements. Furthermore,

$$d(x_m,\ x) \leqslant \frac{\alpha}{1 - \alpha}\, d(x_{m-1},\ x_m) \qquad (9.24)$$

and

$$d(x_m,\ x) \leqslant \frac{\alpha^m}{1 - \alpha}\, d(x_0,\ x_1) \qquad (9.25)$$

The proof of the error estimates follows in an analagous way to the proof of Theorem 3. By now it may be asked how (9.19) relates to the usual form

$$Ax = b \qquad (9.26)$$

of a system of linear equations.

Let $A = I - C$ so that (9.26) becomes $Ix - Cx = b$. The form (9.19) clearly follows. Convergence criteria for Gauss-Seidel iteration may be derived from (9.23) but these are beyond our scope.

Application to differential equations

Consider the initial-value problem

$$\frac{dx}{dt} = f(x, t) \quad \text{with} \quad x(t_0) = x_0 \tag{9.27}$$

An iterative method of solution is sought. The first step is to integrate (9.27) to obtain

$$x = x_0 + \int_{t_0}^{t} f(x, \tau)\, d\tau$$

where x in the integrand is now a function of τ.

The **Picard iteration method** is defined by

$$x_{n+1}(t) = x_0 + \int_{t_0}^{t} f(x_n, \tau)\, d\tau \tag{9.28}$$

Perhaps you begin to see the link with the other applications of Theorem 1. We now quote an existence and uniqueness theorem due to Picard.

Theorem

In the initial-value problem (9.27) suppose $f(x, t)$ is continuous in the region

$$S = \left\{ (x, t) \,\middle|\, x_0 - \beta \leqslant x \leqslant x_0 + \beta,\ t_0 - \alpha \leqslant t \leqslant t_0 + \alpha \right\}$$

(Try to sketch S.) Suppose that on S

$$|f(x, t)| \leqslant M \tag{9.29}$$

and that $|f(x, t) - f(y, t)| \leqslant L|x - y|$ (the **Lipschitz condition**) for all points (x, t) (y, t) in S. Then (9.27) has a unique solution on an interval

$$t_0 - \gamma \leqslant t \leqslant t_0 + \gamma = [t_1, t_2]$$

Outline proof

First, let C be the metric space of real-valued continuous functions on $[t_1, t_2]$ with the metric

$$d(x, y) = \max_{[t_1, t_2]} | x(t) - y(t) | \tag{9.30}$$

C is known to be complete. If D is the subspace of C which comprises those functions in C which satisfy the condition

$$| x(t) - x_0 | \leqslant M\gamma \tag{9.31}$$

where γ is a constant to be determined, then it can be shown that D is complete (not easy, though). We define a mapping T: $D \rightarrow D$ by

$$Tx = x_0 + \int_{t_0}^{t} f(x, \tau) \, d\tau \tag{9.32}$$

Remember that x_0 is a constant function, defined on $[t_1, t_2]$). Strictly, we need to prove that T does map D into D. However, we content ourselves with proving that T is a contraction mapping on D.

By the Lipschitz condition we obtain,

$$| Tx - Ty | = \left| \int_{t_0}^{t} [f(x, \tau) - f(y, \tau)] d\tau \right|$$

$$\leqslant |t - t_0| \, . \max_{[t_1, t_2]} L|x(\tau) - y(\tau)|$$

$$= L\,|t - t_0|\, . \, d(x, y)$$

The interval $[t_1, t_2]$ is in fact the interval $[t_0 - \gamma, t_0 + \gamma]$; then $|t - t_0| \leqslant \gamma$ so that $|Tx - Ty| \leqslant L\,\gamma\, d(x, y)$. However, the r.h.s. does not depend on t so by taking the maximum value of the l.h.s. we obtain

$$d(Tx, Ty) \leqslant L\,\gamma\, d(x, y)$$

Now $L\gamma < 1$ if $\gamma < 1/L$ so that T is a contraction mapping. By Theorem 1 it is unique. Kreysig (1978) gives the proof in more detail and shows that

$$\gamma < \min\left\{\alpha, \frac{\beta}{M}, \frac{1}{L}\right\}$$

This is because γ has to satisfy the conditions

$\gamma < \alpha$ (continuity restriction on t)

$M\gamma < \beta$ (continuity restriction on x)

$L\gamma < 1$ (contraction mapping restriction)

Although the Picard iteration is not really a practical method of solution, its value lies in demonstrating the essential link between basic iteration, matrix iteration and iterative solution of ordinary differential equations. It also leads to an existence and uniqueness theorem.

The essence of what we have done so far lies in the abstraction of certain concepts and the demonstration that the processes examined have essentially the same underlying structure. We may look for parallel results in these processes and gain a clearer understanding of their nature.

Problems

1. Consider X as the set of real numbers $\geqslant 1$ and define $T : X \to X$ by $Tx = \frac{1}{2}x + 1/x$. Show that T is a contraction mapping.

2. If T is a contraction mapping show that T^2 is also a contraction mapping.

3. Discuss the iteration formula $x_{n+1} = \frac{1}{2}(x_n + A/x_n)$.

4. In the solution of simultaneous equations, choose the metric $d(x, z) = \max|\xi_i - \zeta_i|$. Show that in Theorem 4 the required criterion is

$$\sum_{j=1}^{n} |C_{ij}| < 1 \quad \text{for} \quad i = 1, 2, \ldots\ldots, n$$

5. Show that the function $f(x, t) = |x|^{\frac{1}{2}}$ satisfies the Lipschitz condition except in regions including $x = 0$.

9.5 NORMED LINEAR SPACES

First recall the axioms for a vector space.† Let X be a set of vectors over a field of scalars (real line **R** or complex plane **C**) which we denote by K. The elements of X are $x, y, z, \ldots\ldots$ and the elements of K are $\alpha, \beta, \ldots\ldots$ There are two operations: vector addition and scalar multiplication, and the axioms may be stated:

(i) To every pair of vectors $x, y \in X$ there is a unique vector $(x + y) \in X$

(ii) Vector addition is associative: $x + (y + z) = (x + y) + z$

(iii) Vector addition is commutative: $x + y = y + x$

(iv) There is a zero vector $0 \in X$ such that $x + 0 = x$ for all $x \in X$

(v) For each vector $x \in X$ there is a unique vector $-x \in X$ such that $x + (-x) = 0$

† Refer to AEM.

(vi) For each vector $x \in X$ and each scalar $\alpha \in K$ there is a unique vector $\alpha x \in X$

(vii) $(\alpha\beta)x = \alpha(\beta x)$

(viii) $1x = x, \ 1 \in K$

(ix) $\alpha(x+y) = \alpha x + \alpha y$

(x) $(\alpha+\beta)x = \alpha x + \beta x$

} Distributive laws

We shall impose a metric on a vector space in the form of a **norm** to generalise the idea of the length of a vector. Then we shall briefly examine the idea of completeness in the resulting normed space. In the next section we glance at mappings from one normed space to another.

A **norm** on a vector space is a function defined at each $x \in X$ which is real-valued denoted by $\|x\|$ and obeys the rules

(i) $\|x\| \geqslant 0$

(ii) $\|x\| = 0$ if and only if $x = 0$

(iii) $\|\alpha x\| = |\alpha| \, \|x\|$

(iv) $\|x+y\| \leqslant \|x\| + \|y\|$

(This is the triangle inequality.)

Examples

(a) $X = \mathbf{R}, \ \|x\| = |x|$.

(b) $X = \mathbf{R}^n$, so that $x = (\xi_1, \xi_2, \ldots\ldots, \xi_n)$

The Euclidean norm is $\|x\| = [\sum_{i=1}^{\infty} \xi_i^2]^{1/2}$. Another possible norm is

$$\|x\| = \max_{1 \leqslant i \leqslant n} |\xi_i|$$

(c) l^2, Hilbert sequence space: $\|x\| = \left(\sum_{i=1}^{\infty} \xi_i^2\right)^{1/2}$

(d) $C[0, 1] : \|x\| = \max_{0 \leqslant t \leqslant 1} |x(t)|$

Note Every normed space is a metric space where $d(x, y) = \|x - y\|$. It is left to you to verify this.

A complete normed space is called a **Banach space.**

An example of an incomplete normed space is provided by the set of continuous functions on $[0, 1]$ where

$$\|x\| = \int_0^1 |x(t)| \, dt$$

This space can be shown to be incomplete by taking the functions

$$x_m(t) = t^m, \quad 0 \leqslant t \leqslant 1$$

Via metric space theory it can be shown that, for the fundamental sequence of functions $x_m(t)$, convergence takes place to the function

$$x(t) = \begin{cases} 0, & 0 \leqslant t < 1 \\ 1, & t = 1 \end{cases}$$

which, not being continuous, does not belong to the set X. Hence the sequence is not convergent and the space is incomplete. You may argue that the sequence of functions is somewhat artificial, but at this level we must be precise in laying down conditions under which the results we derive do hold true.

Notes

(i) An incomplete normed space can be completed.

(ii) Not every metric on a vector space can be obtained from a norm.

(iii) If an infinite series is defined on a normed space,† then an absolutely convergent series is one for which $\|x_1\| + \|x_2\| + \|x_3\| + \ldots..$ converges. An absolutely convergent series is convergent if and only if the space is complete.

The importance of a normed space, then, is that it permits us to extend naturally the ideas of the calculus into more general contexts.

It is worth remarking that a sequence x_m is convergent if there is an x in X such that

$$\lim_{m \to \infty} \|x_m - x\| = 0$$

† Such a definition is not possible in a general metric space.

Application to Approximation Theory

Let X be a normed space and suppose that Y is a specific subspace of S. We approximate $x \in X$ by some $y \in Y$ and denote the "distance" from x to Y by

$$\Delta(x, Y) = \inf_{y \in Y} \| x - y \|$$

that is, for each $y \in Y$ we form $\|x - y\|$ and take the infimum (or greatest lower bound) of all these values. Should it be the case that there *is* an element $y^* \in Y$ such that $\| x - y^* \| = \Delta$ then y^* is the *best* approximation in Y to x.

As an example, consider the set $C[0, 1]$; a subspace Y of C is the set of polynomials of degree $\leqslant 2$. (Prove this.) Then we seek a polynomial $y^* = a_0 + a_1 t + a_2 t^2 \in Y$ for which

$$\max_{0 \leqslant t \leqslant 1} \| x(t) - y^*(t) \| \leqslant \max_{0 \leqslant t \leqslant 1} \| x(t) - y(t) \|$$

for any $y \in Y$.

That such a "best approximating parabola" exists is guaranteed by the following result.

Theorem 1

If Y is a subspace of finite dimension of a normed space X then for every $x \in X$ there exists a best approximation to x from among the elements of Y.

For the proof we need several preliminaries. First the idea of (sequential) compactness.

A metric space (X, d) is said to be **compact** if every sequence x_n in X has a subsequence which is also convergent.

A subset Z of X is called a **compact subset** if every sequence in Z has a convergent subsequence which has a limit in Z.

Example

The subset of the real line consisting of the open interval $(0, 1)$ is not compact. For instance the sequence $\frac{1}{2}, \frac{1}{3}, \frac{1}{4}$ ..... and any subsequence converge to 0 which is not in the open interval.

We now state two theorems; for their proofs see Kreysig (1978).

Theorem 2

Every compact subset of a metric space is closed and bounded.

Theorem 3

Any subset of a finite dimensional normed space is compact if and only if it is closed and bounded.

Proof of Theorem 1

Suppose we are given $x \in X$. Then consider the closed ball $\overline{B}$ defined as the set of those elements $y \in Y$ for which $\| y \| \leqslant 2 \|x\|$†. It is left to you to verify that this defines a closed ball; we chose 2 fairly arbitrarily. The element $0 \in \overline{B}$ and hence

$$\Delta(x, \overline{B}) = \inf_{y \in \overline{B}} \|x - y\| \leqslant \| x - 0 \| = \| x \|$$

It is a general result that

$$\| x - y \| \geqslant \left| \|y\| - \|x\| \right| \quad \text{(See page 370)}$$

so that if now $y \notin \overline{B}$ and therefore $\| y \| > 2\| x \|$, then

$$\| x - y\| \geqslant \| y \| - \| x \| > 2\| x\| - \| x \| = \| x\| \geqslant \Delta(x, \overline{B}) \tag{9.33}$$

Hence $\Delta(x, \overline{B}) = \Delta(x, Y)$. (Check why.)

The existence of the $>$ sign in (9.33) means that $\Delta(x, Y)$ cannot be assumed if $y \in Y/\overline{B}$. The only hope for a best approximation is that it exists in $\overline{B}$. Since Y is of finite dimension and B is closed and bounded, Theorem 3 assures us that it is compact. Now it can be shown that the norm is a continuous mapping of X into the real line. This is coupled with the result that a continuous mapping of a compact subset of a metric space into the real line assumes its maximum and minimum values. Hence there is an element $y^* \in \overline{B}$ (which is a subset of Y so that $y^* \in Y$) for which $\| x - y\|$ is a minimum. Then y^* is by definition a best approximation to x in Y. (Notice that we say "a best" rather than "the best".)

Uniqueness of approximation

Under what conditions does "a best" approximation become "the best"? First, a

† This is written $\overline{B} = \left\{ y \in Y \,\middle|\, \|y\| \leqslant 2\|x\| \right\}$

efinition is needed. A norm is said to be **strictly convex** if for all x, y with $\|x\| = \|y\| = 1$, $\|x+y\| < 2$ (x and y are distinct).

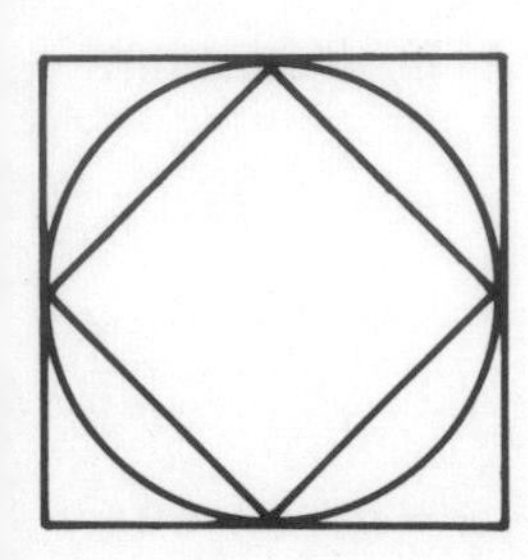

Figure 9.4

Consider Figure 9.4; it depicts the two-dimensional unit sphere (unit circle) under three different norms. The outermost one is under

$$\|x\| = \max\left\{|\xi_1|, |\xi_2|\right\}$$

the centre one is under

$$\|x\| = (\xi_1^2 + \xi_2^2)^{1/2}$$

and the innermost one is under

$$\|x\| = |\xi_1| + |\xi_2|$$

As usual, $x = (\xi_1, \xi_2)$. Verify for yourselves that these unit circles are as depicted.

It is clear that the three regions $\|x\| \leqslant 1$, are convex sets. The centre region as the property, however, that it cannot contain straight line segments on its boundary, nlike the other two; it can be said to be **strictly convex**. Now justify why the orresponding norm may be termed strictly convex.

Theorem 4

Let X be a normed space with a strictly convex norm and Y be any subspace. Then to any $x \epsilon X$ there is at most one best **approximation** from among the elements of Y.

Example 1

It can be shown that the space $C[0, 1]$ with the usual norm is not strictly onvex. With this norm the approximation is known as **uniform approximation**. If the ubspace Y of dimension n satisfies the *Haar condition*, **namely** that every non-zero element of Y has at most $(n-1)$ zeros in $[a, b]$ then the best **approximation** is unique. An example of such a subspace is the set of polynomials of degree $\leqslant n$. (The polynomial 0 is included in this set.) A second example is that of the Chebyshev polynomials on the interval $[0, 1]$.

Example 2 (Least squares approximation)

The vector space comprising all continuous, real-valued functions defined on

$[a, b]$ together with the norm

$$\| x \| = \left(\int_a^b [x(t)]^2 \, dt \right)^{1/2} \tag{9.34}$$

is not complete. The corresponding completed space is denoted $L^2 [a, b]$ and this leads to the idea of least squares approximation.

Problems

1. Verify that $\big| \|x\| - \|y\| \big| \leqslant \|x - y\|$.

2. Verify that the following are norms on ordered pairs of real numbers $x = (\xi_1, \xi_2)$ and $y = (\zeta_1, \zeta_2)$.

   (a) $\|x\| = |\xi_1| + |\xi_2|$

   (b) $\|x\| = (\xi_1^2 + \xi_2^2)^{1/2}$

   (c) $\|x\| = \max [|\xi_1|, |\xi_2|]$

   Sketch the curves $\|x\| = 4$. Interpret.

3. Show that every absolutely convergent series in a Banach space is convergent.

4. Show that a closed subset of a compact metric space is itself compact.

5. Show that the norm $\|x\| = |\xi_1| + |\xi_2|$ is not strictly convex.

6. Let X be the vector space of ordered pairs of real numbers $x = (\xi_1, \xi_2)$. Consider $x_1 = (0, 1)$ and $x_2 = (0, -1)$. Under the norms of Problem 2, determine the intersection of the two spheres $\|x - x_1\| = 1$ and $\|x - x_2\| = 0$.

7. Show that for any two members of a normed space $x \neq 0$, $y \neq 0$ the equation

$$\|x + y\| = \|x\| + \|y\|$$

   implies that $x = \lambda y$ for some $\lambda > 0$.

9.6 LINEAR OPERATORS

In this section a brief look at mappings from one vector space to another is taken. Such mappings are called **operators.** If the operator maintains "structure" in the following way it is of particular importance. Let E_1 and E_2 be vector spaces over the same scalar field.

An operator $T : E_1 \to E_2$ is called **linear** if for all $x,\ y \in E_1$, and α any scalar

(i) $T(x+y) = Tx + Ty$

(ii) $T(\alpha x) = \alpha Tx$

Examples of linear operators are :

(i) the differentiation operator $Tx = x'$ where X is the space of polynomials on $[a,\ b]$ and $T : X \to X$

(ii) the integration operator $Tx = \int_a^t x(\tau)\, d\tau$ where X is the space $C[a,\ b]$,

$T : X \to X$ and $a \leqslant t \leqslant b$

(iii) the scalar product operator $Tx = \xi_1 a_1 + \xi_2 a_2 + \xi_3 a_3$ where a is a fixed vector in $\mathbf{R}^3$ and $T : R^3 \to R$

(iv) the matrix operator. The operator $T : \mathbf{R}^n \to \mathbf{R}^m$ can be defined by $y = Ax + b$ where $x \in \mathbf{R}^n$, A is an $n \times m$ matrix and $b \in \mathbf{R}^m$.

Verify that each of the above examples is a linear operator.

Boundedness

The operator T is said to be **bounded** if a real number M can be found such that

$$\| Tx \| < M \| x \| \tag{9.35}$$

for all $x \in E_1$. Note that $\| Tx \|$ is on E_2.

Ignoring $x = 0$ (why?) the smallest possible value of M is given by

$$\sup \frac{\| Tx \|}{\| x \|}$$

where the supremum is taken over all non-zero vectors in E_1. This value is sometimes denoted $\| T \|$ and this requires justification. First,it should be clear that

(i) $\| T \| \geqslant 0$

(ii) $\| 0 \| = 0$

Second, if $\| T \| = 0$ then $\| Tx \| = 0$ for all $x \in E_1$, so that $Tx = 0$ for all $x \in E_1$ and hence $T = 0$, the zero operator. Third,

$$\sup_{\|x\|=1} \| \alpha T x \| = \sup_{\|x\|=1} |\alpha| . \|Tx\| = |\alpha| . \sup_{\|x\|=1} \| Tx \|$$

If x is restricted by $\| x \| = 1$ so that $\| Tx \| = \| T \| . \| x \| = \| T \|$, we may define $\| T \| = \sup_{\|x\|=1} \| Tx \|$. Hence $\| \alpha T \| = |\alpha| . \| T \|$.

Fourth, it may be shown in like manner that the triangle inequality holds. Note that if the dimension of E_1 is finite then all linear operators on E_1 are bounded.

Examples

(i) The differentiation operator defined on the space of polynomials on $[0, 1]$ with the norm $\| x \| = \max_{[0,1]} | x(t) |$ is not bounded. It should be clear that $x(t) = t^n$ and $x'(t) = nt^{n-1}$ each take their largest value at $t = 1$. Hence $\| x \| = 1$ and $\| Tx \| = n$. Then $\| Tx \| / \| x \|$ can be made as large as required by increasing n.

(ii) The integration operator defined on $C[0, 1]$ by

$$Tx = \int_0^1 K(s, t) \, x(t) \, dt$$

where $K(s, t)$ is a continuous function of s and t on the closed square $E = \{0 \leqslant s \leqslant 1, \ 0 \leqslant t \leqslant 1\}$, is bounded when the norm is $\| x \| = \max_{[0,1]} | x(t) |$.

For

$$\| Tx \| = \max_{[0,1]} \left| \int_0^1 K(s, t) \, x(t) \, dt \right|$$

$$\leqslant \max_{[0,1]} \int_0^1 | K(s, t) | . | x(t) | dt$$

Since $K(s, t)$ is continuous, it is bounded so that $\max_E | K(s, t) | = M$, say. Then $\| Tx \| < M . \| x \|$ as required.

(iii) The matrix operator with norm $\| x \| = \left(\sum_{i=1}^{n} \xi_i^2 \right)^{1/2}$ is bounded. (The proof is beyond our present scope.) The corresponding value of M is

$$\left(\sum_{i=1}^{n} \sum_{j=1}^{n} a_{ij}^2 \right)^{1/2} .$$

Continuous operators

An operator T is said to be **continuous** if given $x^* \epsilon E_1$ then for any $\epsilon > 0$ we can find a $\delta > 0$ such that

$$\| Tx - Tx^* \| < \epsilon \quad \text{whenever} \quad \| x - x^* \| < \delta \tag{9.36}$$

it being assumed that $x \epsilon E_1$.

(In very rough terms, whenever x is close to x^* in E_1, Tx is close to Tx^* in E_2.)

A useful result is that if T is continuous it is bounded and vice-versa.

Applications of norms of operators

Several problems hinge on finding that vector x for which $Ax = b$ where b is a known vector and A is a linear operator. The special case of solution of a set of simultaneous linear equations can be used as an example to carry us through this development.

If A has an inverse, we may write the (unique) solution as $x = A^{-1} b$. It is important, in general, to know whether a linear operator has an inverse.

One result is that if a positive number λ exists such that $\| Ax \| > \lambda \| x \|$ for every $x \epsilon E_1$, then A^{-1} exists and is bounded. But how do we compute the inverse operator?

From the binomial series

$$(1 - x)^{-1} = 1 + x + x^2 + \ldots.. \quad , \quad |x| < 1$$

we may generalise to obtain the following.

Theorem

If $T : E_1 \to E_2$ is a bounded linear operator with $\| T \| < 1$ then $(\mathrm{I} - T)^{-1}$ exists and is equal to

$$\mathrm{I} + T + T^2 + T^3 + \ldots.. \tag{9.37}$$

Here, I is the identity operator defined by $Ix = x$, for all x.

It can be shown that the norm of $(I - T)^{-1}$ is less than or equal to $(1 - \|T\|)^{-1}$.

Now if we are solving a set of equations $Ax = b$ via Gauss elimination we expect round-off errors to occur. In effect, we shall be solving a different set of equations, say $Cx = b$. The value of $\|A - C\|$, which is a measure of the round-off error, can be estimated.

Let the solution of $Ax = b$ be x_1 and the solution of $Cx = b$ be x_2. Then

$$\|x_1 - x_2\| = \|A^{-1}b - C^{-1}b\| = \|(A^{-1} - C^{-1})b\| < \|A^{-1} - C^{-1}\|.\|b\| \tag{9.38}$$

A result which takes us further is given below.

Theorem

If $\|A - C\| < 1/\|A^{-1}\|$ then C^{-1} exists and

$$\|C^{-1} - A^{-1}\| < \frac{\|A^{-1}\|^2 . \|A - C\|}{1 - \|A^{-1}\| . \|A - C\|} \tag{9.39}$$

From an estimate of $\|A - C\|$, together with (9.38) and (9.39) it is possible to obtain an estimate of $\|x_1 - x_2\|$.

It follows that the **relative error** is bounded:

$$\frac{\|x_1 - x_2\|}{\|x_1\|} \leqslant \frac{\|A^{-1}\| . \|A - C\|}{1 - \|A^{-1}\| . \|A - C\|} \tag{9.40}$$

An application of the idea of continuity for operators lies with iterations of the form $x_{n+1} = Tx_n$; if $x_n \to x$ then $x_{n+1} \to x$. Now, x is certainly the solution of the equation $x = Tx$ when T is continuous, but if T is not continuous then we cannot justifiably say anything about the solution.

Problems

1. Consider the operators $T: \mathbf{R}^2 \to \mathbf{R}^2$ defined by

   (i) $T(\xi_1, \xi_2) = (\xi_1, 0)$

   (ii) $T(\xi_1, \xi_2) = (\xi_2, \xi_1)$

   (iii) $T(\xi_1, \xi_2) = (3\xi_1, 3\xi_2)$

Show that they are linear and interpret them geometrically.

The null space of a mapping is the set of all x such that $Tx = 0$. Find the null space for each mapping T in Problem 1.

Express the mappings of Problem 1 via matrices.

A non-zero bounded linear operator T operates on an element x for which $\|x\| < 1$; show that $\|Tx\| < \|T\|$.

.7 FURTHER REMARKS

We have so far not used the term "functional". In fact a **functional** is an operator hich maps into the real line **R** or the complex plane **C**. If the domain E_1 is real, ıen the functional maps into $E_2 = \mathbf{R}$, otherwise it maps into $E_2 = \mathbf{C}$.

Examples of functionals are

) The scalar product

i) The definite integral on $C[a, b]$: $f(x) = \int_a^b x(t)\,dt$. (Note that we use the symbol f for a functional, by analogy with functions in calculus.) In this example f is linear and bounded. Linearity follows from linearity of the integration procedure. Also

$$|f(x)| = \left| \int_a^b x(t)\,dt \right| < \int_a^b |x(t)|\,dt < (b-a) \max_{[a,b]} |x(t)| = (b-a)\|x\|$$

Further it may be shown that $\|f\| = b - a$.

ii) Let $t^* \in [a, b]$ and define

$$f(x) = x(t^*) \qquad x \in C[a, b]$$

Then f is linear, bounded and has norm $\|f\| = 1$.

The theory of linear functionals is very rich and indeed its study initiated the whole ctivity in functional analysis. The interested reader is referred to the Bibliography.

An important concept in mathematics is that of orthogonality. The scalar product n "ordinary" vector algebra is a useful measure of orthogonality. This idea may be eneralised.

An **inner product space** is a vector space X which has an inner product that maps $X \times X$ into the scalar field. It is written $< x, y >$ and defined on all pairs of elements $x, y \in X$ as follows, where $x, y, z \in X$ and α is any scalar:

(i) $< x, y > = \overline{< y, x >}$ †

(ii) $< \alpha x, y > = \alpha < x, y >$

(iii) $< x + y, z > = < x, z > + < y, z >$

(iv) $< x, x > \geqslant 0$

(v) $< x, x > = 0$ if and only if $x = 0$

The inner product leads naturally to a norm

$$\| x \| = \sqrt{< x, x >} \tag{9.41}$$

and a metric

$$d(x, y) = \| x - y \| = \sqrt{< x - y, x - y >} \tag{9.42}$$

Hence any inner product space is a normed space.

Finally, two elements $x, y \in X$ are **orthogonal** to each other if $< x, y > = 0$.

A complete inner product space is called a **Hilbert space.**

A familiar application of inner product spaces is in the theoretical background to Fourier series.

Let X be the inner product space of all real-valued functions continuous on $[-\pi, \pi]$ with the inner product

$$< x, y > = \int_{-\pi}^{\pi} x(t) . y(t) \mathrm{d}t$$

Then consider the particular members of X:

$$1, \quad \cos t, \quad \sin t, \quad \cos 2t, \quad \sin 2t, ...$$

It is easy to show that these functions are orthogonal to each other. This fact lead to the determination of the coefficients in a Fourier series by formulae which give each coefficient independently of the others. To add more terms to a Fourier series

† The bar denotes complex conjugate.

approximation involves merely the determination of the coefficients of the new terms: a direct consequence of orthogonality.

Concluding Remarks

As mentioned at the outset, we have merely trodden one path in the realm of functional analysis. The books in the Bibliography give a wider view of the subject.

Problems

1. Show that the functional defined on $C[a, b]$ by

$$f(x) = \int_a^b x(t)x_0(t)\,dt$$

where x_0 is a fixed element in $C[a, b]$, is a linear functional.

2. Show that if x and y are orthogonal to each other in an inner product space then

$$\|x+y\|^2 = \|x\|^2 + \|y\|^2$$

3. Show, for an inner product space, that if the relationship $\langle x, z\rangle = \langle x, y\rangle$ holds for all x then $z = y$.

4. Prove that a norm on an inner product space satisfies

$$\|x+y\|^2 + \|x-y\|^2 = 2(\|x\|^2 + \|y\|^2)$$

APPENDIX A - FUNCTIONS OF A MATRIX

1. Polynomials

Let $p(\lambda)$ be the characteristic polynomial of a square matrix $\mathbf{A}$; then $p(\lambda) = 0$ is the characteristic equation of $\mathbf{A}$. As is well-known, the Cayley-Hamilton theorem states that $p(\mathbf{A}) = \mathbf{0}$. If we wish to evaluate a polynomial in $\mathbf{A}$ of higher degree than $p(\lambda)$, then the Cayley-Hamilton theorem can help simplify matters.

For example, the matrix $\mathbf{A} = \begin{bmatrix} 0 & 1 \\ -4 & -5 \end{bmatrix}$ has characteristic equation $p(\lambda) = \lambda^2 + 5\lambda + 4 = 0$. Hence $\mathbf{A}^2 + 5\mathbf{A} + 4\mathbf{I} = \mathbf{0}$. If we wish to evaluate $f(\mathbf{A}) = \mathbf{A}^4 + 6\mathbf{A}^3 + 4\mathbf{I}$ notice that

$$\frac{f(\lambda)}{p(\lambda)} = \lambda^2 + \lambda - 9 + \frac{41\lambda + 40}{\lambda^2 + 5\lambda + 4}$$

Hence $f(\mathbf{A}) = (\mathbf{A}^2 + \mathbf{A} - 9\mathbf{I})p(\mathbf{A}) + 41\mathbf{A} + 40\mathbf{I}$. But $p(\mathbf{A}) = \mathbf{0}$, therefore $f(\mathbf{A}) = 41\mathbf{A} + 40\mathbf{I}$.

If $\mathbf{A}$ has repeated eigenvalues, then it may satisfy a polynomial equation of lower degree than the characteristic equation: that equation of lowest degree satisfied by $\mathbf{A}$ is called the **minimum polynomial** of $\mathbf{A}$.

In the example above the minimum polynomial is the characteristic polynomial.

However, the matrix $\mathbf{A} = \begin{bmatrix} 1 & 2 & 2 \\ 2 & 1 & 2 \\ 2 & 2 & 1 \end{bmatrix}$ has characteristic polynomial $\lambda^3 - 3\lambda^2 - 9\lambda - 5 = 0$, but its minimum polynomial is $\Phi(\mathbf{A}) = \mathbf{A}^2 - 4\mathbf{A} - 5\mathbf{I} = \mathbf{0}$.

To find the minimum polynomial, the following strategy may be adopted.

(i) Form adj($\mathbf{A} - \lambda\mathbf{I}$)

(ii) Factorise the elements

(iii) Put $d(\lambda)$ as the greatest common divisor of the elements: if there is no common divisor, put $d(\lambda) = (-1)^n$ where $n \times n$ is the size of $\mathbf{A}$.

(iv) $$\Phi(\lambda) = \frac{|\mathbf{A} - \lambda\mathbf{I}|}{d(\lambda)}$$

In the last example, $d(\lambda) = 2 - \lambda$. Work out the steps yourself.

Matrix Series

Consider the two infinite series

$$\sigma(x) = a_0 + a_1 x + a_2 x^2 + \ldots\ldots + a_k x^k + \ldots\ldots \qquad \text{(A.1)}$$

and

$$\sigma(\mathbf{A}) = a_0 + a_1 \mathbf{A} + a_2 \mathbf{A}^2 + \ldots\ldots + a_k \mathbf{A}^k + \ldots\ldots \qquad \text{(A.2)}$$

By analogy we may say that

$\sigma(\mathbf{A})$ **converges** to a matrix $\mathbf{B}$ if the sequence of

partial sums $\sum_{k=0}^{n} a_k \mathbf{A}^k$ tends to $\mathbf{B}$ as $n \to \infty$.

A useful result is that $\sigma(\mathbf{A})$ converges if $\sigma(x)$ has a radius of convergence $> \rho(\mathbf{A})$ where $\rho(\mathbf{A})$ is the **spectral radius** of $\mathbf{A}$ which is given by

$$\rho(\mathbf{A}) = \max_i |\lambda_i| \qquad \text{(A.3)}$$

the λ_i being the eigenvalues of $\mathbf{A}$.

A related result is that $\sigma(\mathbf{A})$ converges if $\sigma(x)$ converges for each of the values $x = \lambda_i$.

It can be shown that $\sigma(\mathbf{A})$ converges for all $\mathbf{A}$ if $\sigma(x)$ converges everywhere in the complex plane.

Hence we may define

$$e^{\mathbf{A}} = \mathbf{I} + \mathbf{A} + \frac{1}{2!}\mathbf{A}^2 + \frac{1}{3!}\mathbf{A}^3 + \ldots\ldots\ldots \qquad \text{(A.4)}$$

$$\sin \mathbf{A} = \mathbf{A} - \frac{1}{3!}\mathbf{A}^3 + \frac{1}{5!}\mathbf{A}^5 - \ldots\ldots\ldots \quad (A.5)$$

$$\cos \mathbf{A} = \mathbf{I} - \frac{1}{2!}\mathbf{A}^2 + \frac{1}{4!}\mathbf{A}^4 - \ldots\ldots\ldots \quad (A.6)$$

and these series are convergent for all **A**.

However, the series

$$\log_e(\mathbf{I} + \mathbf{A}) = \mathbf{A} - \frac{1}{2}\mathbf{A}^2 + \frac{1}{3}\mathbf{A}^3 - \ldots\ldots \quad (A.7)$$

converges for those matrices **A** whose spectral radius < 1.

A useful concept is that of a **matrix norm**, dealt with more fully in Chapter 9.

In essence, the norm is a measure of the size of a matrix (cf modulus of a vector). It is written $\|\mathbf{A}\|$.

Two examples of matrix norms are

(i) Euclidean norm: $$\|\mathbf{A}\| = \left(\sum_{i=1}^{n} \sum_{j=1}^{n} |a_{ij}|^2 \right)^{1/2} \quad (A.8)$$

(ii) Cubic norm: $$\|\mathbf{A}\| = \max_{i,j} |a_{ij}| \quad (A.9)$$

We state the necessary and sufficient condition for $\sigma(\mathbf{A}) \to \mathbf{B}$ as

$$\lim_{n\to\infty} \left\| \sum_{k=0}^{n} a_k \mathbf{A}^k - \mathbf{B} \right\| = 0$$

for *any* norm.

Note the special case $\mathbf{B} = \mathbf{0}$. Also note that $\rho(\mathbf{A}) < \|\mathbf{A}\|$ for *any* norm.

It is important to note that a sufficient condition for $\mathbf{A}^k \to 0$ is that $\|\mathbf{A}\| < 1$. A necessary and sufficient condition is that $\rho(\mathbf{A}) < 1$.

It follows that

$$(\mathbf{I} - \mathbf{A})^{-1} = \mathbf{I} + \mathbf{A} + \mathbf{A}^2 + \ldots\ldots\ldots \quad (A.10)$$

provided $\|\mathbf{A}\| < 1$.

An important result, following directly from (A.4) is that

$$e^{\mathbf{A}} \, . \, e^{\mathbf{B}} = e^{(\mathbf{A}+\mathbf{B})}$$

provided $\mathbf{AB} = \mathbf{BA}$, i.e. $\mathbf{A}$ and $\mathbf{B}$ *commute.*

The same condition is necessary for the identity

$$\sin(\mathbf{A} + \mathbf{B}) = \sin \mathbf{A} \cos \mathbf{B} + \cos \mathbf{A} \sin \mathbf{B}$$

If $\mathbf{A} = \mathbf{B}$ then it is easy to see that results such as

$$\sin 2\mathbf{A} = 2 \sin \mathbf{A} \cos \mathbf{A}$$

obtain.

A matrix series is **absolutely convergent** if $\sum_{k=0}^{\infty} \| a_k \mathbf{A}^k \|$ converges. An absolutely convergent series is also a convergent series.

3. Closed form for matrix functions

The matrix $\mathbf{A} = \begin{bmatrix} 1 & 2 & 2 \\ 2 & 1 & 2 \\ 2 & 2 & 1 \end{bmatrix}$ has minimum polynomial $\Phi(\mathbf{A}) \equiv \mathbf{A}^2 - 4\mathbf{A} - 5\mathbf{I}$.

It also has eigenvalues -1 (repeated) and $+5$. Now $e^{\mathbf{A}t}$, as expanded in (A.4) can be simplified to a polynomial of the form

$$f(\mathbf{A}) = \alpha_0 \mathbf{I} + \alpha_1 \mathbf{A}$$

since the equation $\Phi(\mathbf{A}) = 0$ can be used to substitute for $\mathbf{A}^2$ in terms of $\mathbf{I}$ and $\mathbf{A}$.

Now $f(\lambda) = \alpha_0 + \alpha_1 \lambda$. Let $\lambda = -1$. Then

$$e^{-t} = f(-1) = \alpha_0 - \alpha_1$$

If $\lambda = +5$, then $e^{5t} = f(5) = \alpha_0 + 5\alpha_1$. Hence

$$\alpha_0 = \frac{5}{6} e^{-t} + \frac{1}{6} e^{5t}$$

and

$$\alpha_1 = -\frac{1}{6}e^{-t} + \frac{1}{6}e^{5t}$$

It follows that

$$e^{\mathbf{A}t} = \alpha_0 \mathbf{I} + \alpha_1 \mathbf{A} = \begin{bmatrix} \frac{2}{3}e^{-t} + \frac{1}{3}e^{5t} & -\frac{1}{3}e^{-t} + \frac{1}{3}e^{5t} & -\frac{1}{3}e^{-t} + \frac{1}{3}e^{5t} \\ -\frac{1}{3}e^{-t} + \frac{1}{3}e^{5t} & \frac{2}{3}e^{-t} + \frac{1}{3}e^{5t} & -\frac{1}{3}e^{-t} + \frac{1}{3}e^{5t} \\ -\frac{1}{3}e^{-t} + \frac{1}{3}e^{5t} & -\frac{1}{3}e^{-t} + \frac{1}{3}e^{5t} & \frac{2}{3}e^{-t} + \frac{1}{3}e^{5t} \end{bmatrix}$$

4. Derivative of a matrix

The **derivative** of a matrix $\mathbf{A}(t)$ is defined by

$$\frac{d\mathbf{A}}{dt} = \lim_{\delta t \to 0} \frac{\mathbf{A}(t + \delta t) - \mathbf{A}(t)}{\delta t} \tag{A.11}$$

provided the limit exists.

In effect, each element of $\mathbf{A}$ is differentiated with respect to t.

Note that $\frac{d}{dt}\mathbf{A}^2 = 2\mathbf{A} \cdot \frac{d\mathbf{A}}{dt}$ only if $\mathbf{A}$ commutes with $\frac{d\mathbf{A}}{dt}$.

APPENDIX B - STATISTICAL TABLES

F distribution - 5% and 1% points

ν_2 \ ν_1	1	2	3	4	5	6	7	8	10	12	24	∞
1	161	200	216	225	230	234	237	239	242	244	249	254
	4052	5000	5403	5625	5764	5859	5928	5981	6056	6106	6235	6366
2	18.5	19.0	19.2	19.2	19.3	19.3	19.4	19.4	19.4	19.4	19.5	19.5
	98.5	99.0	99.2	99.2	99.3	99.3	99.4	99.4	99.4	99.4	99.5	99.5
3	10.13	9.55	9.28	9.12	9.01	8.94	8.89	8.85	8.79	8.74	8.64	8.53
	34.1	30.8	29.5	28.7	28.2	27.9	27.7	27.5	27.2	27.1	26.6	26.1
4	7.71	6.94	6.59	6.39	6.26	6.16	6.09	6.04	5.96	5.91	5.77	5.63
	21.2	18.0	16.7	16.0	15.5	15.2	15.0	14.8	14.5	14.4	13.9	13.5
5	6.61	5.79	5.41	5.19	5.05	4.95	4.88	4.82	4.74	4.68	4.53	4.36
	16.26	13.27	12.06	11.39	10.97	10.67	10.46	10.29	10.05	9.89	9.47	9.02
6	5.99	5.14	4.75	4.53	4.39	4.28	4.21	4.15	4.06	4.00	3.84	3.67
	13.74	10.92	9.78	9.15	8.75	8.47	8.26	8.10	7.87	7.72	7.31	6.88
7	5.59	4.74	4.35	4.12	3.97	3.87	3.79	3.73	3.64	3.57	3.41	3.23
	12.25	9.55	8.45	7.85	7.46	7.19	6.99	6.84	6.62	6.47	6.07	5.65
8	5.32	4.46	4.07	3.84	3.69	3.58	3.50	3.44	3.35	3.28	3.12	2.93
	11.26	8.65	7.59	7.01	6.63	6.37	6.18	6.03	5.81	5.67	5.28	4.86
9	5.12	4.26	3.86	3.63	3.48	3.37	3.29	3.23	3.14	3.07	2.90	2.71
	10.56	8.02	6.99	6.42	6.06	5.80	5.61	5.47	5.26	5.11	4.73	4.31
10	4.96	4.10	3.71	3.48	3.33	3.22	3.14	3.07	2.98	2.91	2.74	2.54
	10.04	7.56	6.55	5.99	5.64	5.39	5.20	5.06	4.85	4.71	4.33	3.91
11	4.84	3.98	3.59	3.36	3.20	3.09	3.01	2.95	2.85	2.79	2.61	2.40
	9.65	7.21	6.22	5.67	5.32	5.07	4.89	4.74	4.54	4.40	4.02	3.60
12	4.75	3.89	3.49	3.26	3.11	3.00	2.91	2.85	2.75	2.69	2.51	2.30
	9.33	6.93	5.95	5.41	5.06	4.82	4.64	4.50	4.30	4.16	3.78	3.36

ν_2 \ ν_1	1	2	3	4	5	6	7	8	10	12	24	∞
14	4.60	3.74	3.34	3.11	2.96	2.85	2.76	2.70	2.60	2.53	2.35	2.1
	8.86	6.51	5.56	5.04	4.70	4.46	4.28	4.14	3.94	3.80	3.43	3.0
16	4.49	3.63	3.24	3.01	2.85	2.74	2.66	2.59	2.49	2.42	2.24	2.0
	8.53	6.23	5.29	4.77	4.44	4.20	4.03	3.89	3.69	3.55	3.18	2.7
18	4.41	3.55	3.16	2.93	2.77	2.66	2.58	2.51	2.41	2.34	2.15	1.9
	8.29	6.01	5.09	4.58	4.25	4.01	3.84	3.71	3.51	3.37	3.00	2.5
20	4.35	3.49	3.10	2.87	2.71	2.60	2.51	2.45	2.35	2.28	2.08	1.8
	8.10	5.85	4.94	4.43	4.10	3.87	3.70	3.56	3.37	3.23	2.86	2.4
24	4.26	3.40	3.01	2.78	2.62	2.51	2.42	2.36	2.25	2.18	1.98	1.7
	7.82	5.61	4.72	4.22	3.90	3.67	3.50	3.36	3.17	3.03	2.66	2.2
28	4.20	3.34	2.95	2.71	2.56	2.45	2.36	2.29	2.19	2.12	1.91	1.6
	7.64	5.45	4.57	4.07	3.75	3.53	3.36	3.23	3.03	2.90	2.52	2.0
32	4.15	3.29	2.90	2.67	2.51	2.40	2.31	2.24	2.14	2.07	1.86	1.5
	7.50	5.34	4.46	3.97	3.65	3.43	3.26	3.13	2.93	2.80	2.42	1.9
36	4.11	3.26	2.87	2.63	2.48	2.36	2.28	2.21	2.11	2.03	1.82	1.5
	7.40	5.25	4.38	3.89	3.58	3.35	3.18	3.05	2.86	2.72	2.35	1.8
40	4.08	3.23	2.84	2.61	2.45	2.34	2.25	2.18	2.08	2.00	1.79	1.5
	7.31	5.18	4.31	3.83	3.51	3.29	3.12	2.99	2.80	2.66	2.29	1.8
60	4.00	3.15	2.76	2.53	2.37	2.25	2.17	2.10	1.99	1.92	1.70	1.3
	7.08	4.98	4.13	3.65	3.34	3.12	2.95	2.82	2.63	2.50	2.12	1.6
120	3.92	3.07	2.68	2.45	2.29	2.18	2.09	2.02	1.91	1.83	1.61	1.2
	6.85	4.79	3.95	3.48	3.17	2.96	2.79	2.66	2.47	2.34	1.95	1.3
∞	3.84	3.00	2.60	2.37	2.21	2.10	2.01	1.94	1.83	1.75	1.52	1.0
	6.63	4.61	3.78	3.32	3.02	2.80	2.64	2.51	2.32	2.18	1.79	1.0

For each value of $F_{\nu_2}^{\nu_1}$, the upper number is the 5% point and the lower one is the 1% point.

Duncan's Multiple Range test

The tables below show values of r_p

$\alpha = 0.05$

ν \ p	2	3	4	5	6
1	18.0	18.0	18.0	18.0	18.0
2	6.09	6.09	6.09	6.09	6.09
3	4.50	4.50	4.50	4.50	4.50
4	3.93	4.01	4.02	4.02	4.02
5	3.64	3.74	3.79	3.83	3.83
6	3.46	3.58	3.64	3.68	3.68
7	3.35	3.47	3.54	3.58	3.60
8	3.26	3.39	3.47	3.52	3.55
9	3.20	3.34	3.41	3.47	3.50
10	3.15	3.30	3.37	3.43	3.46
11	3.11	3.27	3.35	3.39	3.43
12	3.08	3.23	3.33	3.36	3.40
13	3.06	3.21	3.30	3.35	3.38
14	3.03	3.18	3.27	3.33	3.37
15	3.01	3.16	3.25	3.31	3.36
20	2.95	3.10	3.18	3.25	3.30
30	2.89	3.04	3.12	3.20	3.25
40	2.86	3.01	3.10	3.17	3.22
60	2.83	2.98	3.08	3.14	3.20
100	2.80	2.95	3.05	3.12	3.18
∞	2.77	2.92	3.02	3.09	3.15

$\alpha = 0.01$

ν \ p	2	3	4	5	6
1	90.0	90.0	90.0	90.0	90.0
2	14.0	14.0	14.0	14.0	14.0
3	8.26	8.5	8.6	8.7	8.8
4	6.51	6.8	6.9	7.0	7.1
5	5.70	5.96	6.11	6.18	6.26
6	5.24	5.51	5.65	5.73	5.81
7	4.95	5.22	5.37	5.45	5.53
8	4.74	5.00	5.14	5.23	5.32
9	4.60	4.86	4.99	5.08	5.17
10	4.48	4.73	4.88	4.96	5.06
11	4.39	4.63	4.77	4.86	4.94
12	4.32	4.55	4.68	4.76	4.84
13	4.26	4.48	4.62	4.69	4.74
14	4.21	4.42	4.55	4.63	4.70
15	4.17	4.37	4.50	4.58	4.64
20	4.02	4.22	4.33	4.40	4.47
30	3.89	4.06	4.16	4.22	4.32
40	3.82	3.99	4.10	4.17	4.24
60	3.76	3.92	4.03	4.12	4.17
100	3.71	3.86	3.98	4.06	4.11
∞	3.64	3.80	3.90	3.98	4.04

Cochran's test for equal variances

The tables below show $C_{(1-\alpha, n, k)}$

$C_{(0.95, n, k)}$

k \ n	2	3	4	5	6
2	0.9985	0.9750	0.9392	0.9057	0.8772
3	0.9669	0.8709	0.7977	0.7457	0.7071
4	0.9065	0.7679	0.6841	0.6287	0.5895
5	0.8412	0.6838	0.5981	0.5441	0.5065
6	0.7808	0.6161	0.5321	0.4803	0.4447
7	0.7271	0.5612	0.4800	0.4307	0.3974
8	0.6798	0.5157	0.4377	0.3910	0.3595
9	0.6385	0.4775	0.4027	0.3584	0.3286
10	0.6020	0.4450	0.3733	0.3311	0.3029
20	0.3894	0.2705	0.2205	0.1921	0.1735
30	0.2929	0.1980	0.1593	0.1377	0.1237
40	0.2370	0.1576	0.1259	0.1082	0.0968
60	0.1737	0.1131	0.0895	0.0765	0.0682
120	0.0998	0.0632	0.0495	0.0419	0.0371
∞	0	0	0	0	0

$C_{(0.99, n, k)}$

k \ n	2	3	4	5	6
2	0.9999	0.9950	0.9794	0.9586	0.9373
3	0.9933	0.9423	0.8831	0.8335	0.7933
4	0.9676	0.8643	0.7814	0.7212	0.6761
5	0.9279	0.7885	0.6957	0.6329	0.5875
6	0.8828	0.7218	0.6258	0.5635	0.5195
7	0.8376	0.6644	0.5685	0.5080	0.4659
8	0.7945	0.6152	0.5209	0.4627	0.4226
9	0.7544	0.5727	0.4810	0.4251	0.3870
10	0.7175	0.5358	0.4469	0.3934	0.3572
20	0.4799	0.3297	0.2654	0.2288	0.2048
30	0.3632	0.2412	0.1913	0.1635	0.1454
40	0.2940	0.1915	0.1508	0.1281	0.1135
60	0.2151	0.1371	0.1069	0.0902	0.0796
120	0.1225	0.0759	0.0585	0.0489	0.0429
∞	0	0	0	0	0

Bibliography

The following is a selection of books and papers relevant to the material presented in the foregoing chapters. Some of these works are referred to directly in the text.

Barnett, S. (1975), *Introduction to Mathematical Control Theory*, O.U.P., Oxford.

Barnett, S. and Storey, C. (1970), *Matrix Methods in Stability Theory*, Nelson, London.

Bellman, R.E. and Dreyfus, S.C. (1962), *Applied Dynamic Programming*, Princeton University Press, New Jersey.

Chung, Y.K. and Yeo, M.F. (1979), *A Practical Introduction to Finite Element Analysis*, Pitman, London.

Desai, C.S. and Abel, J.F. (1972), *Introduction to the Finite Element Method*, Van Nostrand Reinhold, New York.

Dixon, L.C.W. (1972), *Non-Linear Optimization*, E.U.P., London.

Garrard, W.L. and Kornhauser, A.L. (1973), Design of optimal feedback systems for longitudinal control of automated transit vehicles, *Transportation Research*, Vol 7, 125 - 144.

Hicks, C.R. (1966), *Fundamental Concepts in the Design of Experiments*, Holt, Rinehart and Winston, New York.

Jacobs, O.L.R. (1974), *Introduction to Control Theory*, O.U.P., Oxford.

Kreysig, E. (1978), *Introduction to Functional Analysis with Applications*, Wiley, New York.

McCausland, I. (1969), *Introduction to Automatic Control*, Wiley, New York.

Meditch, J.S. (1964), On the problem of optimal thrust programming for a lunar soft landing, *I.E.E.E. Trans. Automatic Control*, AC–9, 477 - 484.

Milne, R.D. (1980), *Applied Functional Analysis*, Pitman, London.

Ogata, K. (1967), *State Space Analysis of Control Systems*, Prentice Hall Inc., New Jersey.

Rumsey, A.F. and Powner, E.T. (1974), Longitudinal control of automated vehicles in guided transportation systems, *Proc. I.E.E.*, Vol. 121, 11, 1435 - 1440.

Sage, A. (1968), *Optimum Systems Control*, Prentice Hall Inc., New Jersey.

Sánchez, D.A. (1968), *Ordinary Differential Equations and Stability Theory*, W.H. Freeman & Co., San Francisco.

Scheffé, H. (1959), *The Analysis of Variance*, John Wiley, New York.

Sokolnikoff, I.S. (1957), *Tensor Analysis*, John Wiley, New York.

Stark, R.M. and Nicholls, R.L. (1972), *Mathematical Foundations for Design*, McGraw Hill, New York.

Thaler, G.J. and Brown, R.C. (1960), *Analysis and Design of Feedback Control Systems*, McGraw Hill, New York.

Wakely, P.G. (1970), Mathematics in Engineering, *Int. J. Math. Educ. Sci. Technol.*, 1, 392.

Zienkiewicz, O.C. (1971), *The Finite Element Method for Engineering Science*, McGraw Hill, London.

Answers

Chapter 1

Page 7

2. $$\dot{x}_1 = \mu_1(t) + \mu_2(t) - 0.5 \frac{F_0}{V_0} x_1(t)$$

$$V_0 \dot{x}_2 + c_0 \dot{x}_1 = c_1 \mu_1(t) + c_2 \mu_2(t) - 0.5 c_0 \frac{F_0}{V_0} x_1(t) - F_0 x_2(t)$$

Page 17

9. $$\dot{\mathbf{x}} = \begin{bmatrix} -3 & 1 & 0 & 0 \\ 0 & 0 & -2 & 0 \\ -4 & 0 & 0 & 1 \\ 0 & 0 & -3 & 0 \end{bmatrix} \mathbf{x} + \begin{bmatrix} 0 & 2 \\ 1 & 0 \\ 0 & 3 \\ 1 & -3 \end{bmatrix} \mathbf{u}$$

$$\mathbf{y} = \begin{bmatrix} 1 & 0 & 0 & 0 \\ 0 & 0 & 1 & 0 \end{bmatrix} \mathbf{x} + \begin{bmatrix} 0 & 0 \\ 0 & 1 \end{bmatrix} \mathbf{u}$$

10. $$\dot{\mathbf{x}} = \begin{bmatrix} 0 & 1 \\ -e^{2t} & -e^{-t} \end{bmatrix} \mathbf{x} + \begin{bmatrix} 0 \\ 1 \end{bmatrix} u, \quad \mathbf{y} = (1 \quad 0)\mathbf{x}$$

Chapter 2

Page 34

3. $$\begin{bmatrix} 1 & \dfrac{e^{2t}}{1+e^t} - 1 \\ 0 & \dfrac{e^t}{1+e^t} \end{bmatrix}$$

4. $$e^{\tau} \begin{bmatrix} \cosh \tau & \sinh \tau \\ \sinh \tau & \cosh \tau \end{bmatrix}$$

where $\tau = \cos t_0 - \cos t$

5. $$e^{4(t-t_0)} \begin{bmatrix} \cos(e^{-t_0} - e^{-t}) & \sin(e^{-t_0} - e^{-t}) \\ -\sin(e^{-t_0} - e^{-t}) & \cos(e^{-t_0} - e^{-t}) \end{bmatrix}$$

Page 45

1. (a) $\frac{1}{4}\begin{bmatrix} 2e^{t} + 2e^{4t} & 4e^{t} - 4e^{4t} \\ e^{t} - e^{4t} & 2e^{t} + 2e^{4t} \end{bmatrix}$ (b) $\begin{bmatrix} 1 & t & t + t^2 \\ 0 & 1 & t \\ 0 & 0 & 1 \end{bmatrix}$

(c) $\begin{bmatrix} \frac{1}{2} e^{-2t} + \frac{1}{2} e^{-4t} & \frac{1}{2} e^{-2t} - \frac{1}{2} e^{-4t} & 0 \\ \frac{1}{2} e^{-2t} - \frac{1}{2} e^{-4t} & \frac{1}{2} e^{-2t} + \frac{1}{2} e^{-4t} & 0 \\ 0 & 0 & e^{-3t} \end{bmatrix}$

(d) $\begin{bmatrix} 1 - te^{t} & 1 - (t+1)e^{t} & (2t + 1)e^{t} - 1 \\ 1 + (t-1)e^{t} & 1 + te^{t} & (1 - 2t)e^{t} - 1 \\ 1 - e^{t} & 1 - e^{t} & 2e^{t} - 1 \end{bmatrix}$

2. (a) $\frac{1}{s^2 - 9s + 18}\begin{bmatrix} s - 4 & \sqrt{2} \\ \sqrt{2} & s - 5 \end{bmatrix}$ (b) $\frac{\begin{bmatrix} (s^2 + 4s + 4) & 0 & 0 \\ 0 & (s^2 + s) & (4s + 4) \\ 0 & -(s+1) & (s^2 + 5s + 4) \end{bmatrix}}{s^3 + 5s^2 + 8s + 4}$

Page 55

1. (a) $\begin{bmatrix} 1 & 0 & 0 \\ 0 & 2 & 0 \\ 0 & 0 & 3 \end{bmatrix}$ (b) $\begin{bmatrix} -1 & 1 & 0 \\ 0 & -1 & 0 \\ 0 & 0 & -1 \end{bmatrix}$ (c) $\begin{bmatrix} -1 & 1 & 0 \\ 0 & -1 & 1 \\ 0 & 0 & -1 \end{bmatrix}$

(d) $\begin{bmatrix} -1 & 0 & 0 \\ 0 & -1 & 0 \\ 0 & 0 & -1 \end{bmatrix}$ (e) $\begin{bmatrix} 1 & 0 & 0 \\ 0 & 0.6 & 0.8 \\ 0 & -0.8 & 0.6 \end{bmatrix}$

(f) $\begin{bmatrix} 1 & 1 & 0 \\ 0 & 1 & 1 \\ 0 & 1 & 1 \end{bmatrix}$

2. (a) $\begin{bmatrix} 1 \\ 0 \\ 0 \end{bmatrix}, \begin{bmatrix} 1 \\ 0.5 \\ 0 \end{bmatrix}, \begin{bmatrix} 1 \\ 5/16 \\ 1/8 \end{bmatrix}; \begin{bmatrix} 1 & 1 & 0 \\ 0 & 1 & 1 \\ 0 & 0 & 0 \end{bmatrix}$

(b) $\begin{bmatrix}1\\0\\0\end{bmatrix}$, $\begin{bmatrix}1\\1\\-2\end{bmatrix}$, $\begin{bmatrix}0\\0\\1\end{bmatrix}$; $\begin{bmatrix}4&0&0\\0&4&1\\0&0&4\end{bmatrix}$

(c) $\begin{bmatrix}0\\1\\0\\0\end{bmatrix}$, $\begin{bmatrix}0\\0\\0\\1\end{bmatrix}$, $\begin{bmatrix}0\\1\\0\\0\end{bmatrix}$, $\begin{bmatrix}1\\0\\0\\0\end{bmatrix}$

4. $\mathbf{M} = \begin{bmatrix}-2&1&-2\\1&0&-1\\-1&0&3\end{bmatrix}$, $\mathbf{M}^{-1} = \begin{bmatrix}0&1.5&0.5\\1&4&2\\-1&0&3\end{bmatrix}$,

$$\dot{\mathbf{q}} = \begin{bmatrix}-1&0&0\\0&-2&0\\0&0&-3\end{bmatrix}\mathbf{q};\quad \mathbf{q} = \begin{bmatrix}10e^{-t}\\5e^{-2t}\\2e^{-3t}\end{bmatrix} + \mathbf{C};\quad \mathbf{x} = \mathbf{Mq}$$

Page 66

1. neither
2. vector (ii)

Chapter 3

Page 85

1. (a) Saddle point, unstable; 2, −2 (b) Node, unstable; 2, 3
   (c) Centre, stable; ± 4i (d) Focus, asymptotically stable; −1 ± i
   (e) Focus, unstable; 4 ± 2i (f) Node, stable; −5, −1

3. (a) Unstable saddle point (b) Stable node (c) Unstable node

4. (0, 0) centre; (1, 0) and (−1, 0) saddle points

5. (0, 0) and (2, 0) centres; (−1, 0), (1, 0) and (3, 0) saddle points

6. No

7. No

Page 96

1. (a) Unstable (b) Unstable (c) Asymptotically stable (d) Unstable

2. (a) No (b) No

Page 106

3. (a) Unstable (b) stable (c) asymptotically stable

   (d) unstable (e) unstable

Page 115

2. (a) $V = 0.5\,x_1^2 + x_2^2$ (b) $V = \dfrac{x_1^2}{1 - x_1 x_2} + x_2^2$

3. Asymptotic stability at origin.

4. Not a stable focus at origin.

6. Linearised case needs to be asymptotically stable to be of use.

Chapter 4

Page 131

4. (i) 4/3 (ii) 23/15

Page 146

4. Not possible; system unstable.

Page 154

1. ABEFJ

2. ABCG

3. 1, 1, 1

5. $u_0 = -1,\ u_1 = -2$

Chapter 5

Page 173

2. (ii) ,0, $\frac{1}{2}A^2 \cos \omega (t_1 - t_2)$

3. $t_1 t_2 + b^2$

Page 187

3. (i) and (iii)

4. (i) 1/24 (ii) $(3 + 2\sqrt{6})/\left\{(4\sqrt{6})(2 + \sqrt{6})\right\}$ (iii) 1/13.01

Page 191

3.

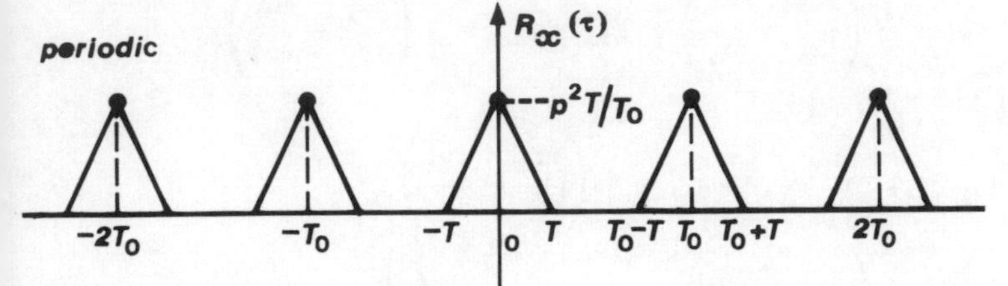

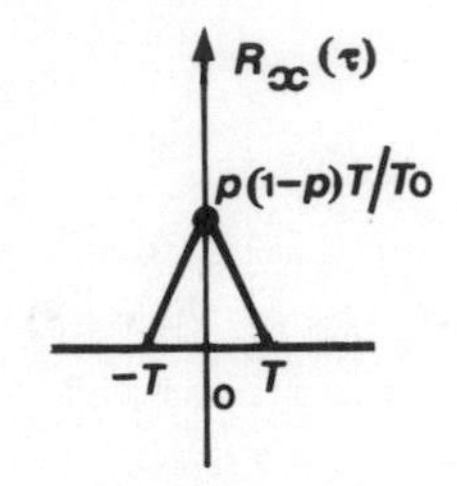

Chapter 6

Page 205

1. $\begin{bmatrix} -1/\sqrt{2} & 1/2 & -1/2 \\ 0 & 1/\sqrt{2} & 1/\sqrt{2} \\ 1/\sqrt{2} & 1/2 & -1/2 \end{bmatrix}$ 2. 1

3. $[\lambda]^{-1} = [\lambda]^{T} = \begin{bmatrix} -1/\sqrt{2} & 0 & 1/\sqrt{2} \\ 1/2 & 1/\sqrt{2} & 1/2 \\ -1/2 & 1/\sqrt{2} & -1/2 \end{bmatrix}$

7. $\begin{bmatrix} 1/\sqrt{3} & 1/\sqrt{3} & 1/\sqrt{3} \\ -2/\sqrt{6} & 1/\sqrt{6} & 1/\sqrt{6} \\ 0 & -1/\sqrt{2} & 1/\sqrt{2} \end{bmatrix}$

Page 213

1. (a) $d\phi = \frac{\partial \phi}{\partial x_i} dx_i$ (b) $a_i x_i x_3$ (c) $\sigma'_{ij} = \lambda_{ik} \lambda_{jl} \sigma_{kl}$

 (d) $\lambda_{ij} \lambda_{jk} = \delta_{ik}$

2. (a) $\begin{cases} a_1 x_1 x_1 + a_2 x_2 x_1 + a_3 x_3 x_1 \\ a_1 x_1 x_2 + a_2 x_2 x_2 + a_3 x_3 x_2 \\ a_1 x_1 x_3 + a_2 x_2 x_3 + a_3 x_3 x_3 \end{cases}$ (b) $\begin{cases} a_{11}x_1 + a_{12}x_2 + a_{13}x_3 = b_1 \\ a_{21}x_1 + a_{22}x_2 + a_{23}x_3 = b_2 \\ a_{31}x_1 + a_{32}x_2 + a_{33}x_3 = b_3 \end{cases}$

(c)
$$\begin{aligned} u'_{11} = {} & \lambda_{11}\lambda_{11}u_{11} + \lambda_{11}\lambda_{12}u_{12} + \lambda_{11}\lambda_{13}u_{13} \\ & + \lambda_{12}\lambda_{11}u_{21} + \lambda_{12}\lambda_{12}u_{22} + \lambda_{12}\lambda_{13}u_{23} \\ & + \lambda_{13}\lambda_{11}u_{31} + \lambda_{13}\lambda_{12}u_{32} + \lambda_{13}\lambda_{13}u_{33} \end{aligned}$$
is a typical example.

(d) $\lambda_{ij}\lambda_{jk} = \delta_{ik}$

(e) $A_{11} = B_{22} + B_{33}$, $A_{21} = -B_{21}$, $A_{31} = -B_{31}$, $A_{12} = -B_{12}$, $A_{22} = B_{33} + B_{11}$, $A_{32} = -B_{32}$, $A_{13} = -B_{13}$, $A_{23} = -B_{23}$, $A_{33} = B_{11} + B_{22}$

(f)
$$e_{11} = \frac{\partial u_1}{\partial x_1}, \quad e_{22} = \frac{\partial u_2}{\partial x_2}, \quad e_{33} = \frac{\partial u_3}{\partial x_3}, \quad e_{12} = e_{21} = \frac{1}{2}\left(\frac{\partial u_1}{\partial x_2} + \frac{\partial u_2}{\partial x_1}\right),$$
$$e_{13} = e_{31} = \frac{1}{2}\left(\frac{\partial u_1}{\partial x_3} + \frac{\partial u_3}{\partial x_1}\right), \quad e_{23} = e_{32} = \frac{1}{2}\left(\frac{\partial u_2}{\partial x_3} + \frac{\partial u_3}{\partial x_2}\right)$$

(g)
$$a_{11} = a_{22} = a_{33} = 0, \quad a_{12} = -a_{21} = \frac{1}{2}\left(\frac{\partial u_1}{\partial x_2} - \frac{\partial u_2}{\partial x_1}\right),$$
$$a_{13} = -a_{31} = \frac{1}{2}\left(\frac{\partial u_1}{\partial x_3} - \frac{\partial u_3}{\partial x_1}\right), \quad a_{23} = -a_{32} = \frac{1}{2}\left(\frac{\partial u_2}{\partial x_3} - \frac{\partial u_3}{\partial x_2}\right)$$

4. (i) $\left(\frac{2}{\sqrt{3}}, \frac{5}{\sqrt{6}}, \frac{3}{\sqrt{2}}\right)$ (ii) (8/3, −1/3, 4/3)

Page 226

1.
$$\begin{bmatrix} 10 & 5 & 2.5 \\ 5 & 7 & -2 \\ 2.5 & -2 & 3 \end{bmatrix} \begin{bmatrix} 0 & 3 & -3.5 \\ -3 & 0 & 7 \\ 3.5 & -7 & 0 \end{bmatrix}$$

2. $B_{ij}U_iV_j = B_{ij}V_iU_j = 378$

Scalars, equal since B_{ij} is symmetric.

4.
$$\begin{bmatrix} \tau & 0 & 0 \\ 0 & \tau & 0 \\ 0 & 0 & \tau \end{bmatrix}$$

5. $$\begin{bmatrix} 600 & 0 & 0 \\ 0 & 400 & 0 \\ 0 & 0 & 100 \end{bmatrix} \begin{bmatrix} \dfrac{1100}{3} & \dfrac{300}{\sqrt{6}} & \dfrac{-700}{\sqrt{18}} \\ \dfrac{300}{\sqrt{6}} & 250 & \dfrac{300}{\sqrt{12}} \\ \dfrac{-700}{\sqrt{18}} & \dfrac{300}{\sqrt{12}} & \dfrac{1450}{3} \end{bmatrix}$$

6. $-2,\ 1,\ 4$ with directions $(0,\ 1/\sqrt{2},\ -1/\sqrt{2})$, $(1/\sqrt{3},\ -1/\sqrt{3},\ -1/\sqrt{3})$, $(-2/\sqrt{6},\ -1/\sqrt{6},\ -1/\sqrt{6})$.

7. $3\tau,\ 0,\ 0$ with direction $(1/\sqrt{3},\ 1/\sqrt{3},\ 1/\sqrt{3})$ and two other directions $(a,\ b,\ c)$ which are perpendicular and such that $a + b + c = 0$.

8. Principal stresses 6, 2 with principal axes at 30° to Ox_1x_2. Max shear stress = 2.

9. $$e_{ij} = \begin{bmatrix} 1 & -3 & \sqrt{2} \\ -3 & 1 & -\sqrt{2} \\ \sqrt{2} & -\sqrt{2} & 4 \end{bmatrix}$$

10. $$\begin{bmatrix} 3x_2^2 & 3x_1x_2 + x_3 & -\tfrac{1}{2}x_2 \\ 3x_1x_2 + x_3 & 0 & \tfrac{1}{2}x_1 \\ -\tfrac{1}{2}x_2 & \tfrac{1}{2}x_1 & 2x_3 \end{bmatrix}$$; six equations are of two types:

$$\frac{\partial^2 e_{11}}{\partial x_2^2} + \frac{\partial^2 e_{22}}{\partial x_1^2} = 2\,\frac{\partial^2 e_{12}}{\partial x_1\,\partial x_2} \quad \text{and} \quad \frac{\partial}{\partial x_1}\left(\frac{-\partial e_{23}}{\partial x_1} + \frac{\partial e_{31}}{\partial x_2} + \frac{\partial e_{12}}{\partial x_3}\right) = \frac{\partial^2 e_{11}}{\partial x_2\,\partial x_3}$$

11. $$\begin{bmatrix} 12 & 2 & -1 \\ 2 & 0 & 0 \\ -1 & 0 & 4 \end{bmatrix}, \quad \begin{bmatrix} 0 & \tfrac{1}{2} & 0 \\ \tfrac{1}{2} & 0 & 0 \\ 0 & 0 & 1 \end{bmatrix}$$; $(1,\ \tfrac{1}{2},\ -\tfrac{1}{2})$, directions $(0,\ 0,\ 1)$, $(1,\ 1,\ 0)$, $(1,\ -1,\ 0$

12. $a_1 + a_2 + a_3$; $a_1a_2 + a_2a_3 + a_3a_1$; $a_1a_2a_3$

13. $0,\ 6 \pm 2\sqrt{6}$

Page 234

4. A_{ik}, 11

Page 243

2. $$\frac{1}{E}\begin{bmatrix} \sigma_1 - \nu(\sigma_2 + \sigma_3) & 0 & 0 \\ 0 & \sigma_2 - \nu(\sigma_3 + \sigma_1) & 0 \\ 0 & 0 & \sigma_3 - \nu(\sigma_1 + \sigma_2) \end{bmatrix},$$

$$\frac{J_1}{3E}\begin{bmatrix} 1-2\nu & 0 & 0 \\ 0 & 1-2\nu & 0 \\ 0 & 0 & 1-2\nu \end{bmatrix}$$

3. $$\frac{(1+\nu)}{E}\begin{bmatrix} \sigma_1 - \sigma & 0 & 0 \\ 0 & \sigma_2 - \sigma & 0 \\ 0 & 0 & \sigma_3 - \sigma \end{bmatrix}$$

4. It remains a cube with change of volume.

5. 0; gives rise to a change of shape, but not volume.

Chapter 8

Page 305

1. $F = 11.3$, differences in metals significant; estimated difference $\bar{x}_A/\bar{x}_B = 1.23$. Homogeneity breaks down at $p = 3$ level.

2. $F = 20.7$, significant; $\hat{\mu} = 251.8$, $\hat{\alpha}_1 = -4.6$, $\hat{\alpha}_2 = -2.2$, $\hat{\alpha}_3 = 6.8$, $\sigma_0{}^2 = 13.9$

3. $F = 29.3$, effect of flow rate significant; $E_5 = -1.08\dot{3}$ grains/m³; $E_{10} = -0.25$ grains/m³; $E_{15} = -0.08\dot{3}$ grains/m³; $E_{20} = 1.417$ grains/m³. (Note that $\Sigma E_n = 0$.)

4. $F = 11$; A, C, D > B, E

5. $F = 12$; 1 and 3 grouped, 2 and 4 grouped.

Page 313

1. $F = 34$, effect of shifts significant.

2. $F = 5.57$, effect of machines significant at 5% level; A and B differ.

3. $F = 4.03$, effect of treatments not significant at 5% level.

4. Times are significant at 5% risk.

Page 316

1. Positions $F = 32$, Lines $F = 6.83$, Compositions $F < 1$.

2. Months $F = 8$, Cars $F = 5.4$, Oils $F = 25$.

3. Days $F = 22.6$, Threads $F = 1.41$, Technicians $F = 1.76$.

4. Nozzles $F = 30.7$, Temperatures $F = 8.3$.
Nozzle A results significantly higher than others.

Page 330

1. Temperatures $F = 3.3$, Volts $F = 4.5$, Interaction $F = 7.3$.

2. Fuels $F = 25.7$, Burners $F = 37.2$, Interaction $F = 5.8$

3. All main effects, Temperatures-Plastic and Speeds-Plastic significant at 1%.

4. All main effects highly significant; temperature x time very significant, temperature x detergent significant at 1% level, time x detergent not significant.
$\bar{D}_1$, $\bar{T}_1$, $\overline{\text{II}}$, $\bar{D}_3$ unlike other means, I, D_2, T_2 are the same, III and T_3 are the same.

5. At the 5% level neither batch nor interaction is significant; method is highly significant; A, C and B are similar but D is definitely greater than the others.

Page 338

1. A and C are significant.

2. Mn; C, Ni, Ni x C, Mn x Ni x C are all significant at 5%.

3. A, C, E, $B \times C \times E$ are significant at 1%; D, $B \times E$, $A \times B \times D$ are significant at 5%.

Page 343

1. a, b, c, abc, d, abd, acd, bcd; (1), ab, ac, bc, ad, bd, cd, $abcd$

2. AB

3. B significant, $A \times B$ significant, Blocks significant.

4. No interaction, A & B significant.
$\hat{\mu} = 20$, $\hat{\alpha}_1 = 6$, $\hat{\alpha}_2 = -6$, $\hat{\beta}_1 = -4$, $\hat{\beta}_2 = 4$

Page 346

1. $2.66 < \mu_0 < 5.85$

2. $10.25 < \mu_0 < 31.35$

Chapter 9

Page 351

3. Yes

4. (1.2, 2.4), $1.2\sqrt{5}$; (0, 3), 3; (2, 2), 2

5.

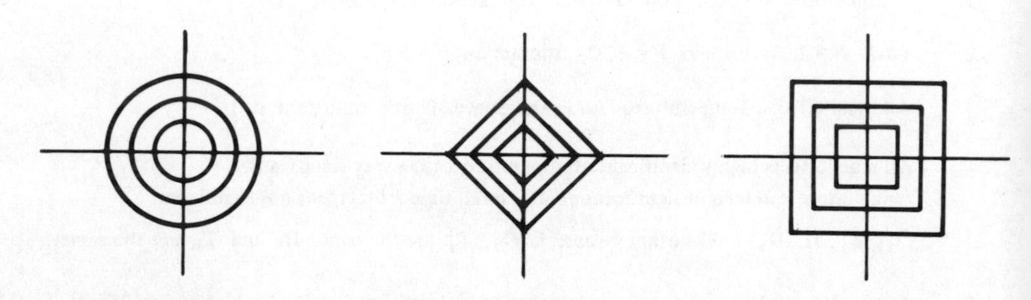

Page 370

6. (0, 0); (0, 0); $\{(x, 0) \mid |x| < 1\}$

Page 374

1. (i) Projection on to ξ_1 axis; (ii) reflection in line $\xi_1 = \xi_2$;
   (iii) enlargement, centre at the origin, scale factor 3.

2. Only (i) has non-trivial null space and this is $\xi_1 = 0$.

3. (i) $\begin{bmatrix} 1 & 0 \\ 0 & 1 \end{bmatrix}$ (ii) $\begin{bmatrix} 0 & 1 \\ 1 & 0 \end{bmatrix}$ (iii) $\begin{bmatrix} 3 & 0 \\ 0 & 3 \end{bmatrix}$

Index